T0295208

Chipless RFID Printing Technologies

For a listing of recent titles in the
Artech House Electromagnetic Analysis Library,
turn to the back of this book.

Chipless RFID Printing Technologies

Santanu Kumar Behera
Durga Prasad Mishra

ARTECH
HOUSE
BOSTON | LONDON
artechhouse.com

Library of Congress Cataloging-in-Publication Data
A catalog record for this book is available from the U.S. Library of Congress.

British Library Cataloguing in Publication Data
A catalog record for this book is available from the British Library.

ISBN-13: 978-1-63081-999-6

Cover design by Joi Garron

685 Canton Street
Norwood, MA 02062

10 9 8 7 6 5 4 3 2 1

Contents

Foreword

The rapidly evolving landscape of technology constantly presents us with new challenges and opportunities. In this age of innovation, radio frequency identification (RFID) technology has stood as a beacon of progress, enabling us to connect with and understand the world around us. Within the RFID domain, chipless RFID printing technologies have emerged as a promising frontier that holds vast potential and exciting possibilities.

This book is a comprehensive exploration of chipless RFID printing technologies, offering readers a deep dive into the intricacies and promise of this cutting-edge field. The ten insightful chapters provide a panoramic view of this technology, comparing it to existing solutions and paving the way for future advancements. The authors, experts in their respective fields, have strived for technical precision, practical insights, and the pursuit of excellence, making this book an invaluable resource.

With each chapter and appendix, the book uncovers the versatility and importance of chipless RFID, from its fundamental principles to its cutting-edge applications. The future of chipless RFID is boundless, urging us to embark on a journey of creativity and discovery. Climate monitoring employs chipless RFID-based mobile climate sensors to address the pressing issue of understanding climate changes and patterns. Traditional climate monitoring methods are limited by fixed weather stations, while chipless RFID technology offers a technical solution with the flexibility to collect data at various altitudes and locations. This technical approach paves the way for a deeper understanding of climate-related challenges and serves as an example of how chipless RFID technology can revolutionize environmental monitoring. The book, rich in its technical depth, offers a comprehensive exploration of a technology that promises to reshape the way we interact with the world. It highlights the transformative potential, explores applications across diverse sectors, and identifies future research directions that will further enhance the capabilities of chipless RFID technology.

In conclusion, I hope that this book serves as a guiding light for researchers, engineers, students, and industry stakeholders, inspiring them to delve deeper into the fascinating world of chipless RFID printing technologies. The transformative journey is already underway, and it is through our collective efforts that we can harness the full potential of this technology, drive innovation, and usher in a new era of technological progress.

Nemai Chandra Karmakar
Monash University Clayton, VIC, Australia
March 2024

Preface

It is with great pleasure and a sense of accomplishment that we present this comprehensive work. This book represents the culmination of years of research, dedication, and collaboration among experts in the field of radio frequency identification technology. Our journey through the intricate world of chipless RFID has led us to explore its multifaceted aspects, delving into its uses, challenges, and promising horizons. We provide here a brief overview of the goals, topics, and journey that this book will take you on.

Understanding the evolution of RFID. The landscape of RFID technology has evolved significantly in recent years, and chipless RFID has emerged as a groundbreaking innovation. While traditional RFID systems rely on integrated circuits (chips) for data storage, chipless RFID introduces a paradigm shift by utilizing patterns and resonators for identification. This revolutionary approach offers increased flexibility, reduced costs, and broader potential applications. The motivation behind this book is to shed light on the developments in chipless RFID, from its fundamental principles to its real-world applications.

We cordially encourage you to turn the pages of this book into a voyage of exploration. Chipless RFID technology is reshaping the landscape of various industries, offering innovative solutions and expanding the boundaries of human knowledge. It is a field driven by passion, curiosity, and dedication from researchers and inventors worldwide. Together, we embrace the challenges and opportunities that lie ahead, influencing the future of chipless RFID technology and driving positive global change. We extend our heartfelt thanks to the contributors, experts, researchers, and practitioners who have shared their knowledge and insights, making this book a valuable resource for both novices and experts in the field of RFID. It is our hope that this book inspires innovation and opens doors to the limitless possibilities of chipless RFID technology.

Acknowledgments

I would like to thank Casey Gerard, assistant editor, Artech House, for the invitation to write a book on chipless RFID printing technologies. Our sincere gratitude goes to the reviewers of the book proposal and chapters. Enormous support from the research scholars and their timely support for submission of chapters are highly acknowledged. Our special thanks to former and current students Tanmay Kumar Das, Priyabrata Sethy, Subhasish Pandav, Khushbu Patel, and G. Ashish Kumar for their generous support and artwork for the chapters. We must acknowledge Natalie McGregor, acquisitions editor, and Leona Crawford, assistant editor, Artech House, for their continuous support and patience throughout the editing and writing process of the manuscript. I would also like to offer special thanks to my former student Amrutha Subrahmannian for the artwork, writing the front matter, and arranging all the figures and tables for each chapter of the book. Special thanks also go to Mahesh Kumar Sahoo, senior technical assistant, for his help collating the electronic copies of the chapters of the book.

The book project was my ambition. I guided my student coauthor to fulfill this ambition throughout the whole project. Therefore, the book is a pinnacle of hard work of my dedicated coauthor Dr. Durga Prasad Mishra. Without his continuous motivation, dedication, and perseverance, the book would not have taken this shape.

My heartfelt thanks go to my family members for their moral support during editing the book.

Santanu Kumar Behera

I am profoundly grateful to my esteemed family members for their unwavering support in my journey as a writer. Special appreciation goes to my mother, Mrs. Prasovini Devi, for her boundless love and guidance, and to my heavenly father, Mr. Dinabandhu Mishra, whose blessings continue to inspire me. My beloved wife, Anisha Mohapatra, and my son, Sriram Mishra, have been my pillars of strength with constant encouragement and understanding. Heartfelt thanks to my two elder sisters, Priyadarsana Mishra and Sanghamitra Mishra, for their unwavering support and guidance. Sincere gratitude extends to extended family members, Diptikanta Mishra, Deviprasad Mishra, Amrita Sarangi, Manas Kumar Mishra, and the younger members, Anrulipta, Mehul, Aditya, Archana, and Anshuman, whose collective support has played a crucial role in shaping my journey. Lastly, I express my heartfelt devotion to my guide Prof. Santanu Kumar Behera and Lord Jagannath with the spiritual invocation, "Oam Namah Shibaya," as an integral part of his divine presence throughout this literary endeavor.

Durga Prasad Mishra

Introduction

The twenty-first century has witnessed an unprecedented surge in technological advancements, reshaping the landscape of various industries and, in many ways, revolutionizing the way we interact with the world. Among these groundbreaking technologies, RFID has emerged as a transformative force, enabling the seamless exchange of information and data capture across a spectrum of applications. RFID, at its core, utilizes radio waves to identify and track objects, individuals, and assets, offering unparalleled efficiency, accuracy, and convenience. While traditional RFID systems, which are reliant on integrated circuits or chips, have enjoyed considerable success and widespread adoption, a new horizon beckons—the realm of chipless RFID printing technologies.

The journey into chipless RFID, characterized by its departure from conventional chip-based solutions, promises an array of benefits that hold the potential to reshape industries and create innovative applications. In this comprehensive exploration, we delve into this burgeoning field, unraveling the myriad facets, challenges, and opportunities it presents. Through the lens of 10 distinctive chapters, this book embarks on a journey that encompasses RFID technology's evolution, the significance of chipless RFID, smart materials, and antennas, as well as the implications for applications spanning healthcare, logistics, and environmental monitoring.

Chapter 1: RFID Revolution Unveiled

Our odyssey begins with this chapter serving as a foundational guide, introducing readers to the diverse universe of RFID technology. We embark on this journey by first understanding the core principles of RFID, a technology that relies on radio waves for automatic data capture and identification. These principles lay the groundwork for comprehending the fundamental differences between chip-based and chipless RFID systems. Chip-based RFID, with its reliance on integrated circuits, has been a mainstay in numerous industries, while chipless RFID has emerged as an exciting alternative. It offers a more cost-effective and flexible approach to RFID, reducing manufacturing expenses while enhancing scalability. We explore the radar cross section (RCS) of a chipless RFID tag and delve into the nuances of chipless RFID printing technology. The innovation in printing techniques offers a transformative approach, with the potential for large-scale deployment in various industries. As we close this chapter, a preview of the chapters ahead reveals a broader perspective on the transformative potential of chipless RFID technology.

Chapter 2: The Present and Beyond: Chipless RFID Research

Chapter 2 delves into the contemporary research and breakthroughs that define the chipless RFID landscape. We discover a broad spectrum of scholarly work that lays the foundation for understanding the current state of the art. This chapter's content is enriched by exploring traditional RFID systems and their uses across multiple industries. The distinction between active and passive RFID sensors becomes evident, impacting their suitability for diverse scenarios. We encounter the emergence of fully printable smart sensing materials and their profound impact on the flexibility and cost-effectiveness of RFID tag fabrication. The concept of multiple parameter sensing intrigues us, opening doors to enhanced data capture capabilities. In exploring the importance of reader antennas in RFID applications, we realize their critical role in ensuring efficient tag reading and data exchange. This chapter reaffirms that chipless RFID technology holds immense potential for innovation and scalability, transforming how we perceive and utilize RFID in a myriad of applications.

Chapter 3: Smart Materials: The Bedrock of Chipless RFID

Chapter 3 unravels the pivotal role of smart materials in the context of chipless RFID printing technologies. These materials, including conductive inks, nanomaterials, and functional polymers, possess the remarkable ability to sense and respond to environmental changes. Such characteristics make them ideal candidates for RFID sensing applications. We categorize these smart materials based on their ability to sense temperature, humidity, pH levels, gases, strain, and cracks. A spotlight on graphene unveils its exceptional properties and potential applications in RFID sensing and beyond. In summary, this chapter reveals that smart materials empower a new era of RFID sensing applications across diverse industries, encompassing healthcare, agriculture, environmental monitoring, and more.

Chapter 4: The Art of Characterization in Chipless RFID

Chapter 4 addresses the critical process of characterizing smart materials for printing chipless RFID tags. Precise material characterization is paramount for successful chipless RFID tag fabrication. We explore various characterization techniques such as X-ray diffraction, transmission electron microscopy, scanning electron microscopy, and ultraviolet (UV)–visible spectrophotometry, and how they contribute to understanding structural, optical, electrical, and microwave properties. These techniques empower researchers and engineers to optimize smart materials for specific sensing applications, further advancing RFID technology across various industries.

Chapter 5: Pioneering Chipless RFID in Biomedical Applications

Chapter 5 shines a spotlight on the transformative influence of chipless RFID technology in the healthcare sector. We investigate the design principles and geometries of passive chipless RFID tags, particularly those based on multisection split ring resonator (MSRR) and rectangular resonator. The chapter also delves into the importance of comparing simulated and measured RCS data, as well as analyzing current distribution patterns. These insights help in understanding the performance of chipless RFID tags and their potential for innovative biomedical applications.

Chapter 6: Wireless Body Area Networks and Chipless Sensors

Chapter 6 underscores the significance of chipless RFID technology in advancing personalized healthcare solutions. We begin with an introduction highlighting the role of passive printable chipless RFID tags in body area networks, which facilitate nonintrusive, efficient, and seamless data capture for healthcare monitoring. We explore chipless sensors based on modified spiral resonator, high bit density based on MSRR, and modified complementary split ring resonator (CSRR). The comparison of findings with existing literature validates the originality and contributions of chipless sensors in the context of body area network applications. Simulated and measured RCS data and current distribution patterns further enrich our understanding of chipless sensor performance in biomedical scenarios.

Chapter 7: Resonators and Signal Processing Techniques

Chapter 7 delves into the design and optimization of resonators to enhance tag sensitivity and reliability. Signal processing techniques for data extraction and tag authentication are also explored. The chapter categorizes various chipless sensors and investigates their diverse applications in industries such as healthcare, retail, logistics, and environmental monitoring. Signal processing techniques, including the convolutional encoder, Viterbi algorithm, windowing techniques, and wavelet transform, emerge as critical components in enhancing data accuracy and reliability in RFID sensing. Noise reduction algorithms further contribute to the accuracy of RFID data. This chapter reinforces the vital role of resonators and signal processing techniques in chipless RFID systems, empowering researchers and engineers to select the most suitable techniques for specific applications.

Chapter 8: Antennas for RFID Reader Applications and Printing Techniques

Chapter 8 delves into the significance of reader antennas in RFID reader systems. Reader antennas play a pivotal role in enabling efficient and reliable communication between RFID readers and tags. We gain insights into the basic architecture of an RFID reader and explore linearly polarized (LP) and circularly polarized (CP) antennas, which cater to various requirements and scenarios in RFID systems. Antenna arrays enhance signal strength, directionality, and coverage, ensuring robust performance in diverse environments. However, the chapter also highlights the limitations and challenges in RFID applications, guiding researchers and engineers in addressing these hurdles to further enhance RFID technology.

Chapter 9: Challenges in Chipless RFID Technologies and Future Research Directions

Chapter 9 addresses the challenges encountered in chipless RFID systems implementation and deployment. Hurdles related to material selection, tag readability, and signal processing are studied, leading to a visionary outlook on potential research areas. These include security enhancements, advanced manufacturing techniques, data encoding techniques, reader sensitivity optimization, extended reading range, energy-efficient tags, and improved manufacturing processes. Recognizing the limitations faced in the current state of chipless RFID, we acknowledge areas that require attention to overcome hurdles in commercial adoption. Addressing these challenges promises to pave the way for the seamless integration of chipless RFID systems in diverse applications, revolutionizing industries and enriching human experiences.

Chapter 10: Concluding Remarks

In our final chapter, we reflect on the captivating journey through the world of chipless RFID printing technologies. From understanding the fundamental principles to exploring cutting-edge applications, this book has unwrapped the versatility and significance of chipless RFID across a spectrum of industries. We acknowledge the transformative potential of this technology, which holds the power to change industries such as healthcare, logistics, and consumer electronics. However, the journey does not end here; it ushers in a new era of chipless RFID technology, sustained by the passion, curiosity, and dedication of researchers and inventors worldwide. By embracing challenges, seizing opportunities, and pushing the boundaries of knowledge and innovation, we can anticipate significant advancements in the field. The future of chipless RFID technology is limitless, encouraging us to embark on a collective journey of creativity and discovery, influencing the technology's future and driving positive global change.

In conclusion, the chapters of this book collectively chart the course of chipless RFID printing technologies, from its inception to its revolutionary applications. The significance of RFID technology is reaffirmed, and chipless RFID emerges as a cost-effective, scalable, and flexible alternative with immense potential. The critical roles of smart materials, antennas, and signal processing techniques are highlighted, showcasing their contributions to RFID innovation. The chapters discussing biomedical applications, body area networks, resonators, and printing techniques underline the versatility of chipless RFID across healthcare and various industries. The challenges in chipless RFID and future research directions serve as a call to action for researchers and industry stakeholders to overcome limitations and enhance RFID technology. As we close this book, the future of chipless RFID technology beckons, promising innovation, transformation, and positive global change.

CHAPTER 1

Introduction to Chipless RFID

Currently, radio frequency identification (RFID) has become in great demand for current research. Surprisingly, the development of this technology dates back to World War II, when military forces utilized it to identify hostile aircraft after realizing its potential. However, the optical bar code is the greatest opponent of RFID [1]. Due to their low-cost execution and advantages over RFID technologies, it became a great success and is predominant to date. Harry Stockman [1] first demonstrated wireless communication with the help of reflected power in 1948 and filed a patent for the passive tag in 1950. However, from 1970 onward, due to the volume of business and the increase in complications, the requirement for new technology came to the limelight, and the journey of RFID started. The actual research in RFID technology was enhanced with the installation of an auto ID center at the Massachusetts Institute of Technology in Cambridge, Massachusetts, in 1999.

RFID is an emanate automatic identification (auto-ID) technology that is used to detect and monitor products or different items with the help of low-power wireless data transmission [1]. It is a contact-free data-capturing technique that usually receives radio frequency (RF) signals for its operation. This technique has a wide range of applications in healthcare, supply chain management, logistics, community service, transportation, counterfeit notes, the Internet of Things (IoT), and smart cities [2, 3]. Many times these systems are not visible to the outside world. The RFID system generally consists of a reader (or interrogator) and a tag (or transponder). The reader is an important block that consists of middleware and hardware. The middleware deals with a control unit, whereas the hardware consists of the reader circuits and the antenna. The reader transfers the electromagnetic energy to the passive tag. It sends and catches the RF signal from and to the tag. The tag is a data-capturing unit that can be placed on the items or products to be tracked. In a special case, it can be possible to measure many tags at a time and receive their information. RFID systems are of two types: (1) near-field RFID, and (2) far-field RFID. A schematic of the RFID system is displayed in Figure 1.1. The first category of RFID is formed with Faraday's magnetic induction technique. In this case, both the tag and reader are made up of metallic coils. The chip-based tag takes energy from the reader, and it can be rewritten. This technology is used for short distances (i.e., $\lambda/2\pi$ for tracking purposes, where λ is the wavelength, which is typically a few centimeters long). These tags are used only for smart cards and access control due to range limitations [4].

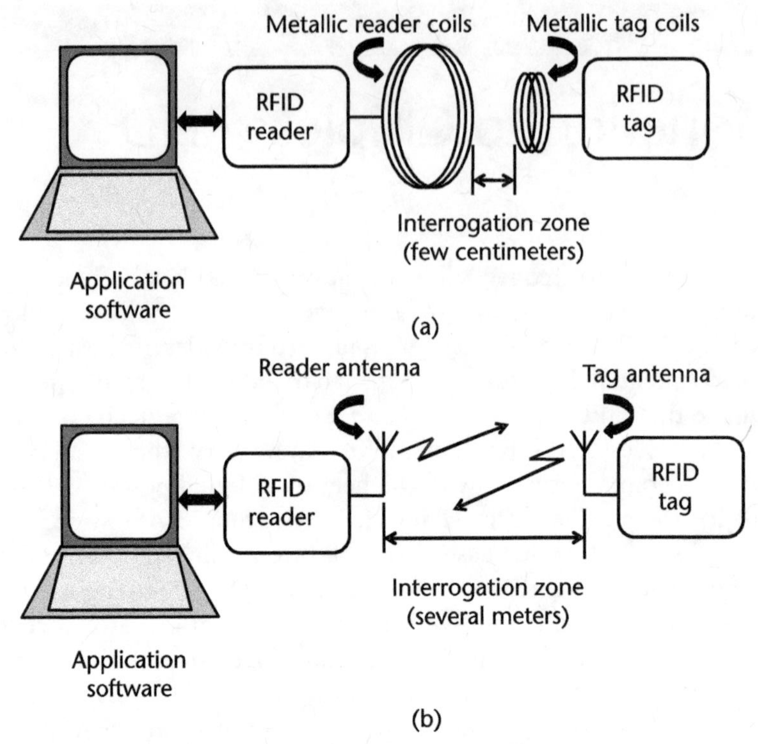

Figure 1.1 Basic RFID system: (a) near-field RFID, and (b) far-field RFID.

In the case of a far-field RFID system, the transponder receives electromagnetic (EM) energy from the reader antenna. The reader may be used in the monostatic or bistatic mode of operation using one or two antennas for transmitting and receiving the signals to and from the tag. The reader transmits the EM waves to the tag and also receives the backscattered signal from the tag. The distance between the reader and the transponder is known as the interrogation zone, which is a few meters long.

The tags in the far-field system may be categorized into three types: (1) active tag, (2) semipassive tag, and (3) passive tag [5, 6]. The active RFID tags operate with an onboard power supply to energize the application-specific integrated circuit (ASIC) chip to transfer data from the tagged items to the reader. The onboard battery makes the system bulky and needs continued maintenance. Therefore, these tags are more suitable for tracking large-size products over a long distance. These tags operate at various frequencies like 455 MHz (ultrahigh frequency (UHF)), 2.4 or 5.8 GHz (industrial, scientific, and medical (ISM)), or 8–12 GHz (X-band), based on the application and the read range requirements. The read range of these tags is around 20–100 meters [3, 4]. The semipassive tags have the power supply to operate the IC and receive EM signals from the interrogation zone to generate information. Similarly, the passive tags do not need any onboard battery. Therefore, it depends on the interrogation signal to backscatter the captured signal to the reader. During the absence of radio waves at the interrogator, the tag is treated as dead [3, 4]. Passive tags are cheaper than active ones due to the absence of ASIC chips and onboard batteries. It can be easily integrated into common materials and items based on the latest technology. These tags are compact in size apart from the cost, and the present antenna technology helps the tag for further miniaturization. But the greater the tag sizes, the more the reading range. Further, the operating range can be improved by considering an interrogation wave with more power. A passive tag receives the EM energy from the reader with the help of the transponder's antenna. It stores the EM energy in an inductive-coupling capacitor [5]. Further, when the capacitor is charged sufficiently, it discharges to the tag's circuitry and as a result, the signal can be transmitted back to the reader [3–5].

1.1 Tracking Technologies

In the present era, auto-ID is a prevalent technology that is popularly used in logistics, industries, airports, e-health, e-commerce, and many more fields [5]. In these cases, error-free tracking of items and monitoring of patients in the hospitals are important tasks. Barcodes are well-known auto-ID systems. These are the cheapest identification systems available commercially. The greatest drawback of barcodes is having less storage capacity, line-of-sight (LOS) contact, and lack of reprogrammable facility.

The best possible solution for the storage of captured data is the smart card with an integrated circuit (IC) chip. But the smart cards need to touch the product for the operation of data transfer. This technique is only sometimes feasible for data tracking. A contactless non-line-of-sight data capturing technique is a radio frequency identification system. Recently, contactless identification has played a vital role in the interdisciplinary fields. It deals with various fields: RF technology, electromagnetics, semiconductor devices, cryptography, smart materials, and manufacturing technology areas. Various auto-ID techniques are outlined in Figure 1.2 for easy understanding of the research [2, 4].

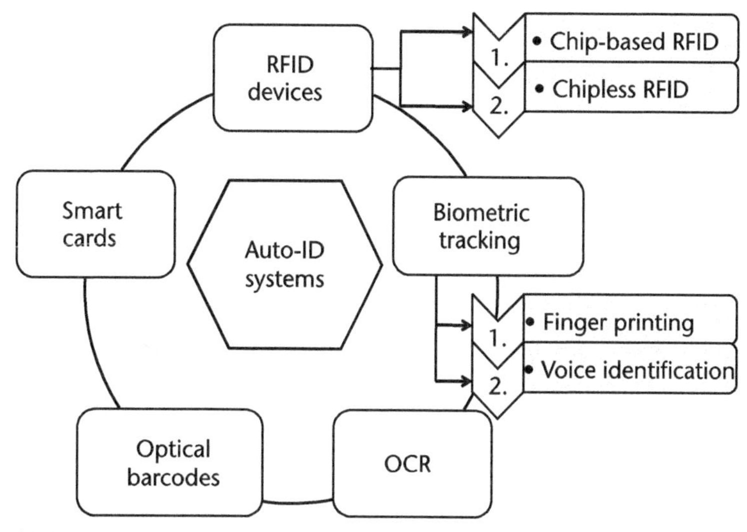

Figure 1.2 Various auto-ID systems.

1.1.1 Smart Cards

A smart card is just like a credit card, which has data storage capacity for automatic identification [4]. It can be placed in a reader unit that makes a dynamic link to the contact place of the smart card. The card gets EM energy from the reader through the contact surface. The data can be transferred from the reader to the card with the help of a bidirectional serial input/output port. The smart card is of two types: (1) memory card, and (2) microprocessor card. The greatest advantage of the smart card is that the stored data may be protected from unwanted access and interpretation. This card makes the tracking process easy, safe, and less costly. However, contact-based cards are prone to corrosion and dust. The devices are very costly for maintenance during any fault.

1.1.2 Biometric Tracking

Biometric is a counting and measurement technique for living beings. It is generally an identifying device that recognizes a human being by sensing its physical characteristics. Fingerprinting, voice identification, face recognition, and retina identification are the common procedures under this technique [2–5].

The fingerprinting technique is based on the verification of papillae and dermal ridges of the fingertips. The device counts the data recorded from the pattern and compares it with the recorded reference patterns. It takes a few seconds to recognize and verify a fingerprint in the present day. Most advanced fingerprint devices are developed to prohibit counterfeit cases and also to accurately and correctly detect the fingerprints of human beings [2–4].

Today, devices are also available to identify the human voice through speaker recognition. In this case, the person talks into a microphone, which must be connected to a personal computer. The sound signal is converted into digital data by the device and identified by the software installed in it. This is simply a speech recognition technique that can be compared with an existing reference speech pattern [2–4].

1.1.3 Optical Character Recognition

Optical character recognition (OCR) was invented for the application of a specialized character that can be read automatically by a machine and normally by the public. This device has high data storage capability and can read the data visually. The application of this device is in the areas of industries, banking sectors (checkbooks), and administrative fields. It is not widespread due to its high cost and the requirement of a complex reader system [2–4].

1.1.4 Barcode

The barcode system is a very old automatic identification system. It is basically a binary code consisting of a series of bars and spaces organized in a parallel sequence. The arrangement is prepared with a prefixed pattern and represents data in terms of symbols. The wide and narrow spaces in the barcode represent numerical and alphanumerical codes [2–4]. This barcode can be read with optical laser scanners. The details of the barcode are discussed in the next section.

1.1.5 RFID Devices

RFID devices are based on contactless technology. These are very close to barcodes and smart cards. The power supply to the chipless transponder/tag can be accomplished without any contact, instead using inductive coupling phenomena for near-field and EM waves for far-field propagation. This is not possible in the case of barcode and smart card sensing technology. These devices cannot be deployed in all the items used in daily life due to the price of tags, including ASIC, battery, antenna, and accessories. Therefore, barcodes are still available in the market with all consumer products, although RFID devices have many advantages over them [6, 7].

RFID technology relies on the backscattering EM wave by the tag, which is radiated by the interrogator antenna. The signal-capturing techniques of a reader antenna are well understood by the normal antenna parameters, but their scattering properties are more complicated. The RFID properties [8–10] may be more complicated than traditional radar models, which assume targets in free space and the far field of the transmit/receive antennas. These assumptions are normally not considered in real RFID systems. Further, almost all the techniques of RFID system design rely on the radar equation [11–13].

There are two main techniques related to the scattering principles of antennas. The first one comprises an EM operation of the backscattered signal relying on a full three-dimensional (3D) EM numerical technique. The second technique is to model the backscattering signal relying on an equivalent circuit through the coupling network, the reader, and transponder antenna ports. In 1960, this technique was introduced for electromagnetic field measurements with the help of the modulated scatterer technique (MST) [11–13].

The actual cost of the tag comes from the ASIC installed in it. This cost can be reduced by using a chipless tag where ASIC is absent, and it does not need any battery to energize the tag. Many applications use this technology due to the practical advantages in terms of wireless and batteryless features. Chipless RFID tags will be discussed in detail in Section 1.4.

1.2 Barcoding and RFID

Although barcodes are less expensive, they provide acceptable hindrances in terms of their small reading range and manual tracking. These restrictions increase the maintenance cost of a large industry by millions of dollars per annum [7, 14]. Today, much research is going on to replace the barcode with RFID tags with distinct ID codes for each item, which can be tracked from a long distance. Therefore, the difficulty of tracking range and automatic identification may be resolved with RFID tags. The price of a tag is the main reason for not replacing the barcode. A schematic diagram of the optical barcode and the chipless RFID tag is shown in Figure 1.3. A barcode may be realized by placing marks and spaces within nonconductive strips. It is portrayed in Figure 1.3(a). Similarly, a chipless RFID tag can be created by designing conductive and dielectric (substrate) strips of dissimilar dimensions

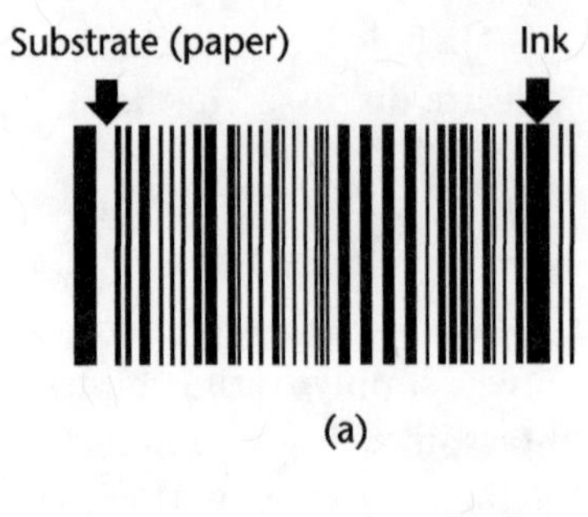

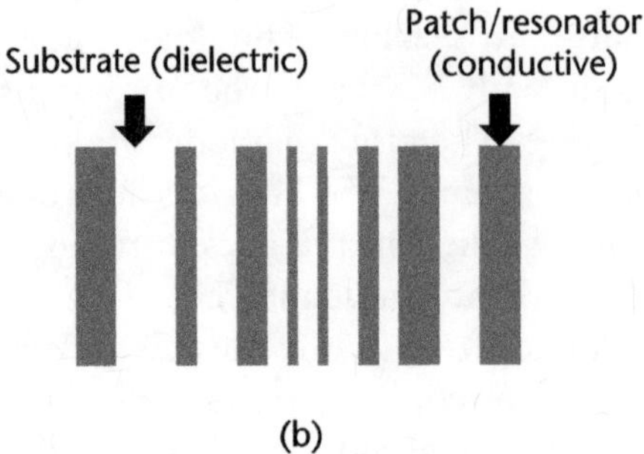

Figure 1.3 Development of tracking ID technology: (a) optical barcode label, and (b) chipless RFID tag.

near each other [7, 14]. The characteristics of barcodes and chipless RFID tags are detailed in Table 1.1.

1.3 Chipless RFID

The RFID tag can replace optical barcodes owing to its multifarious benefits (e.g., good read range capability, non-line-of-sight reading, huge data handling capability, and flexible operations). The cost of the tag restricts the technology from becoming a competition to the optical barcode. The chipless tag is less expensive in comparison to the barcode as it does not have any chip and takes EM energy from the reader for data encoding [5, 6]. Due to the absence of an ASIC chip, the microelectronic manufacturing technique is not required. Further, the complex connections between the antennas on the chip are also not needed. As a result, the tag becomes stronger and more low-cost. The future market for these tags can be anticipated to be the maximum [6, 7]. Many characteristics of chipless RFID tags are

Table 1.1 Characteristics of Auto-ID Systems

Auto-ID Systems	*Characteristics*
Barcodes	• Manual and line-of-sight tracking ID
	• Short tracking range
	• No security
Chipless RFID tags	• Automatic and non-line-of-sight read/tracking ID
	• Long-distance tracking range
	• Security feature available

superior to the chip-based ones; hence, these are very much preferred in the present day. The features of chipless RFID and chip-based RFID tags are listed in Table 1.2.

The read distance of a chipless RFID tag is nearly 3m or more. As the ASIC chip is absent in a chipless tag, the realization process can be adjusted with the manufacturing of tags. Generally, these tags can be manufactured on any product directly, lowering the cost of the tag below that of any optical barcode. Chipless tags identify the product when the EM wave intercepts the item. The signature can be analyzed to recognize the identity of the product.

1.3.1 Chipless RFID System

In the last few years, RFID tags have had excellent item tracing and identification features. Passive RFID tags are less costly than active tags, as they are batteryless and have a longer lifetime. Much research is going on to reduce the cost of tags very near to the cost of barcode. The accepted cost of a chipless tag is $0.01. The principle of operation of a chipless RFID device is shown in Figure 1.4.

The various features of chipless RFID tag categories are noted in Figure 1.5. The chipless tag can be divided into three categories. Category 1 is subdivided into retransmission type or time-domain reflectometry (TDR) type tags. Antennas required for these types should have large bandwidths and omnidirectional radiation patterns. Similarly, category 2 may be either a backscattering or millimeter-wave imaging type [7]. Category 3 can be designed as a special type tag with a near-field printed coil antenna. The features of the antennas used for categories 2 and 3 are narrow bandwidth, high Q-resonance, and unidirectional pattern.

1.4 Radar Cross Section Principle

The radar cross section (RCS) of any target may be associated with polarization of the incident wave, incidence angle and observation angle, shape of the target, operating frequency, and electrical properties of the target. The unit of RCS is m^2, and it can be expressed in decibels as dBsm.

1.4.1 RCS of a Tag

In a tag, the reflected power in the direction of the target can be calculated by its RCS. RCS of a chip-based tag can be given by [7]

Table 1.2 Chipless RFID and Chip-Based RFID Tags

Features	*Chipless RFID Tag*	*Chipped RFID Tag*
ASIC chip	Absent	Present
Tracking distance	Short (3m or more)	Long (20m or more)
Reader power transmission	Low (10 mW)	High (3–4W)
Cost	Less expensive (1 cent)	Expensive (10 cents)
Tag activation	Interrogation energy	Battery/interrogation energy
Tag antenna design	Complicated	Effortless
Physical structure	Flexible and printable	Not flexible and printable

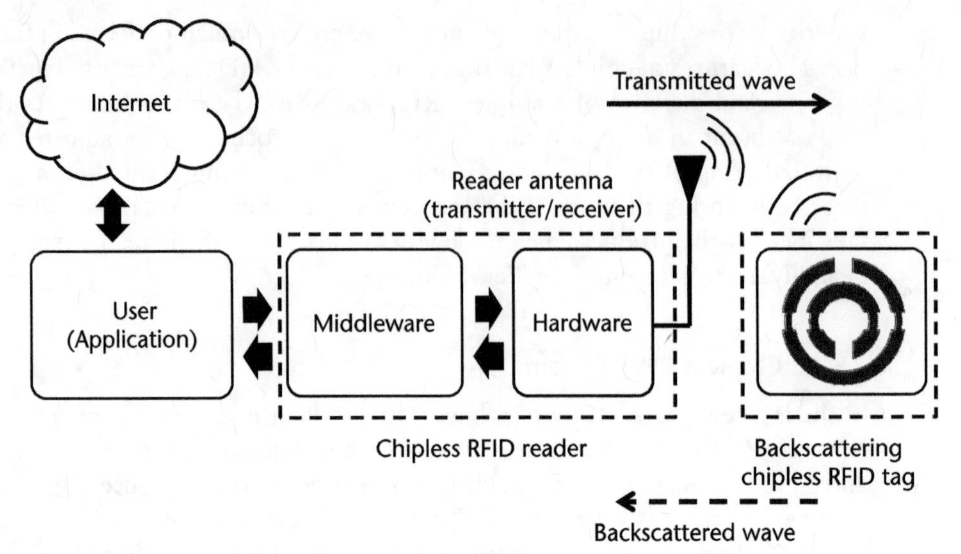

Figure 1.4 Principle of operation of a chipless RFID system.

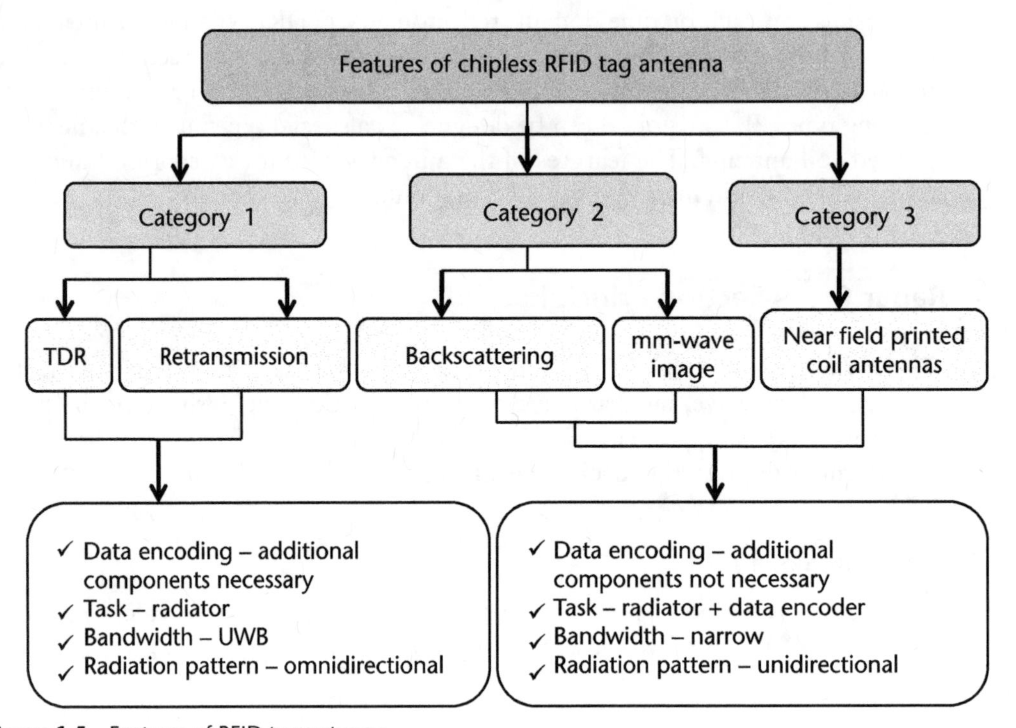

Figure 1.5 Features of RFID tag antenna.

$$\sigma_{tag} = \frac{\lambda^2 R_A G_{tag}^2}{\pi |Z_A + Z_C|} \tag{1.1}$$

where $R_e\{Z_A\}$; Z_A is the input impedance of antenna, Z_C is the impedance of a chip-based RFID tag, and G_{tag} is the gain of the tag.

In the case of a chipless tag, Z_C is assumed to be 50Ω due to the absence of a chip. At the frequency of resonance, (1.1) can be reduced to

$$\sigma_{tag} = \frac{\lambda^2 R_A G_{tag}^2}{\pi |2Z_A|} = \frac{\lambda^2 G_{tag}^2}{2\pi} \quad (\text{Since, } Z_A = R_A) \tag{1.2}$$

Therefore, the RCS of a chipless tag may be stated as the ratio of backscattering power per unit solid angle in the direction of the receiver to the power density that is captured by the target. For a spherical target, RCS may simply be defined as the product of three terms as given in (1.3).

$$\sigma = A|\Gamma|^2 D \tag{1.3}$$

where

$|\Gamma|^2$= the reflectivity;
D = directivity;
A = cross sectional area of the target.

Similarly, when the target is a patch antenna,

$$\sigma_{patch} = 4\pi W^2 h^2 \lambda^2 \tag{1.4}$$

where

W = the patch width;
h = the height of the patch;
λ = the wavelength.

Based on radar theory, additional design formulas are needed for the RCS of the antenna if the target is an antenna. There are two modes used for chipless RFID tags: (1) antenna mode RCS (σ_A), and (2) structural mode RCS (σ_S). An equivalent circuit for antenna mode RFID tag is shown in Figure 1.6.

The antenna mode RCS follows antenna theory for its design and analysis. It provides the designer with a command of the RCS performance of the chipless tag [15–18]. The antenna mode RCS relies on the reflected power initiated by load mismatch in the antenna. Similarly, the structural mode RCS (σ_S) depends on the antenna orientation and physical size. The chipless tag RCS can be given by (1.5) and (1.6).

$$\sigma = \lim_{r\to\infty} 4\pi r^2 \frac{\left|\vec{E_r}\right|^2}{\left|\vec{E_i}\right|^2} = \frac{\lambda^2}{4\pi} G^2 \left|\Gamma - A_s\right|^2 \tag{1.5}$$

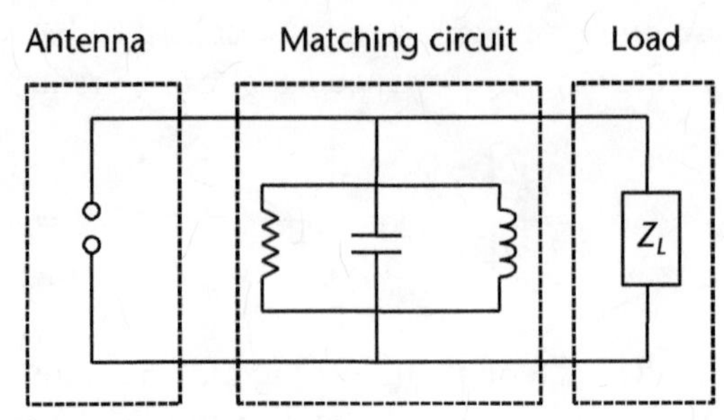

Figure 1.6 Chipless RFID tag in antenna mode.

$$\Gamma = \frac{Z_L - Z_A}{Z_L + Z_A} \tag{1.6}$$

where

σ = combination of structural and antenna mode RCS;
$\vec{E_i}$ = incident electric field intensity;
$\vec{E_r}$ = reflected electric field intensity;
Γ = reflection coefficient;
r = distance between the transponder and interrogator antenna;
A_s = a constant due to structural mode backscatter.

Under the matching condition (i.e., load impedance (Z_L) and antenna impedance (Z_A) are matched), Γ becomes zero. It is clear that no antenna mode component to backscatter waves produces the RCS (σ). If the mismatch occurs with a change in load impedance, the antenna mode reflected signal (Γ) will increase.

Further, as the structural mode component of the RCS is fixed, the variation in the RCS is directly proportional to the change in the antenna mode component.

1.5 Chipless RFID Printing Technology

At present, flexible electronic devices are in great demand due to the increasing market for miniaturized and lightweight products. In order to manufacture all these flexible products, namely RFID tags and readers, wireless sensors, and touch pads for keyboards with a low price need the idea of additive manufacturing techniques and proper materials. Chipless RFID tags are still costlier than optical barcodes. However, it is a difficult task to reduce the price of the tag through printing. The ability of chipless tags to get rid of the restrictions of barcodes has motivated broad scientific and commercial interest. These tags have non-line-of-sight reading and direct printing capabilities on packaging materials. Further, researchers have designed and fabricated chipless tags, creating huge prospects for analysis of printing techniques for cheap direct printing tags and sensors.

Currently the organic materials fabricated with conductive inks are patterned using several printing methods like inkjet, flexography, screen, thin film, and gravure. All these printing techniques have their own advantages and disadvantages. The latest developments in cellulose-based substrates (CBSs) like plastics, paper, and wood are flexible and renewable. These materials may be printed along with conductive polymer and efficient inks, and they can be recyclable.

1.5.1 Challenges for Chipless RFID Tag Printing

At present, researchers are trying to decrease the price of a tag below 10 cents ($0.10) so that many tags can be printed on the cheapest items in the commercial market [14–16]. Therefore, the challenges for chipless tag printing are elaborated in this section. The main challenge is the mass fabrication of direct printed tags with the proper substrate material. It is due to the frequency response of tags printed on various materials varying based on the substrates' EM behaviors. Furthermore, if the tag is placed inside the layers of a product, the resonant frequency can be computed considering the permittivity of both layers (substrate and superstrate).

1.5.2 Tag Printing Process

The tag printing process can be divided into three parts: (1) the effect of different parameters on tag printing, (2) printing techniques, and (3) errors in printing. The effects of different parameters can be ink conductivity, dielectric permittivity (ε_r), loss tangent (tan δ), and substrate materials [14–16]. An overview of the chipless RFID tag printing process is portrayed in Figure 1.7. Printing techniques like inkjet printing (paper, wood, brush painting), doctor blading, thermal printing, screen printing (3D printing), gravure, aerosol jet, and thin film (thin film transistor, organic thin film transistor, CBS), are discussed in detail in the subsequent sections. The flaws in printing due to shapes, edges, and the aging effect of tags are also discussed in later sections.

1.6 Applications of RFID Technology

Today RFID technology plays a vital role in the day-to-day life of human beings. RFID tags are implemented in every product that is available in the world. Tags are used in logistics, supply chain, manufacturing, transport, hospitals, railways, airports, and many other areas, as well as in document management, sports, pharmaceuticals, livestock, baggage handling, marketing, and access control [16–21].

Marketing and supply chains are the two highest application areas of this technology. In a manufacturing process, the raw materials collected from the suppliers may be tagged at the initial stage. The raw materials are tracked by the reader before passing through several marketing processes. This tracking process is performed with various sensors (pressure, temperature, humidity, etc.), and then the database records everything at each moment of the process. In this process, the quality of the product can be maintained.

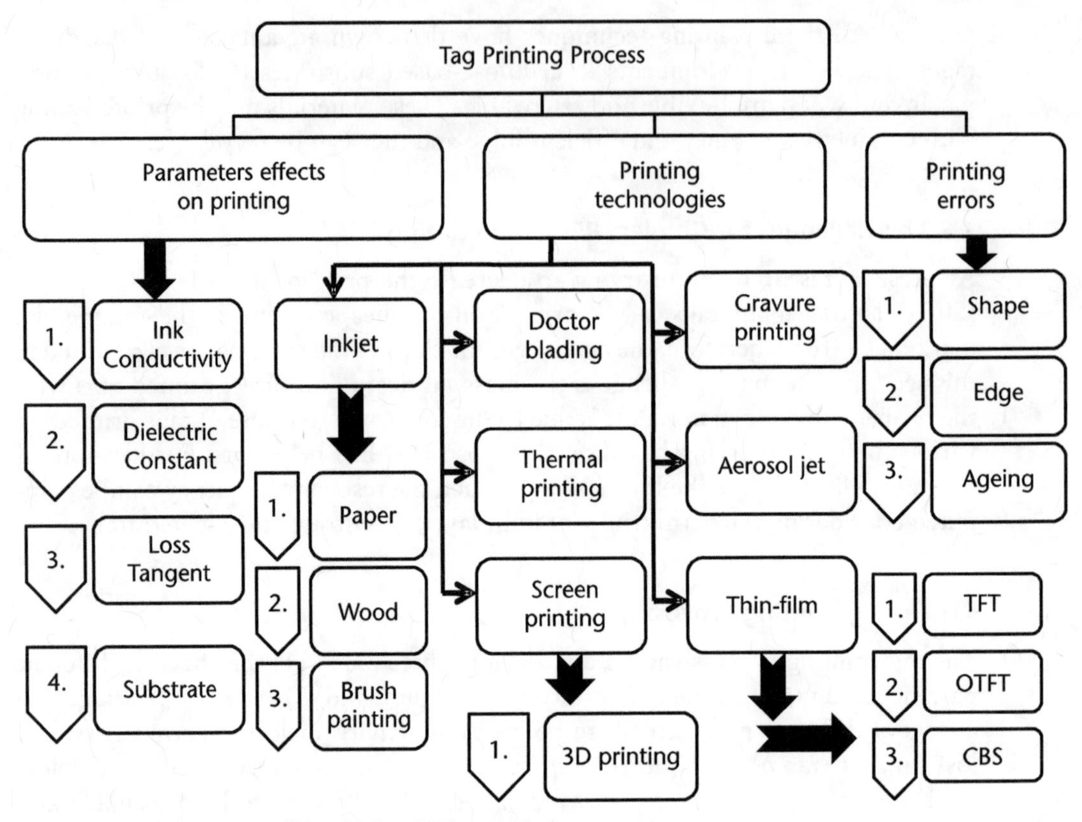

Figure 1.7 An overview of the chipless RFID tag printing process.

1.6.1 Supply Chain Management

Normally, the original equipment manufacturer (OEM) supplies its products to the market through airways or sea routes. At the OEM's end, the items are tagged with item level or packaging level with a distinctive identification code. At the airport or dock, the information about the date, time, and place are stored against the distinctive identification code in the database of the OEM and the supply chain personnel [16, 17]. The OEM sends the information to the supply chain personnel in advance related to the item's serial number of the product to be tagged. The information can be used to protect against loss or theft during transportation. At the retailer end, a similar process should be carried out for item level and packaging level. In this process, the information can be read in a particular interval so that any inconsistency can be reported.

1.6.2 Security

Personal identification and security are very important applications of RFID. Accessing office rooms, classrooms, laboratories, and hotel rooms in a building through RFID cards is a general use of this technology. In advanced banking systems, credit cards, American Express cards, and other payment cards are RFID tags. RFID tags are used in certain payment cards for automatic travel payment in mass-transit systems like the Smart Trip card for the Washington DC subway and bus terminal [16, 17]. These cards replace the magnetic stripes, allowing a much more reliable technique to store data since the magnetic strips are prone to wear out and lose data over time. The keys for a four-wheeler (car/taxi) also use a chipless tag, which works with a reader near the ignition switch. The interrogator only accepts the particular code stored in the keys. The car will not start if the code in a key does not match with the reader, which prevents theft.

1.6.3 Movement Tracking

RFID tags can keep track of the movement of people and related data associated with them. Modern hospitals utilize the tags to keep track of newborn babies to confirm identification and warn hospital staff of any unauthorized removal of a baby to the outside. In libraries, books are tagged in order to track them on shelves, protect against theft, and carry out automatic checkout procedures. In schools, children can wear tag-implemented bracelets or wristbands to monitor attendance and to detect lost pupils.

1.7 Chapter Overview

This section presents a brief summary of the chapters in this book.

This chapter presented an introduction to RFID systems, details of tracking technologies, barcodes, and RFID. Chipless RFID, radar cross section, chipless RFID systems, and printing technologies were discussed in brief. The important applications of RFID technology were presented, followed by a chapter overview.

Chapter 2 focuses on conventional RFID, active and passive RFID sensors, passive RFID resonators in chipless RFID (both frequency and time domain), and fully printable tags. Smart sensing materials, along with multiple parameter sensing and reader antennas for RFID applications, are also presented.

Chapter 3 discusses smart sensing materials and the classification of smart materials used for sensing. The details of temperature, humidity, pH, gas, strain and crack sensing, and light-sensing materials followed by graphene are also presented.

Chapter 4 discusses material characterization techniques followed by X-ray diffraction, Raman scattering, secondary ion mass spectrometer, transmission electron microscopy, scanning electron microscope, atomic force microscopy, and UV-visible spectrometers. The electrical properties of materials, such as conductivity, microwave-material interaction aspects, and smart material characterization, are also discussed briefly.

Chapter 5 details the design of the passive resonator-based tag, along with design steps, MSRR-based chipless sensor, and their results.

Chapter 6 presents the design of the modified spiral resonator-based chipless sensor, high bit density chipless sensor based on MSRR, and modified CSRR-based chipless sensor and their results.

Chapter 7 focuses on the comparison of chip-based and chipless sensors, categories of chipless sensors, and applications. Signal processing techniques in chipless RFID containing convolution encoders, Viterbi algorithm, windowing techniques, and wavelet transform are detailed. As well, noise reduction algorithms and results due to different signal processing techniques are explained.

Chapter 8 presents the importance of reader antennas in RFID systems, reader architecture, LP and CP antennas, and fractal antennas (Koch and Minkowski). Reader antennas for biomedical applications, antenna arrays, and limitations in RFID applications are addressed. As well, printing techniques and their classifications are presented.

Chapter 9 addresses the challenges and future research directions of printing techniques.

Chapter 10 focuses on the important findings of the design of sensors and highlights the future works for printing tags efficiently.

References

[1] Karmakar, N. C., *Handbook of Smart Antennas for RFID Systems,* Hoboken, NJ: Wiley, 2010.

[2] Finkenzeller, K., *RFID Handbook,* Second Edition, John Wiley & Sons, 2003.

[3] Preradovic, S., and N. C. Karmakar, "RFID Transponders—A Review," *2006 International Conference on Electrical and Computer Engineering,* 2006, pp. 96–99.

[4] Ramos, A., A. Lazaro, D. Girbau, and R. Villarino, *RFID and Wireless Sensors Using Ultra-Wideband Technology*, Netherlands: Elsevier Science, 2016.

[5] Weinstein, R., "RFID: A Technical Overview and Its Application to the Enterprise," *IT Professional*, Vol. 7, No. 3, May-June 2005, pp. 27–33.

[6] Behera, S. K., and N. C. Karmakar, "Wearable Chipless Radio-Frequency Identification Tags for Biomedical Applications: A Review [Antenna Applications Corner]," *IEEE Antennas and Propagation Magazine*, Vol. 62, No. 3, June 2020, pp. 94–104.

[7] Karmaker, N. C., "Tag, You're It Radar Cross Section of Chipless RFID Tags," *IEEE Microwave Magazine*, Vol. 17, No. 7, July 2016, pp. 64–74.

[8] Bolomey, J. C., S. Capdevila, L. Jofre, and J. Romeu, "Electromagnetic Modeling of RFID-Modulated Scattering Mechanism. Application to Tag Performance Evaluation," in *Proceedings of the IEEE*, Vol. 98, No. 9, September 2010, pp. 1555–1569.

[9] Hu, M., "On Measurements of Microwave E and H Field Distributions by Using Modulated Scattering Methods," *IRE Transactions on Microwave Theory and Techniques,* Vol. MTT-8, No. 3, May 1960, pp. 295–300.

[10] Harrington, R., "Small Resonant Scatterers and Their Use for Field Measurements," *IRE Transactions on Microwave Theory and Techniques,* Vol. MTT-10, No. 3, May 1962, pp. 165–174.

[11] Harrington, R., "Electromagnetic Scattering by Antennas," *IEEE Transactions on Antennas and Propagation,* Vol. AP-11, No. 5, September 1963, pp. 595–596.

[12] Harrington, R., "Theory of Loaded Scatterers," *Proceedings of the Institution of Electrical Engineers,* Vol. 111, No. 4, 1964, pp. 617–623.

[13] Bolomey, J.- C., and F. E. Gardiol, *Engineering Applications of the Modulated Scatterer Technique,* Norwood, MA: Artech House, 2001.

[14] Karmakar, N. C., E. M. Amin, and J. K. Saha, *Chipless RFID Sensors,* Hoboken, NJ: John Wiley & Sons, 2015.

[15] Subrahmannian, A., and S. K. Behera, "Chipless RFID: A Unique Technology for Mankind," *IEEE Journal of Radio Frequency Identification*, Vol. 6, 2022, pp. 151–163.

[16] Behera, S. K., and N. C. Karmakar, "Chipless RFID Printing Technologies: A State of the Art," *IEEE Microwave Magazine*, Vol. 22, No. 6, June 2021, pp. 64–81.

[17] Preradovic, S., and N. C. Karmakar, "Chipless RFID: Bar Code of the Future," *IEEE Microwave Magazine*, Vol. 11, No. 7, December 2010, pp. 87–97.

[18] Cook, B. S., et al., "RFID-Based Sensors for Zero-Power Autonomous Wireless Sensor Networks," *IEEE Sensors Journal,* Vol. 14, No. 8, August 2014, pp. 2419–2431.

[19] Dey, S., J. K. Saha, and N. C. Karmakar, "Smart Sensing: Chipless RFID Solutions for the Internet of Everything," *IEEE Microwave Magazine,* Vol. 16, No. 10, 2015, pp. 26–39.

[20] Aminul Islam, M., and N. C. Karmakar, "Real-World Implementation Challenges of a Novel Dual-Polarized Compact Printable Chipless RFID Tag," IEEE Transactions on Microwave Theory and Techniques, Vol. 63, No. 12, December 2015, pp. 4581–4591.

[21] Athauda, T., and N. Karmakar, "Chipped Versus Chipless RF Identification: A Comprehensive Review," *IEEE Microwave Magazine,* Vol. 20, No. 9, September 2019, pp. 47–57.

CHAPTER 2

Literature Review

2.1 Introduction to the Chipless RFID

The identification (ID) of an RFID tag can be found using either time-domain (TID) reflectometry-based or frequency-domain (FRD) spectral-signature approaches [1] [2]. TID-based tags fall into three categories: surface acoustic wave (SAW), thin-film (TF) transistor circuits based, and delay line (DL)-based tags. These tags backscatter the signal to the reader with varying delays thanks to resonators on the tag. Maximum frequency-domain tags have numerous resonators, each with its own coding capabilities. Frequency-based hybrid resonators (HR) tags, which combine two or more kinds of resonators, are frequently employed in chipless tags because they offer the additional advantage of accumulating more bits in a comparatively smaller footprint [1–3].

Due to its numerous uses in automatic identification, item tracking, security, healthcare, and authentication with the least amount of human intervention, chipless RFID (CHLRFID) has the capacity and ability to dominate the auto-ID market. Low-power EM broadcasts are used to gather data from a remote-located device. A transponder/tag and an RFID interrogator/reader make up the two main parts of a physical layer of RFID, as presented in Figure 2.1. The 3D area called the interrogation zone is where the reader antenna locates the tag using the backscattered wave that the tag has reflected. The reading distance is also known as the operating range during the detection process between the tag and reader antenna.

Meanwhile, the backscattered signal strength is very low to be detected by the reader; it is very difficult to retrieve the spectral sign and further process the information to improve the system performance. Effective signal processing and other coding methods must be implemented to do the aforementioned process. The signal that is sent as an interrogation from the reader is backscattered in different directions through the tag. The middleware understands the data before sending it in the appropriate format to the host computer. RFID has not been adopted widely despite its many benefits, including the need for direct LOS procedure, longer or farther tracking distances, and higher encoding capacities [1]. The CHLRFID tags' service abilities, like public, communication, financial, and biomedical applications, are shown in Figure 2.2.

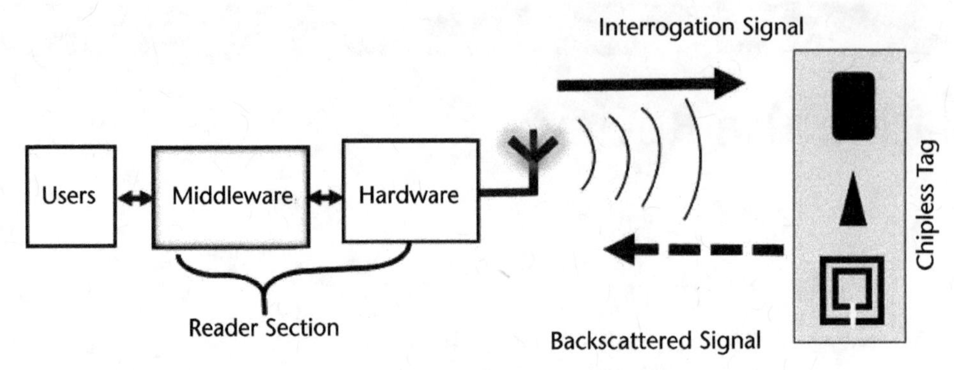

Figure 2.1 Block diagram with the architecture of the CHLRFID system.

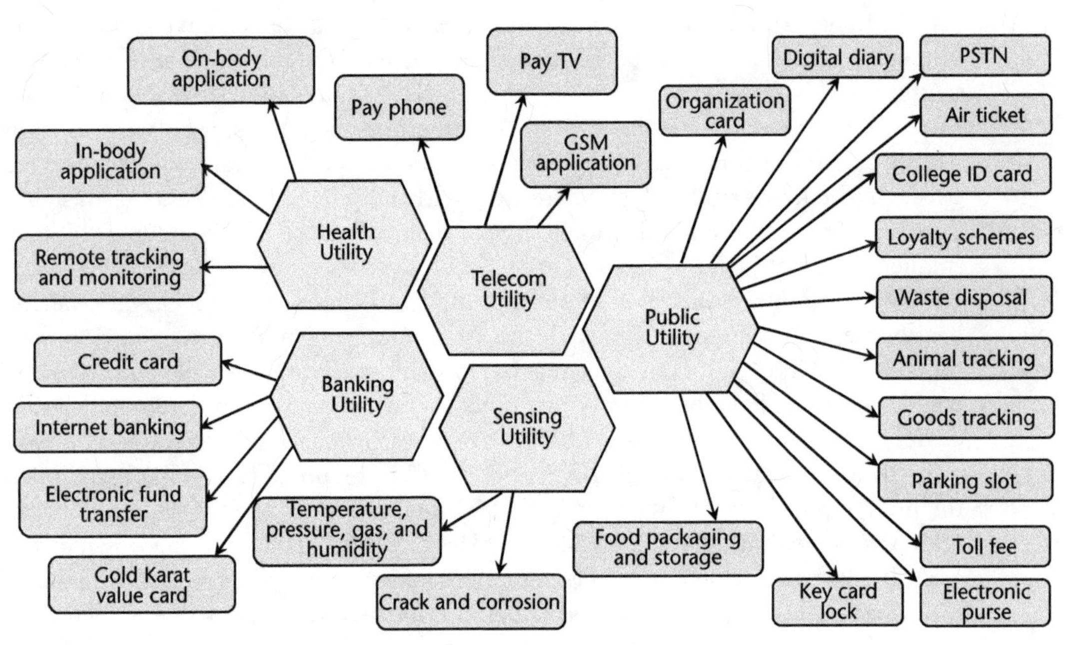

Figure 2.2 CHLRFID applications.

Due to their unique attributes, microwave resonators are exceptionally well suited for material characterization, offering real-time measurement capabilities. This is especially pertinent in challenging environments where sensitivity and selectivity are crucial in resonator-based sensing. In the development of a high-resolution RFID sensor, integrating a passive planar resonator into a tag becomes essential to optimize performance. This inclusion aligns with the goal of achieving noncontact sensing and enabling EM power emission into the surroundings, facilitating interactions with objects of interest. Despite their limited resolution and a moderate quality factor (Q-factor), passive resonators exhibit the ability to detect subtle fluctuations in highly sensitive measurements. Moreover, the utilization of resonators in chipless RFID tags helps minimize the physical footprint of both the tag and the reader. This reduction simultaneously amplifies the RCS and enhances sensitivity, showcasing the advantageous potential of resonators in the context of chipless RFID technology [1, 4].

The quality factor (Q) of a resonator is a critical parameter that characterizes the efficiency and performance of the resonant system. It is an indicator of the energy storage and dissipation characteristics of the resonator. The Q factor is influenced by various factors, including the resonator material, and it plays a crucial role in determining the behavior of resonators in chipless RFID technology. Here are some points that show how the resonator material can affect the Q factor:

- *Dielectric properties:* The dielectric properties of the material, such as permittivity and loss tangent, significantly impact the Q factor. Materials with lower loss tangents and higher permittivities often lead to higher Q factors. Lower loss tangents mean that the material dissipates less energy, contributing to a higher Q factor.
- *Conductivity:* The electrical conductivity of the material affects the loss in the resonator. Materials with higher conductivity tend to have higher losses, resulting in a lower Q factor. Conversely, materials with low conductivity, especially insulating materials, tend to have higher Q factors.
- *Homogeneity and purity:* The homogeneity and purity of the resonator material are crucial. Impurities or variations in composition can introduce additional losses, reducing the Q factor. High-purity, homogeneous materials tend to yield higher Q factors.
- *Loss mechanisms:* Different materials have various loss mechanisms, such as dielectric losses, magnetic losses, and surface losses. The dominant loss mechanism in a particular material will affect the Q factor of the resonator.
- *Temperature stability:* The temperature stability of the material is also important. Materials that exhibit stable dielectric properties over a wide temperature range can help maintain a consistent Q factor across different operating conditions.

The way the material is used or processed in the resonator design can also affect Q. Different materials may have optimal Q factors at different frequency ranges. The resonator material needs to be selected considering the desired operating frequency range to maximize Q. The resonator material significantly impacts the Q factor of the resonant structure in chipless RFID technology. Choosing an appropriate material with favorable dielectric properties, low conductivity, high homogeneity, and suitable loss mechanisms is crucial to achieve high Q factors, which ultimately enhance the performance and efficiency of the resonator in the RFID system.

In chipless RFID technology, the choice of frequency has a significant impact on the dimensions and complexity of the resonators and the overall system. Specifically, when using low frequencies, the dimensions of the resonators become relatively large, posing a challenge in terms of size and practicality. On the other hand, employing high frequencies makes the system more complex and expensive. At low frequencies, the wavelengths are longer, requiring larger resonator dimensions to achieve resonance. This can be impractical for applications where compact and small-sized RFID tags are preferred. The design and implementation of large resonators can also be technically challenging. Conversely, high-frequency RFID systems offer the advantage of smaller wavelengths, allowing for smaller resonator dimensions and, consequently, smaller tag sizes. However, this advantage comes with increased complexity and cost. High-frequency systems require precise engineering and specialized components, leading to a more intricate design and elevated production costs. Balancing the choice of frequency is essential in chipless RFID design. It involves considering factors such as desired tag size, system complexity, manufacturing costs, and the intended application. Achieving an optimal tradeoff between resonator dimensions, system complexity, and cost is a critical aspect of effective chipless RFID implementation. Balancing the choice of frequency and associated design considerations in chipless RFID involves a thoughtful and strategic approach. Here are steps that may help achieve a balance:

1. *Understand application requirements:* Begin by thoroughly understanding the specific requirements of the application. Consider factors such as tag size, reading range, environment, and data storage capacity.
2. *Evaluate frequency trade-offs:* Assess the tradeoffs between low and high frequencies. Low frequencies may require larger resonators but offer longer read ranges, while high frequencies enable smaller tag sizes but may have shorter read ranges and higher costs.
3. *Analyze size constraints:* Evaluate the available space and size constraints for integrating the RFID tag. Consider the dimensions of resonators at different frequencies and how well they fit within the intended application.
4. *Assess cost implications:* Consider the cost implications associated with choosing a particular frequency. Evaluate the costs of components, manufacturing, and system complexity for both low- and high-frequency options.
5. *Optimize for read range and sensitivity:* Balance the choice of frequency to optimize read range and sensitivity, considering the application's requirements. Aim for a frequency that provides an adequate read range while maintaining a reasonable tag size.
6. *Consider manufacturing feasibility:* Evaluate the feasibility of manufacturing resonators at different frequencies. Consider the ease of fabrication, material availability, and complexity of manufacturing processes.
7. *Model and simulate:* Use modeling and simulation tools to simulate the behavior of resonators at different frequencies. This can provide insights into the expected performance and behavior, aiding in the decision-making process.
8. *Iterative design process:* Adopt an iterative design process that involves prototyping and testing resonators at different frequencies. Learn from each iteration to fine-tune the design based on practical results.
9. *Engage with experts and suppliers:* Seek advice and guidance from RFID experts and suppliers specializing in chipless RFID technology. They can provide valuable insights and recommendations based on their experience and knowledge.
10. *Consider future scalability:* Anticipate future needs and scalability of the RFID solution. Choose a frequency that aligns with future advancements in technology and potential scaling requirements.

11. *Conduct cost-benefit analysis:* Perform a comprehensive cost-benefit analysis that considers all relevant factors, including size, read range, cost, and complexity. Use this analysis to make an informed decision.

By carefully evaluating these factors and taking a well-informed approach, one can strike a balance between frequency selection and design considerations, ultimately achieving an optimized chipless RFID system for your specific application.

The core idea and key information regarding the importance of the passive resonators (PR) in CHLRFID technology in this section are discussed, and resonators are the basic components of contemporary CHLRFID industries that improve the transponder's usability by enhancing its encoding capability, information storage capacity, and performance of detection in a variety of retail and healthcare domains. From the technological perspective, smart materials and metamaterials that make use of resonators are critical. This section looks at recent advancements in RFID resonators and the issues they invariably cause, like sensitivity, selectivity, interference, collision, and so on. Resonators are characterized as sensors with classification and attributes in Section 2.3. In Section 2.4, the use of resonators to enhance system performance using various research works and a comparison table is examined. The development of resonators into metamaterials is summarized in Section 2.5. Section 2.6 describes smart materials as sensors. In Section 2.7, a generic FRD reader is used to demonstrate the FRD tag detection. The implementational difficulties and future directions of research are discussed in Section 2.8, which is followed by Section 2.9's conclusion and the chapter's citations.

2.2 CHLRFID Resonators

Analogous/single resonance resonators operate at a unique resonant frequency (f_r), while hybrid/multiresonance resonators produce multiple resonant frequencies. This is how RFID resonators are structurally classified and are presented in Figure 2.3. Multiple resonances depend heavily on the resonator's geometry [1, 2].

2.2.1 Half- and Quarter-Wavelength Microstrip Resonators

A resonator having a half-length of a wavelength by open circuits at both ends is a half- and quarter-wavelength resonator. The essential resonance frequency (f_0) is determined through (2.1).

$$f_0 = \frac{c}{2l\sqrt{\epsilon_{eff}}} \tag{2.1}$$

where l is the length of the microstrip line, ε_{eff} is the microstrip's effective dielectric constant, and c is the light speed. This resonance occurs for $n = 2$, 3, and so on at $f = nf_0$. A quarter-wavelength stub with an open circuit end and a grounded end is referred to as a quarter-wavelength resonator.

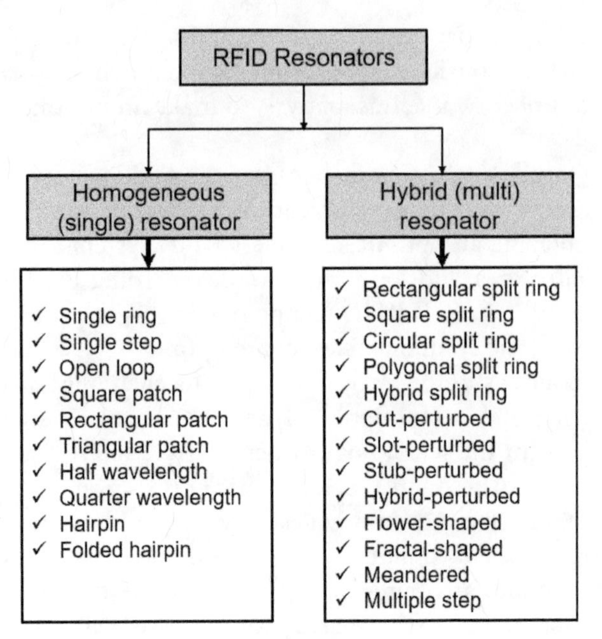

Figure 2.3 The structural classification of CHRFID resonators (frequency domain).

$$f_0 = \frac{c}{4l\sqrt{\epsilon_{eff}}} \tag{2.2}$$

For $n = 2, 3$, and so on, this resonator resonates at $f = (2n-1)f_0$ for $n = 2, 3, \ldots, n$.

2.2.2 Patch Resonators

Patch resonators play a crucial role in chipless RFID technology. These resonant structures, known for their compact size and versatility, are employed to encode data in chipless RFID tags. By altering the dimensions and layout of the patch resonators, unique frequency responses can be generated, allowing for data encoding. The resonant frequencies of the patches serve as distinct identifiers, akin to the functionality of traditional microchips in RFID. Their compactness makes them ideal for integration into various objects, packaging, or labels, offering a promising avenue for innovative and discreet RFID applications. Using a resonator, such as a patch resonator, is chosen to enhance the filter's ability to handle power due to its efficient and compact design. Resonators, like patch resonators, can effectively tune and control the filter response, optimizing power handling capabilities within a limited physical footprint. The patch resonator typically comes in several shapes, like rectangular, triangular, hexagonal, and square configurations (Figure 2.4). Due to its widespread applicability in CHLRFID applications as a crucial component to achieving the necessary RCS, it has attained the highest level of popularity [1].

Planar microwave sensors are preferred when related to other microwave-cavity resonators because of their compact dimension, easy fabrication, low profile, and lightweight features [5, 6]. Half-wavelength (open-ended) resonators are transformed

Figure 2.4 Hexagonal, rectangular, triangular, and circular-structure-based resonator.

into split-ring-resonators (SRR), which are broadly used as detecting apparatuses because of their excellent quality factor and compact dimension [1, 7, 8].

2.2.3 Ring Resonator

A single guided-wavelength communication line in a closed ring is a ring resonator (Figure 2.5). For $n = 2$, 3, and so on, its peak resonant modes also exist at $f = nf_0$. For simplicity of coupling in the filter design, the shape of the ring can be curved into different shapes like triangular, hexagonal, or square form [9, 10]. In addition, a split structure created by cutting the ring (such as a rectangular split ring (SR), square SR, circular SR, etc.) produces many resonances with small diameters.

2.2.4 Hairpin Resonators

The half-wavelength resonator (HWR) can be folded into another structure, like a hairpin, folded hairpin, or meandering structure, to increase its compactness. A U-structured, OC half-wavelength (HW) microstrip transmission line is referred to as a hairpin resonator (Figure 2.6(a)). This resonator resonates at frequencies wherever its physical dimension (of size 2l) is an integral numerous of the guided wavelength (λ_g). The primary frequency is provided by (2.1). The resonators are used to ensure adequate coupling when creating filters. The Butterworth and Chebyshev filter structures are extremely adaptable to a square open-loop resonator [1, 11]. The HWR can be bent into a variety of shapes, including a hairpin, a folded hairpin, an open shape, and a meandering shape, to decrease the dimension of the transponder. An HW microstrip open-transmission line that has been twisted into a U shape is a hairpin resonator (Figure 2.6(a)). This resonator can operate at frequencies when its dimension (2l) is an integer several of half the guided wavelength λ_g. The goal of hairpin resonators is to make it simple to connect the resonators for the filter structure. A hairpin open-loop resonator can be used with the Butterworth and Chebyshev filter designs with exceptional ease [1, 11].

Figure 2.5 Common resonator types: hexagonal-shaped, triangular-shaped, rectangular-shaped, and ring-shaped.

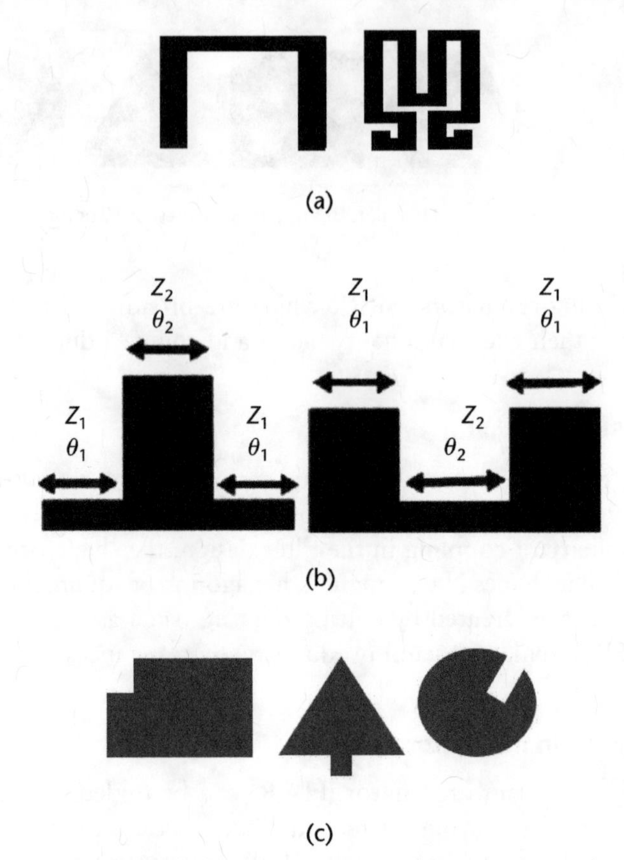

Figure 2.6 (a) Hairpin and rounded hairpin resonators, (b) stepped impedance resonators, and (c) cut, stub, and slot perturbations in resonators.

2.2.5 Stepped Impedance Resonator

Multiple microstrip line parts by variations in impedance magnitudes (Figure 2.6(b)) can be framed as a stepped impedance resonator (SIR) [12]. This resonator is less than an HWR in the form of a physical footprint and also operates at an identical frequency and may be determined as:

$$\tan\theta_1 \tan\theta_2 = \frac{Z_2}{Z_1} \tag{2.3}$$

The two microstrip lines' electrical lengths, represented by the letters θ_1 and θ_2, along with their corresponding characteristic impedances, are Z_1 and Z_2, respectively. For different integer values of n (2, 3, etc.), the resonances occur at $f = nf_0$. To create a small structure and make coupling easier with another resonator in the filter design, the resonator is rounded into an open-loop structure or other structures. The spurious response can be managed by properly combining several stepped impedance resonators [12].

2.2.6 Resonators with Perturbation

Multiple resonances can result from the perturbation. It can be accomplished with the help of adding several patches with a consistent geometrical shape or by creating empty spaces (slots) or cuts (Figure 2.6(c)). They may be of the cut, stub, or slot varieties. The four fundamental types of the dual-mode resonator (DMR) shapes are (i) radiator resonators by stub perturbations (SP), (ii) radiator resonators by cuts, (iii) rings by SP, and (iv) rings by a notch. The use of a ring of arrows resonator or a meandering line resonator by stub perturbation and a fractal-geometry-based resonator (FSR) can reduce the size of the resonator [13–15]. Dual-band bandpass filters are made by stacking two ring resonators with SP stacking like one on another in a multilayer configuration. A DMR that uses a combination of right and left transmission line with two sets of etched holes alters the resonator's fundamental mode frequencies as well [16]. To realize the innovative triple-mode resonator, multiple branch lines are connected along with a ring resonator analyzed in [17]. To display different resonant frequencies, capacitive stubs can be placed into the dual-mode resonator. The following are several different ways to introduce the capacitive stubs: To the circular radiator center is connected at the lowest part, which is centred on a metal, as well as via the radial-line stubs (RLS). The five-part stepped impedance resonator proposed [19] employs a quadruple-mode resonator. Each segment is composed of quarter-wavelength microstrip wires through both low and high impedances.

The numerous resonances provide numerous bits by enhancing the storage capability of CHLRFID tags, which is an astonishing property in detecting applications. The use of resonators as sensors is discussed in the following sections.

2.3 RFID Sensor Classifications

The CHLRFID sensor (Figure 2.7) does not require an active component, such as an embedded chipset, to track and collect sensing data from a far-off object. The chipless sensor is significantly cheaper because there is no integrated chip present. The tag's cost performs a key role in deciding the overall price of the RFID system. The chipless sensor (CS) also has a number of benefits over chip-based sensors due to its durability, adaptability, affordability, and low power consumption [20, 21].

An interrupt signal is sent to the CS by the reader. Both the ID information and the detecting outputs are backscattered by the sensor tag [22]. There are two groups of resonators on the tag, out of which one is only used for ID answers and another for detecting the information. Physical properties like humidity, blood pH, blood pressure, body temperature, and glucose level are translated into EM characteristics via chipless sensors [3]. The Q factor, relative permittivity (ε_r), loss tangent (tan δ), and conductivity of the sensing material, as well as variations in the resonator's electrical and physical footprint, and other factors, also cause variations in the EM characteristics. Replicated electromagnetic waves provide a frequency signature that carries sensory information about the surroundings [23]. According to the method used to encode data, CS can be categorized into three groups: FM, TDR, and phase modulation (PM)-based chipless sensors [3, 21].

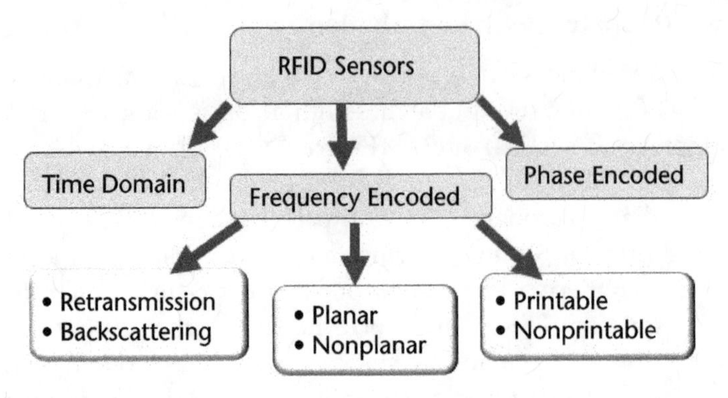

Figure 2.7 CHLRFID sensor types.

TDR-based chipless RFID sensors offer several advantages, including high resolution, the capability to sense various physical parameters (e.g., moisture, temperature, and pressure), and the potential for integration into a wide range of applications such as smart packaging, healthcare, environmental monitoring, and industrial processes. The tag reacts by a sequence of pulses which contain sensor information when it receives a UWB signal from an interrogator. Hybrid TDR tags feature greater storage capacity and greater tracking ranges. The information is encoded into the frequency domain in the mode of sensors based on frequency modulation. This family of chipless sensors that encrypt data using resonant structures has the highest coding capacity. They are also smaller and more reasonably priced [24, 25]. Detection is significantly impacted by background noise, however. The third class of chipless sensors that have problems due to bandwidth limitations are phase-encoded tags [3]. The section focuses on FRD-based tags, several kinds of resonators, and use in CHLRFID classifications from a different collection of works described in the following section.

2.4 Resonators in Chipless RFID System Performance Improvement

SIRs possess a unique ability to independently and automatically adjust their main harmonic frequencies. This adjustment is achieved by changing two specific factors: the length ratio (α), which represents the proportions of the resonator, and the impedance ratio (K), which involves modifying the ratio of resistive and inductive components within the resonator. Essentially, these resonators can adapt and fine-tune their frequencies, which makes them incredibly flexible and efficient in various applications. The capability to vary the frequencies by varying either the length ratio (α) or impedance ratio (K) is the key benefit of SIR over other resonators. By using the resonator's important frequency approach, the majority of tags are encrypted with information. As an outcome, the current frequency range is constrained to the lesser resonant frequency and its likely harmonic. The three–bit chipless tag proposed in [26] has worked on the SIR of Substrate-Rogers 4350. The impedance ratio (IR) can be used to alter the dimension and harmonic frequency. The SIR-based chipless tags (RT/duroid, $\varepsilon_r = 2.2$ and $\tan\delta = 0.0009$, 8 – bit, and C-MET LK, $\varepsilon_r = 4.3$, and $\tan\delta = 0.0018$, 2 – bit tag) are capable to code 2^{2N}

number of bits with N resonators [27]. The integration of DBR in CHLRFID tags is proposed in [28]. The operational bandwidth improvement is achieved using a four-resonator tag. Background subtraction and time-gating techniques are used to improve detection and improve the read range. A compact CHLRFID tag ($\varepsilon_r = 4.4$ and tan $\delta = 0.0018$, tag dimension: $80 \times 60 \times 1.6$ mm^3) taking a coding size of 8 bits utilizing open stubs is designed to work in 2–4 GHz [29].

Using hybrid resonators, which combine several resonator types, is another option to increase encoding capability. This method can minimize dimension and enhance the RCS response. Compactness, lightweight, multifunction, wideband or multiband, optimization of all system expenses, many adjustable tunable ranges, flexibility, low loss, and finally, strong radiator performance are just a few of the numerous advantages of HRs. Orientation-insensitive resonant structures are of great significance in the field of chipless RFID. These structures, which come in diverse shapes like floral, triangular, strip-coupled lines, leaf-shaped, hairpin, and spiral, are designed to have resonant responses that are not affected by their orientation in space. This is crucial in RFID applications, as tags can be oriented differently during reading. For instance, a tag with a floral or spiral shape will resonate at the same frequency whether it is placed horizontally, vertically, or at any angle in between. Such orientation insensitivity ensures a reliable and consistent reading of the tag regardless of its position, enhancing the overall efficiency and robustness of chipless RFID systems. It simplifies tag placement and usage, making chipless RFID technology more practical and user friendly across various applications. Various orientation-insensitive resonant structures, including those with a floral shape (Figure 2.8), a triangular shape, a strip-coupled line, a leaf shape, a hairpin, and a spiral, are discussed, and their RCS is analyzed [30]. There is a peak RCS response of 11 dB. The resonant geometries are put into practice on a Taconic TLX-8 substrate material, which operates between 4 and 4.4 GHz.

In addition to the resonator form, other variables that affect how effectively the resonator functions include the characteristics of the substrate and the properties of the conductor material. These resonators produce a succession of notches in the frequency response range of the tag. The tag ID is encoded using these peaks and notches. The FDSS-based CHLRFID tag's coding uses the deepness of a peak, which is defined by the tag's RCS, to represent a bit. Contingent to the deeper peak of the RCS, the tag may more effectively show the required bit, and the reader may be able to identify the bit more rapidly.

Another novel alternative to increase the robustness of CHLRFID systems is the development of cross-polar and orientation-insensitive CHLRFID tags. The orientation-insensitive tag is especially helpful when the tag's orientation is unsure during measurement. This kind of tag has a lot of potential for applications that include item-level tagging. Basically, two varieties of frequency-based CHLRFID tags are copolar and cross-polar tags. The polarization of the emitted wave can be altered by changing the geometry of the cross-polar resonators. In a cross-polar tag structure, the transmitter and receiver radiators are essential and, therefore, have perpendicular polarizations. Stub and slot resonators can vary the polarization of the transmitted wave and are effective for numerous RCS peaks. Platform-tolerant characteristics are demonstrated via a hybrid resonator idea that combines two three-step impedance parts of shorted radiating elements with a slot [31].

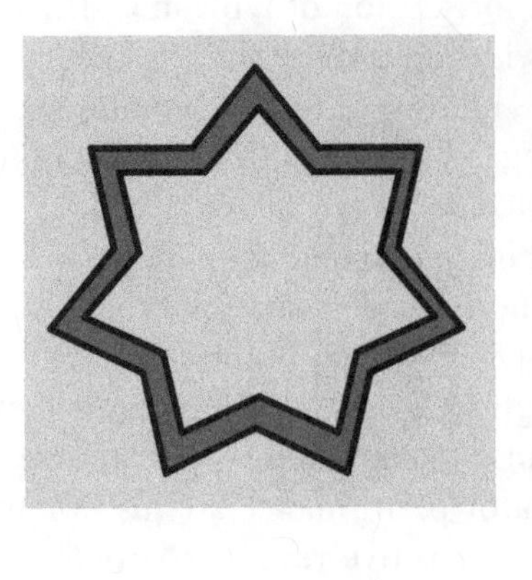

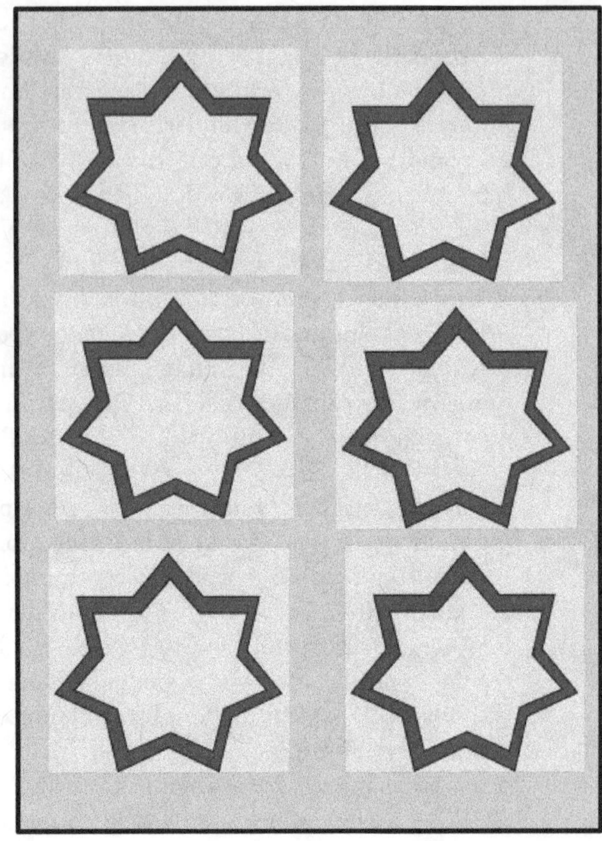

Figure 2.8 Cross-polarized tag (flowers and lines).

With polarization analysis, several square, circular, and octagonal CHLRFID tags were designed. Identification of the tag is normally exciting due to less intensity of the reflected wave; hence, various signal dispensation methods consisting of short-time Fourier transform (STFT), windowing, and interpolation can be utilized at the receiver [32]. These tags are designed on an FR4 material ($\varepsilon_r = 4.4$ and tan $\delta = 0.003$). To achieve the needs of varied applications, it is quite difficult to attain reasonable RCS and considerable bit capacity within the appropriate band of frequencies. HR-based tags can be designed to get around these limitations. They play a crucial role in offering a multibit encoding miniature tag structure [33, 34].

Loss tangent, dielectric constant, and substrate width are three crucial attributes of a substrate. Substrate loss accounts for the RCS level of the tag. Figure 2.9 shows how the depth and actual permittivity of the substrate are responsible for the frequency shift. A flower-geometry-based resonator (Figure 2.8) with a fixed size was designed on a Taconic substrate through an effective permittivity of 2.55 and a tan δ of 0.0019 in order to explore the thickness effect. A frequency shift to the lower frequency band is observed due to the growth in the dielectric constant and the width of the substrates. The tag size of the low-frequency resonators is larger. However, expensive thick substrate material might not be optimal for chipless RFID item labeling [30].

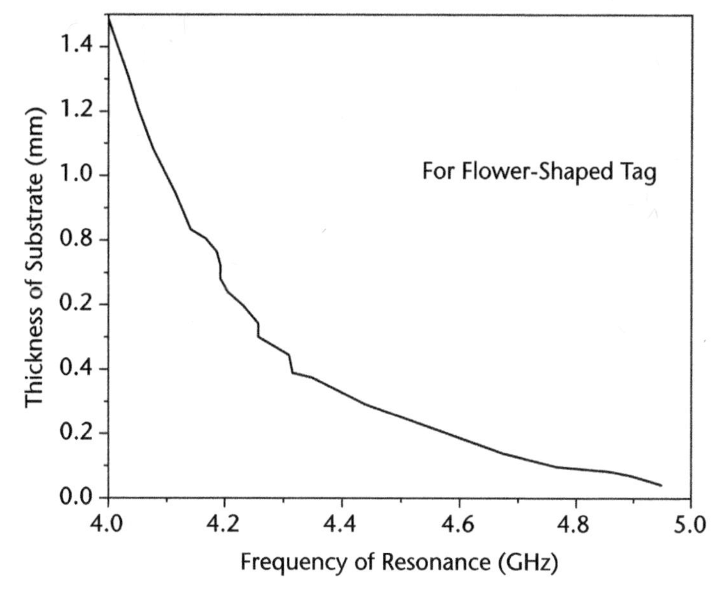

Figure 2.9 Substrate thickness and dielectric constant vs frequency.

The affordable, versatile, and easily accessible substrates for fabricated tags applicable in industrial settings are reedy plastic and paper. The printed version should react similarly because taconic TLX-8 has a dielectric material like that of polyethylene terephthalate (PET) and paper. RCS behavior is affected by the conductive ink's quality. In [3, 30, 35], it is examined how the properties of the resonator substrates and the ink used in the process of printing disturb the performance of resonance.

For minimal sensitivity to orientation, the unit cell with a flower geometry and compact dimension executes superiorly. A design for implanting a chipless RFID tag using quick-response (QR) codes is described and displayed [36]. The chipless RFID tag used to create the QR code is indicated as supporting a minimum of nine distinct resonant frequencies for elementary FSK coding, giving multiple-bit coding capability. The resonator in the QR code will be formed by keeping the modules for tag metallization on copper. Conversely, the gray modules in the QR code printed using other nonconductive materials [36]. If one module is metalized and taken for operation to improve the RCS and coding capacity, then it is called loading. The objective is to choose the correct modules with metallization (by proper loading of the resonator module) to improve the RCS as well as coding capacity. The dominant resonant frequency of 1.246 GHz, as displayed in Figure 2.10, has a strong peak with a 3-dB highest width of roughly 10 MHz, which is sharper than the ideal ring resonator. The sharpness of the peak determines how much information can be encoded using FSK coding. The supplement image of Figure 2.10 reveals a further noteworthy peak at a resonant frequency of 3.01 GHz.

By lowering the density of loading for RFID coding, the f_r in the frequency range from 1.24 to 1.790 GHz will be modified. By altering the loading density, the resonance frequency can be tuned to fall from 1.246 to 3.0 GHz. Nine separate resonant frequency peaks are obtained here using FSK coding, spread out spanning the coding frequency range covering from 1.2 to 1.9 GHz [36]. It is challenging to

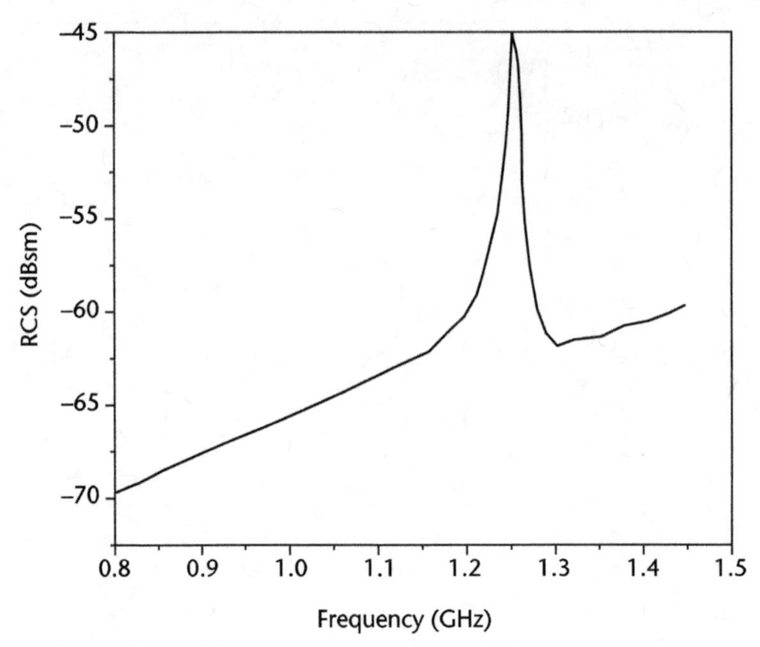

Figure 2.10 Simulated RCS of the resonator.

create an only tag with many detecting abilities, including pH, pressure, moisture, and temperature. This kind of problem can be lessened by utilizing the multisensing ability of metamaterials, which can notice a range of physical elements from the atmosphere and is enclosed in the next section. Diverse resonator types with different properties and applications are described in Table 2.1.

Table 2.1 Comparison of Literature

Ref.	*Type of Resonator*	*Operating Frequency (GHz)*	*Dimension (mm²)*	*No. of Bits*	*Substrate Material*	*RCS (dBsm)*	*Reading Antenna (cm)*	*Spatial Bit Density (Bits/mm²)*	*Applications*
[2]	C-type	2–5	30 × 40	15	FR4(ε_r = 4.6)	–28	60	0.0125	High-capacity tags
[29]	Open-stub	1.9–4.5	80 × 60	8	FR4 (ε_r = 4.4)	—	40	0.001	Low-cost tags
[30]	Flower-shaped	3–8	38 × 38	4	Taconic TLX-8 (ε_r = 2.55)	–26	15	0.002	Orientation independent identification
[33]	Square split ring	1.5–6.7	24 × 24	3	FR4 (ε_r = 4.4)	–22.89	100	0.005	Retail and biomedical applications
[34]	L-shaped slot	3–6	20 × 20	8	Rogers RT5880 substrate (ε_r = 2.2)	–10	45	0.02	Smart retail applications
[37]	Elliptical-based slot	3.5–15.5	22.8 × 16	10	Rogers RT/duroid/5880 (ε_r = 2.2)	–21	3.2	0.027	Moisture sensing
[38]	SIR	1.83–2.15	28.7 × 24.7	3	Rogers 4350 (ε_r = 3.66)	—	100	0.004	High-capacity tags

2.5 Metamaterial-Inspired Advanced Chipless RFID

With important applications in the healthcare, monitoring, and retail industries, metamaterial (MTM) research is an integrative field that has been growing for years. The intriguing properties of metamaterials have stimulated several research in disciplines as varied as optical systems, microwave equipment, radiators, and RFID technology. These structures exhibit various electromagnetic properties that are not commonly found in nature. Negative permittivity, extensive phase alternation, Doppler effect inversion, negative permeability, and Snell's law are some examples of these qualities.

This section aims to provide a brief introduction to geometry strategy and analysis using chipless RFID resonators made of metamaterials for various communications. Analyses of the major properties, including radar cross-section, radiation resistance, read range, surface current, and coding capacity, are presented together with relevant literature. The fundamental component of two-dimensional MTM-based chipless RFID tags and reader radiators are resonators. Because of their unique EM characteristics, like group, phase velocity, and phase fluctuations, metamaterials have the potential to have engineerable permittivity (ε), permeability (μ), and an extremely less index of refraction. This has made them potentially useful in a range of microwave and optical electromagnetic applications, mainly in emitted-wave devices like reader fabrications [39].

A type of engineered media known as metamaterials has unique EM properties that are not present in usual materials. The term "metamaterial" derives from the Greek word meta, which means above or beyond. In the 1960s, a Russian scientist named Victor Georgievich Veselago was credited with suggesting that a substance can be extremely useful in physics. According to Maxwell's equation, the parameters that describe how plane waves move through such medium are (1) the electric field strength E, (2) the magnetic field intensity H, and (3) the wave vector k. In contrast to the right-handed triplet created by a typical right-handed media (RHM), E, H, and k make a left-handed triplet (Figure 2.11).

The modeling of the inductive and capacitive coupling can be done by a coupling capacitor of capacitance C and a transformer through a transforming ratio of n. The single-ring circuit network is described with an RLC resonator [40, 41] by a resonant frequency of ω_0 can be determined as:

$$\omega_0 = \frac{1}{\sqrt{LC}} \tag{2.4}$$

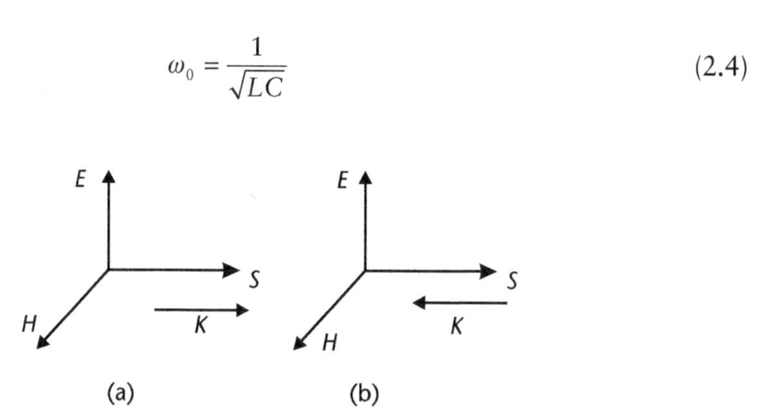

Figure 2.11 (a) Right- and (b) left-handed media coordinates.

The mutual coupling capacitance and inductance are denoted by *C* and *L*. High-current density explains the development of a peak magnetic moment produced.

For chipless CHLRFID tags, a compact MTM-formed resonator along a high coding capacity may be useful. The many resonators inspired by MTM are composed of connected microstrip geometries with various forms and divided (*K*-1) legs (where k is the number of resonators used in the tag). This makes it possible to change the frequency of each resonator to any *K*-resonance frequency. As an outcome, numerous tags are developed and applied to different dielectric material types. In a field crudely equivalent to that of a typical 5-GHz two-state resonator, a passive resonator is applied [42] as an eight-state variant (the printed design is shown in Figure 2.12. The results of the simulated and empirical *S*-parameters correspond well.

To avoid the loading effect, which, if it exists, causes a frequency alteration when greater than a single arm is joined to the connected line, the resonating parts are positioned appropriately. Figure 2.13 displays the produced configuration's simulated and actual return loss (S_{11}). For this code, resonance frequencies of 4.42, 4.66, 5.30, 5.67, 6.41, and 6.91 GHz were noted. In this situation, a spectral density boost is seen between 1 bit/290 MHz and 1 bit/200 MHz. Combining the methods yields an estimated spectral density of more than 1 bit/120 MHz [42].

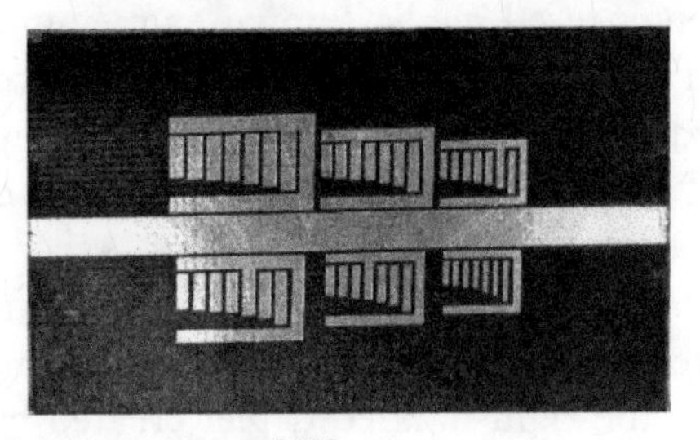

Figure 2.12 MTM-based high-capacity tag [42].

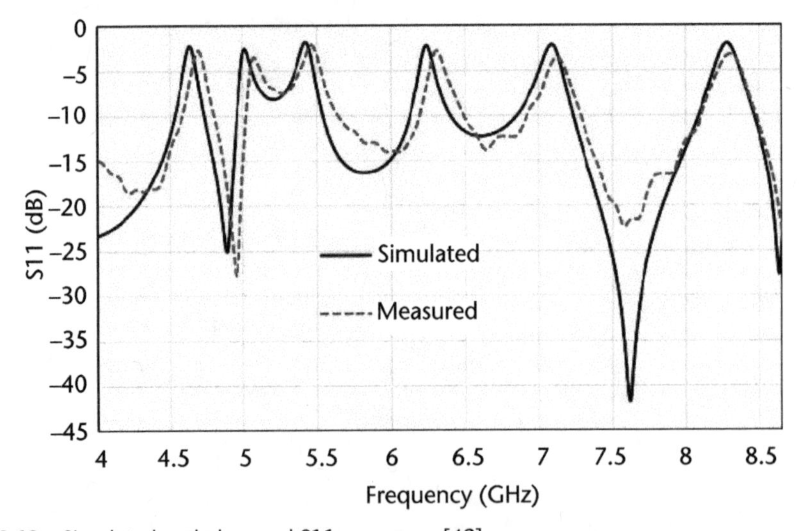

Figure 2.13 Simulated and observed S11 responses [42].

Wearable sensors with characteristics like compact, flexible, and robustness are helpful for handling a simple and secure body area network. RFID sensors are increasingly widely used now because of their sophisticated characteristics and body-centric applications [1, 3]. Temperature sensors are sensitive to graphene, and humidity and temperature sensors are sensitive to polyvinyl alcohol (PVA). Smart materials are preferred in the RFID market because of their advantageous qualities, which are covered in the next section.

2.6 Smart Materials

In RFID technology, smart materials are crucial for precise sensing. Table 2.2 lists a few smart materials that can be used as RFID sensor tags. Polyaniline (PANI), phenanthrene, graphene, hydrophilic polymer, plastic crystals, poly (3,4-ethylenedioxythiophene)-polystyrene sulfonic acid also identified as PEDOT: PSS, Kapton, metallic oxides, PVA, single-walled carbon nanotubes (SWCNTs), and conductive polymers are among the cutting-edge materials widely used in RFID sensors. Meanwhile, there are environmental variations such as temperature, moisture, strain, and pressure, which can alter the electrical characteristics of a sensor. These resources are thought to be excellent for smooth detection. Passive millimeter wave and microwave equipment is typically constructed from extremely conductive resources. Therefore, sensor circuits benefit more from these materials than passive microwave and millimeter wave circuits do. Below is a discussion of a few smart resources used as sensors.

N-methyl-n-butylepyrrolidinium-hexaflurophosphate ($P_{14}PF_6$): Due to its crystal structure, it is frequently an excellent option for temperature-sensing material. Starting with crystal to liquid, the temperature rises in three transit phases from −15°C to 17°C. The crystal conductivity varies as a result of the crystal's molecular mobility [3].

Nanostructure semiconducting metal oxides (ITO and ZnO): These are the two metal oxides with nanostructures that are semiconducting and are extra sensitive to ecological variations such as temperature, pressure, and EM radiation. Due to its strong heat conductivity, better melting point of 2248°K, and widespread market availability, ZnO is better matched to detecting applications. According to

Table 2.2 Smart Sensing Materials

Ref.	*Sensing Materials*	*Dielectric Constant (ε_r)*	*Tag Size (mm^2)*	*Frequency of Operation*
[43]	PVA/XG Hydrogel	1.60	50×50	8.40–9.60 GHZ
[44]	Textile	1.18	60×40	1.9–2.7 GHz
[45]	Kapton film	3.5	158×16	0.78–0.94 GHz
[46]	Textile	1.1	30×30	3.1–10.6 GHz
[47]	Rogers (Ultralam) 3850	2.9	25.6×1	1.5–4.5 GHz
[48]	Textile	1.45	64×49	2.4 GHz
[49]	IgG-AuNP	2.5×10^5	2×6	925 MHz
[50]	Polyethylene foam	1.2	7.5×88	700 MHz–1 GHz

authors in [55], the wavelength of this substance changes with temperature. There is a linear connection between temperature and the material's energy band gap.

Graphene: In sensing devices, graphene is in great demand because of its good mechanical stability, electrical conductivity, and charge mobility. The similar metallic characteristics got in solar exfoliated decreased graphene flakes and graphene oxide (SrGo) can be used to create an on-body temperature sensor. This encourages the creation of widely applicable, adaptable, low-power technologies enthused by graphene high-volume manufacturing in numerous applications. Wearable human body temperature sensors that operate well at 350°C can be built using graphene flakes and SrGo materials [56, 57].

PEDOT: PSS stands out as an excellent choice for sensor applications due to its exceptional characteristics. These include outstanding stability, heightened conductivity, and remarkable transparency, especially in contexts relevant to sensor technology. These materials stand out among all the simply accessible conducting polymers for actual applications since they are strong and have better conductivity. PEDOT film has a conductivity of 5×10^4 S/m and can disperse in water. The ratio of the concentration of PEDOT to PSS molecules has a substantial impact on sheet resistance and electrical conductivity. Therefore, PEDOT film might be a suitable substance for pH detection. The Miller Nelson controller's preset moisture levels are changed to test the transmission coefficient (S_{21}) of the tag under several ecological conditions (Figure 2.14). The temperature throughout the research was kept continuous at around 22.5°C. The relative humidity (RH) inside the chamber however, rose from 60% to 80%. The ambient humidity can be adjusted using this frequency alteration because the electric field coupled indutor (ELC) resonator's operational frequency has been shown to be significantly moved in the direction of a lesser frequency [58].

Single-walled carbon nanotubes (SWCNTs): SWCNTs are used as parts of gas-detecting systems. Owing to their sensitivity to chemical situations, comparatively

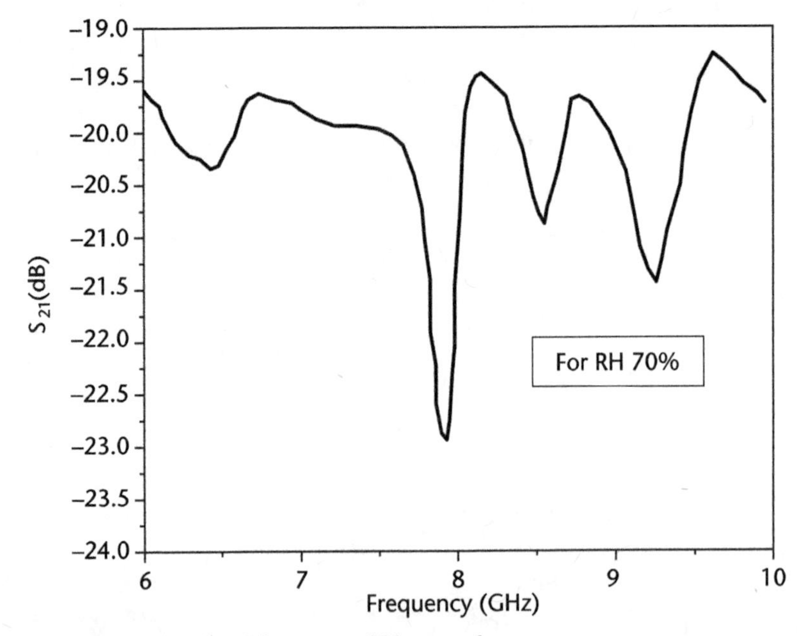

Figure 2.14 Chipless RFID humidity sensor: S21 versus frequency.

limited flexibility, and slant absorption, SWCNTs may be described as having more transparency and electrical conductivity. These films can be employed as crystal-clear conducting materials in the infrared range [59].

Polytetrafluoroethylene (PTFE): It is possible to use PTFE, a fluorocarbon solid along an excessive molecular mass, as a gas-detecting element. It has a relative permittivity of 2.1, a high electrical conductivity, and a tan δ of 0.0005. According to researchers in [60] and [49], PTFE can be employed as a monitoring material for strain and cracking. It has been demonstrated that PTFE resources display a large shift in resonant frequency along with temperature variations because of their high relative permittivity fluctuations.

Kapton (C_{12} H_{12} N_2O) polymer: Kapton polymer is commonly used for humidity monitoring since it is affordable and accessible [51]. It can be combined with passive RFID tags for omnidirectional sensing. At 23°C and 25% humidity, the relative permittivity of Kapton film is 3.25. At room temperature, the Kapton dielectric constant (r) changes linearly with relative moisture.

Polyvinyl alcohol: This is a material used for comparative moisture detection that is extremely sensitive. It is a synthetic polymer with a glass transition temperature of about 700°C, a density of between 1.19 and 1.31 g/cm3, and water solubility. PVA-based sensors are insensitive to some gases (NH_3, NO_2, CO), while they are sensitive to vapors (ethanol, methyl) with OH categories. Relative humidity can be detected using PVA as a sensing material [51, 61]. The many sensor applications for RFID technologies are shown in Table 2.3.

2.7 A Typical Frequency Domain Reader for Transponder Detection

A typical CHLRFID reader is a transceiver that comprises three essential components, as shown in Figure 2.15 [62]. All the interrogation signals that are sent and received between tags and readers are transmitted and received via a single or more radiator. An extensive examination of RFID system antennas is given in [63]. Filters, mixers, gain/phase detectors (GPD), low-noise amplifiers (LNA), power

Table 2.3 Applications of RFID Sensor

Ref.	*Sensing Material*	*Chemical Formula*	*Dielectric Constant*	*Conductivity*	*Sensing Application*
[58]	PEDOT: PSS	Poly (3,4-ethylene dioxythiophene) polystyrene sulfonate	1690	5×10^4 Sm^{-1}	pH
[61]	Cork	$C_{123}H_{182}O_{56}N$	1.7	1.80×10^{-9} S cm^{-1}	Humidity
[62]	Nanostructured metal oxide	ZnO	8.65	1.58 S cm^{-1}	Temperature
[68]	Graphene oxide	$C_{140}H_{42}$ O_{20}	3.20	1.0×10^{-7} S cm^{-1}	Ammonia gas
[69]	Kapton	C_{12} H_{12} N_2O_5	3.50	1.0×10^{-17} S m^{-1}	Humidity
[70]	PTFE	$(C_2F_4)n$	2.020	5.10×10^{-17} S m^{-1}	Strain
[71]	PVA	$(C_2H_4O)x$	328.9	1.63×10^{-12} S cm^{-1}	Humidity
[72]	CNT	C_8H_7N	44	10^7 Sm^{-1}	Ammonia gas

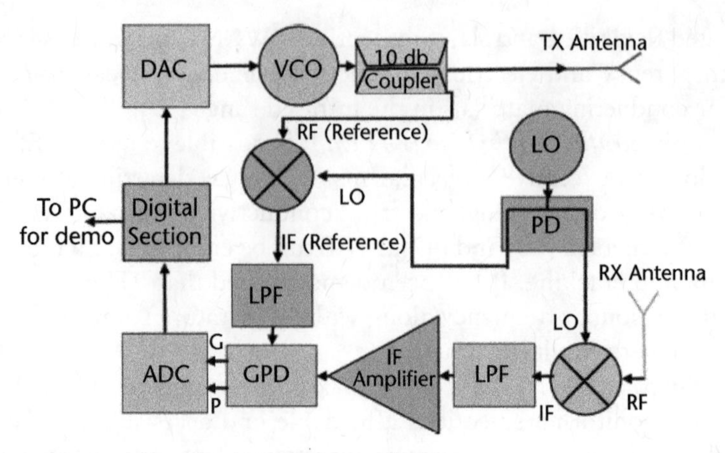

Figure 2.15 A generic chipless reader architecture [62].

dividers, and other RF elements are included in the RF section. The digital section (2.3) contains anticollision algorithms, signal processing techniques, and monitoring of interrogation signals [62–64]. Costly laboratory devices, like vector network analyzers (VNA), is employed to extract the spectral signatures of chipless tags. The ID of the tag can be determined from the spectral signature as long as the RCS fluctuation is recognized by the algorithm and reader.

A typically allowed RCS variance for reading systems is between –15 and –20 dB. The matrix pencil method [66] and adaptive wavelet method [65] can be used for postprocessing. An independent postprocessing technique should be used to discover the ID of the tag regardless of the reading range and RCS level. Here, a series of continuous wave (CW) signals at distinct frequencies, like those transmitted by linear stepped frequency continuous wave (LSFCW) or frequency modulated continuous wave (FMCW) transmitters, are used for interrogation. An oscillator with voltage control creates CW excitation signals [voltage controlled oscillator (VCO)]. A settling period is continuously required for this approach before it can begin generating UWB signals. The VCO's work has a significant influence on the settling time. Designers are thus limited to choosing between a costly, better-performance VCO or a cheap VCO. GPDs typically operate at frequencies below 2.70 GHz. Phase detection can be made simpler if the frequency of the received signal is lowered by a mixer. In [67], it has been provided more specifics regarding the reader's qualities.

The signal sent by the reader antenna is received by the tag, which must be inside the reader's examination region. Data is encoded into chipless tags using two methods: the TID Approach (TIDA) and the FRD Approach (FR-DA) (as explained in Section 2.1). TIDA tags are based on time delays, however, FRDA tags are based on resonant geometry that generates a range of frequency resonances. Meanwhile, chipless tags do not require a time-delay element, and the information is directly determined and implanted in the tag; FDA is easy to implement. The reader's reception radiator receives the signal from the backscattered chipless tags with resonators. This signal, identified as a radar cross-section signal, is essential to tag recognition. The technique used to locate the backscattered base tags is the RCS. The RCS pattern is frequently described using the terms "tag-sign" or "electromagnetic signature." Finding any crests or valleys in the tag's RCS is crucial (because of interfering, some of the peaks may dissolve). As a result, several error-correcting algorithms and coding techniques are used to extract the information content. Majority-rule-based multiple interrogations, Bose-Chaudhuri-Hochquenghem (BCH), and Golay are some of the error-correcting codes that are developed to address the CT encoding or decoding problems. When using a binary-coding scheme to encode bit 0 or 1, the presence or absence of the notch is taken into account [73–75].

In the capacity-increase method, the tag's bits are determined using the notch amplitude. However, the techniquc required both the antenna of the reader and the structure of the tag. Frequency domain CHLRFID tags are used in conjunction with the notch-position modulation (NOPM) technology and medium access control (MAC) protocol to handle multitag identification scenarios [73]. A tag-tag collision occurs when numerous tags all reply to a similar reader at once. Collisions increase the time needed for identification, reduce the efficiency of the system, and hinder the collection of data [76]. There are other implementation-related challenges in the physical world, which are defined in further features in the following section.

A superior RFID reader antenna design is essential for accurate tag detection in any setting. Based on a Monte Carlo analysis, the authors of [77] suggested a way for an RFID antenna's superior detection performance. The tag detection was proven by the authors at various reader antenna orientations and distances from the antenna. Plots are made of the antenna performance's probability distribution curves. The 2.45-GHz wideband radiator monopole RFID reader radiator was designed by the authors in [78] for a vehicle detection application in the underground parking lot next to the curb. The radiating element is made of an FR4 epoxy dielectric material and has dimensions of 34 mm in length, 16 mm in breadth, and 1.6 mm in thickness. The antenna's S-band bandwidth is 280 MHz (2.27 MHz- 2.55 MHz). A dual-frequency circularly polarized UHF-RFID/WLAN circular radiating element radiator for RFID readers was created by the authors in [79]. Slits in the ground plane and slots on the patch of the antenna structure enabled dual band operability and circular polarization at both the UHF-RFID (915 MHz) and WLAN bands (2.45 GHz). The planned radiator is placed on an FR4 epoxy dielectric material with dimensions of 84 mm by 84 mm by 1.6 mm. The authors attempted to provide a small circularly polarized antenna for RFID scanners in [80]. The antenna is constructed with two cross-dipole structures that have dipole pairs in the substrate's upper and lowest layers. The radiator is mounted on an FR4 epoxy dielectric material and has the following dimensions: 29 mm by 29 mm by 1.6 mm. A circularly polarized broadband coplanar waveguide (CPW) fed radiator for a 2.45 GHz RFID reader was suggested by the authors in [81]. The 60 mm by 60 mm by 1.5-mm CPW-fed monopole antenna is constructed on an FR4 epoxy base. The planned antenna has a 40.2% impedance bandwidth in the 2.18–3.28 GHz frequency. A ring-structured triband faulty ground reader radiator for RFID applications was presented by the authors in [82]. The antenna is 83 mm by 55 mm by 3.2 mm and contains two copper surfaces sandwiched between an FR4 epoxy base. In this design, the writers used a flawed ground structure. An aperture-linked UWB radiator array for mm-wave chipless RFID reader was shown by the authors in [83]. Operating between 21 and 27 GHz, the planned antenna array. Wideband miniaturized CPW-fed circularly polarized radiator for universal UHF RFID reader was presented by the authors in [84]. A feed line in the shape of an L supplies the intended antenna. In the ground plane, two L-structured strip lines are introduced for effective impedance matching and wideband circular polarization operation.

2.8 Challenges and Scope for Further Research

Chipless RFID technology needs to overcome the problems listed below in order to contend not only with barcodes but also with chipped RFID. The reader in a passive CHLRFID structure needs to be capable of moving beyond the challenges listed below in order to contest with optical barcodes and CHLRFID. The difficulties are presented in the right order.

2.8.1 Size and Price

Despite the fact that chipless tags (CT) are more affordable than chip-based tags (CBT), readers are more expensive since they require a more expensive antenna and associated components that operate in the UWB spectral band. One strategy to reduce reader costs and footprint is the adoption of metamaterial technology. The two-antenna reader can alternatively be substituted with a single-antenna topology. The use of metamaterial technology is one method to lower reader costs and footprint. Another strategy is to use strategies along with rarer wideband microwave components [63, 66]. Price and reader bandwidth are significantly traded off. An excellent reader who appeals to business is efficient and performs well. A low-priced scanner that can operate across the entire UWB frequency range is consequently more popular. The price of the reader is suggestively obstructed when a novel chipless RFID stage is found [70].

2.8.2 Tag Alignment and Sensitivity

In the majority of tag systems, the research signals' polarization inclination regulates the tagging behaviors. As a result, the system performance is impacted by the chipless tag orientation. To read the tags in dissimilar positioning, consideration must be given to the reader design and orientation-insensitive chipless tags [85, 86].

2.8.3 Reading Speed and Accuracy

The sensing time (more specifically, the window of time during which a moving object may be identified) is crucial in various applications. The primary factors of analysis time and clarifying correctness in frequency area readers are hardware design, operating frequency, and resolution [87]. Transponders use high-quality factors made of low-loss material to decrease the possibility of identifying faults at a lower cost. Employing various resonators and software to increase reading correctness is more practical than using an expensive material to make tags. For this cause, it is recommended that the reader's structure geometry a signal-processing algorithm [88].

2.8.4 Sensitivity to the Tag Alignment

In most of the tag configurations, the tagging behavior is controlled by the polarization inclination of the inquiry signal. The outcomes of the system are impacted by the positioning of the chipless tag; as a result, the reader design must be altered to receive the backscattered signal from the tag in a variety of directions. Employing a reader with a circular-polarized radiator and two dual-polarized radiators may be preferred in this case [86, 89].

2.8.5 Reading Distance and Interferences

If the range can be increased, the chipless RFID technology can be applied in place of the chipped version for a number of long-term applications. However, a variety of factors limit the system's reading range. The broadband noisy band, nearly chipless tags, UWB regulatory limitations, and lack of signal processing units on the tag are just a few examples. EM signal leaks from transponders and investigators, as well as reflected waves from further sources, are causes of interference. In compact-reading chipless RFID arrangements, which are thought to relate in both the time-based and frequency areas, it is difficult to discern between the broadcasted and reflected signals. As a result, the effective detection zone can be increased by raising the system's recognition boundary and adjusting reader-side signal jamming. For reducing the constant sources of interferences, a UWB adjustable module can be used, and signal processing techniques are used to reduce the dynamic interferences [91].

2.8.6 Body Area Network Challenges

Because it surpasses the elastic limit of the structure, the static bio implant is unable to function properly and produces unexpected results when utilized for diagnostic procedures. Using a variety of patient observations, the objective is to spot implant deformation throughout patient recovery. It is also possible to dictate the phases of activities that can be connected to implant carrying capacities and bioimplants to study the effects of various on-body RFID applications on tissues.

2.8.7 Sensing Materials

The effects of chipless sensors on society and the environment are tremendously positive. Numerous industries, consisting of food protection, smart homes, and dynamic symbol monitoring, heavily rely on these sensing materials. Sensitive tags can be made from a variety of materials, including PEDOT, PVA, metallic oxide, SWCNTs, and phenanthrene. By characterizing the material with XRD, AFM, SEM, and TEM, in addition to employing a vector network analyzer to look at microwave characteristics, including tan δ, ε_r, μ, and Q factor, it is possible to find out the RF sensitivity. A possible alternative is by using a variety of fabricated methods on low-cost substrates such as wood, paper, cloth, and plastic to decrease the price of printing. The charge of printing can also be reduced by using the gravure printing technique [21] in contrast to screen and inkjet printing. Utilizing the fabrication method as one another may be helpful in decreasing the price of wearable sensors [92].

2.8.8 Compatibility with Regulations

The reader's optimal effective isotropic radiated power (EIRP) should be accepted by the UWB protocol specification. Because of the short-term examination pulse (rare nanoseconds), this requirement does not influence the performance of time-domain readers; nevertheless, it does lower the actual reading range of frequency-domain readers. There are several approaches in the works for increasing the EIRP level while still abiding by UWB standards [63, 93–95].

2.8.9 Enhancing Data Capacity Using PRs

Tags built on resonators might be a wise solution to boost bit capacity. The development of tags with huge data storage capacities inspired by metamaterials is a crucial field of research. The additional hardware essential is the factor that must be occupied into consideration when setting the pricing and printing challenges of the interrogator structure in order to enhance the coding capacity of the transponder and strike stability among the price and the information managing capacity. MTM-inspired hybrid tags must be carefully built to reduce the cost and complexity for the reader.

2.8.10 Feature Extraction

In chipless sensors, changes to outward situations could lead to variations in the resonant frequency, quality factor, or signal amplitude. The properties of the backscattered signal can also be changed by Environmental interferences. Interference can be caused by other things: background noise, tag and reader spacing, tag and reader antenna polarization incompatibilities, and tag orientation. To offer noise-free sensing data in this approach, feature extraction algorithms are needed [96]. Independent component analysis (ICA) and principal component analysis (PCA) are the strategies for feature extraction [97]. Structural health monitoring using ultra-high frequency (UHF) RFID, these techniques offer better precision [98, 99]. Feature extraction is a key element in CHLRFID sensors to give ruggedness in the process of detection.

2.8.11 Collision Avoidance and Recovery

A collision occurs while reading more tags simultaneously because each tag responds to the inquiry signal simultaneously. Anticollision techniques are employed in the reader's middleware to be rid of contradictory responses from additional tags. A few techniques, like anticollision, were employed, and each one required the usage of pricey laboratory equipment (particularly for moving tags) [73, 100]. Further, the method used in this technique uses background subtraction, which is inapplicable to moving tags and other real-world applications. As a result, more investigation is required in this field of inquiry. The fact that some of these methods call for a hardware design adds to the reader's complexity. The Slotted Termination

Adaptive Collection (STAC) protocol and tree walking algorithm (TWA), which are examples of EPC and IEC standards [93, 101], are some of the anticollision algorithms that have been created for chipped devices. However, because passive tags are unable to perform complex functionalities, chipless devices may not be able to use these protocols. In this circumstance, improving the chipless system's efficiency is challenging. Multiple CT detection can be achieved using notch position modulation (NOPM), fractional Fourier transform (FrFT), and linear frequency modulation (LFM) [31, 73, 102].

2.9 Summary

This chapter explored chipless RFID technology, delving into its components and applications. We began with an introduction to chipless RFID, outlining its principles and purpose. Within chipless RFID, various resonators play a crucial role, including patch, ring, hairpin, and stepped impedance resonators, each with distinct characteristics and applications. The discussion extends to RFID sensor classifications and the integration of resonators for system performance enhancement. Metamaterial-inspired advanced chipless RFID and smart materials were explored as innovative approaches in the field. For material characterization and other sensing applications, this chapter discussed numerous significant characteristics of chipless RFID planar resonators. In this chapter, some of the most significant advantages and drawbacks of planar sensors were discussed along with suitable remedies. Different RFID applications were explored along with interference and collision issues, smart materials, and metamaterials using resonators. The chapter described how to get over problems with sensitivity and selectivity that come with microwave resonator-based sensing along with different smart materials. With the right resonators, selectivity and RCS can be increased while the reader and CHLRFID tag's size are greatly reduced. Several passive resonator characteristics were examined and assessed for how well they perform in terms of footprint, RCS, bit-capacity, polarization, bit-density (bits/cm^2), and printing abilities in a range of applications. Additionally, a typical frequency domain reader for transponder detection was highlighted, shedding light on practical implementations. This chapter concluded by addressing existing challenges and proposing avenues for future research in this evolving domain.

References

[1] Karmakar, N. C. (ed.), *Handbook of Smart Antennas for RFID Systems,* Hoboken, NJ: Wiley, 2010.

[2] Rance, B. O., R. Siragusa, P. Lemaitre-Auger, and E. Perret, "Toward RCS Magnitude Level Coding for Chipless RFID," *IEEE Transactions on Microwave Theory and Techniques,* Vol. 64, No. 7, July 2016, pp. 2315–2325.

[3] Behera, S. K., "Chipless RFID Sensors for Wearable Applications: A Review," *IEEE Sensors Journal,* Vol. 22, No. 2, January 15, 2022, pp. 1105–1120.

[4] Karmakar, N. C., E. M. Amin, and J. K. Saha, *Chipless RFID Sensors,* Hoboken, NJ: John Wiley & Sons, 2016, Chapter 2, pp. 13–28.

[5] Lee, Y.- H., E.- H. Lim, F.- L. Bong, and B.- K. Chung. "Bowtie-shaped Folded Patch Antenna with Split Ring Resonators for UHF RFID Tag Design," *IEEE Transactions on Antennas and Propagation.* Vol. 67, No. 6, 2019, pp. 4212–4217.

[6] Deif, S., and M. Daneshmand, "Multiresonant Chipless RFID Array System for Coating Defect Detection and Corrosion Prediction," *IEEE Transactions on Industrial Electronics,* Vol. 67, No. 10, 2019, pp. 8868–8877.

[7] Athauda, T., and N. C. Karmakar, "The Realization of Chipless RFID Resonator for Multiple Physical Parameter Sensing," *IEEE Internet of Things Journal,* Vol. 6, No. 3, 2019, pp. 5387–5396.

[8] Tariq, N., M. A. Riaz, and H. Shahid, , et al., "Orientation Independent Chipless RFID Tag Using Novel Trefoil Resonators," *IEEE Access,* Vol. 7, 2019, pp. 122398–122407.

[9] Hsieh, L.- H., and K. Chang, "Equivalent Lumped Elements G, L, C, and Unloaded Q's of Closed- and Open-Loop Ring Resonators," *IEEE Transactions on Microwave Theory and Techniques,* Vol. 50, No. 2, 2002, pp. 453–460, IEEE Xplore.

[10] Mo, S.- G., Z.- Y. Yu, and L. Zhang, "Design of Triple-Mode Bandpass Filter Using Improved Hexagonal Loop Resonator," *Progress in Electromagnetics Research,* Vol. PIER96, 2009, pp. 117–125.

[11] Matthaei, G. L., N O., Fenzi, R. J. Forse, and S. M. Rohlfing, 'Hairpin-Comb Filter for HTS and Other Narrow-Band Application," *IEEE Transactions on Microwave Theory and Techniques,* Vol. 45, No. 8, 1997, pp. 1226–1231, IEEE Xplore.

[12] Makimoto, M., and S. Yamashita, "Bandpass Filters Using Parallel Coupled Stripline Stepped Impedance Resonators," *IEEE Transactions on Microwave Theory and Techniques,* Vol. 28, No. 12, December 1980, pp. 1413–1417.

[13] Ma, K., K. C. B. Liang, R. M. Jayasuriya, and K. S. Yeo, "A Wideband and High Rejection Multimode Bandpass Filter Using Stub Perturbation," *IEEE Microwave and Wireless Components Letters,* Vol. 19, No. 1, January 2009, pp. 24–26.

[14] Tu, W.- H., and K. Chang, "Compact Microstrip Bandstop Filter Using Open Stub and Spurline," *IEEE Microwave and Wireless Components Letters,* Vol. 15, No. 4, April 2005, pp. 268–270.

[15] Baik, J.- W., L. Zhu, and Y.- S. Kim, 'Dual-Mode Dual-Band Bandpass Filter Using Balun Structure for Single Substrate Configuration," *IEEE Microwave and Wireless Components Letters,* Vol. 20, No. 11, 2010, pp. 613–615, IEEE Xplore.

[16] Serrano, A. L. C., and F. S. Correra, "A Miniaturized Bandpass Filter wth Two Transmission Zeros Using a Novel Square Patch Resonator," in *Proceedings of the Microwave and Optoelectronics Conference-IMOC, SBMO/IEEE MTT-S Int,* Brazil, October 29–November 1, 2007, pp. 941–945, IEEE Xplore.

[17] Cho, C. S., J. W. Lee, and J. Kim, "Dual- and Triple-Mode Branch-Line Ring Resonators and Harmonic Suppressed Half-Ring Resonators," *IEEE Transactions on Microwave Theory and Techniques,* Vol. 54, No. 11, 2006, pp. 3968–3974, IEEE Xplore.

[18] Zhang, L, Z.- Y. Yu, and S.- G. Mo, "Design of Compact Triple-Mode Bandpass Filter Using Double Center Stubs-Loaded Resonator," *Microwave and Optical Technology Letters,* Vol. 52, No. 8, 2010, pp. 2109–2111, IEEE Xplore.

[19] Cai, P., "Design of Ultra-Wideband Bandpass Filter Using Step-Impedance Four-Mode Resonator and Aperture-Enhanced Coupled Structure," *Microwave and Optical Technology Letters,* Vol. 50, No. 3, 2008, pp. 696–699, IEEE Xplore.

[20] Raju, R., and G. E. Bridges, "A Compact Wireless Passive Harmonic Sensor for Packaged Food Quality Monitoring," *IEEE Transactions on Microwave Theory and Techniques,* Vol. 70, No. 4, April 2022, pp. 2389–2397.

[21] Dey, S., E. M. Amin, and N. C. Karmakar, "Paper Based Chipless RFID leaf Wetness Detector for Plant Health Monitoring," *IEEE Access,* Vol. 8, No. 1359, 2020, pp. 191986–191996.

[22] Preradovic, S., and N. Karmakar, "Chipless RFID Tag with Integrated Sensor," *Proc. IEEE Sensors,* November 2010, pp. 1277–1281.

[23] Borgese, M., F. A. Dicandia, F. Costa, S. Genovesi, and G. Manara, "An Inkjet Printed Chipless RFID Sensor for Wireless Humidity Monitoring," *IEEE Sensors Journal,* Vol. 17, No. 15, August 1, 2017, pp. 4699–4707.

[24] Gorur, A. K., E. Dogan, G. Ayas, C. Karpuz, and A. Gorur, "Multibit Chipless RFID Tags Based on the Transition Among Closed- and Open-Loop Resonators," *IEEE Transactions on Microwave Theory and Techniques,* Vol. 70, No. 1, January 2022, pp. 101–111.

[25] Fathi, P., J. Aliasgari, and N. C. Karmakar, "Analysis on Polarization Responses of Resonators for Frequency-Coded Chipless RFID Tags," *IEEE Transactions on Antennas and Propagation,* Vol. 70, No. 2, February 2022, pp. 1198–1210.

[26] Wang, Y., H. Wu, and Y. Zeng, "Capture-Aware Estimation for Large-Scale RFID Tags Identification," *IEEE Signal Processing Letters,* Vol. 22, No. 9, September 2015, pp. 1274–1277, doi: 10.1109/LSP.2015.2396911.

[27] Nijas, C. M., et al., "Low-Cost Multiple-Bit Encoded Chipless RFID Tag Using Stepped Impedance Resonator," IEEE Transactions on Antennas and Propagation, Vol. 62, No. 9, September 2014, pp. 4762–4770.

[28] Girbau, D., J. Lorenzo, A. Lazaro, C. Ferrater, and R. Villarino, "Frequency-Coded Chipless RFID Tag Based on Dual-Band Resonators," *IEEE Antennas and Wireless Propagation Letters,* Vol. 11, 2012, pp. 126–128.

[29] Nijas, C. M., et al., "Chipless RFID Tag Using Multiple Microstrip Open Stub Resonators," *IEEE Transactions on Antennas and Propagation,* Vol. 60, No. 9, September 2012, pp. 4429–4432.

[30] Babaeian, F., and N. C. Karmakar, "Development of Cross-Polar Orientation-Insensitive Chipless RFID Tags," *IEEE Transactions on Antennas and Propagation,* Vol. 68, No. 7, July 2020, pp. 5159–5170.

[31] Li, Z., Y. Lan, G. He, S. He, and S. Wang, "Optimal Window Fractional Fourier Transform Based Chipless RFID Tag Anti-Collision Algorithm," in *2018 Chinese Automation Congress (CAC),* 2018, pp. 4237–4242, doi: 10.1109/CAC.2018.8623437.

[32] Mishra, D. P., and S. K. Behera, "Modified Rectangular Resonators Based Multi-Frequency Narrow-Band RFID Reader Antenna," *Microwave and Optical Technology Letters,* Vol. 64, No. 3, 2022, pp. 544–551, doi:10.1002/mop.33116.

[33] Mishra, D. P., and S. K. Behera, "A Novel Technique for Dimensional Space Reduction in Passive RFID Transponders," in *2021 2nd International Conference on Range Technology (ICORT),* 2021, pp. 1–4.

[34] Sharma, V., S. Malhotra, and M. Hashmi, "Slot Resonator Based Novel Orientation Independent Chipless RFID Tag Configurations," *IEEE Sensors Journal,* Vol. 19, No. 13, July 1, 2019, pp. 5153–5160.

[35] Shrestha, S., R. Yerramilli, and N. C. Karmakar, "Microwave Performance of Flexo-Printed Chipless RFID Tags," *Flexible and Printed Electronics,* Vol. 4, No. 4, November 2019.

[36] Wang, X., Y. Tao, J. Sidén, and G. Wang, "Design of High-Data-Density Chipless RFID Tag Embedded in QR Code," *IEEE Transactions on Antennas and Propagation,* Vol. 70, No. 3, March 2022, pp. 2189–2198.

[37] Jabeen, I., A. Ejaz, and M. A. Riaz, et al., "Miniaturized Elliptical Slot Based Chipless RFID Tag for Moisture Sensing," *ACES Journal,* Vol. 34, No. 9, September 2019.

[38] Feng, C., W. Zhang, L. Li, L. Han, X. Chen and R. Ma, "Angle-Based Chipless RFID Tag with High Capacity and Insensitivity to Polarization," *IEEE Transactions on Antennas and Propagation,* Vol. 63, No. 4, April 2015, pp. 1789–1797.

[39] Veselago, V. G., "The Electrodynamics of Substances with Simultaneously Negative Values of € and μ," *Physics-Uspekhi,* Vol. 10, No. 4, 1968, pp. 509–514.

[40] Pendry, J. B., A. J. Holden, D. J. Robbins, and W. J. Stewart. "Magnetism from Conductors and Enhanced Nonlinear Phenomena," *IEEE Transactions on Microwave Theory and Techniques,* Vol. 47, No. 11, November 1999, pp. 2075–1084.

[41] Costa, F., A. Monorchio, and G. Manara, "Efficient Analysis of Frequency-Selective Surfaces by a Simple Equivalent-Circuit Model," *IEEE Antennas and Propagation Magazine*, Vol. 54, No. 4, August 2012, pp. 35–48.

[42] Abdulkawi, W. M., and A.- F. A. Sheta, "K-State Resonators for High-Coding-Capacity Chipless RFID Applications," *IEEE Access*, Vol. 7, 2019, pp. 185868–185878.

[43] Occhiuzzi, C., A. Ajovalasit, M. A. Sabatino, C. Dispenza, and G. Marrocco, "RFID Epidermal Sensor Including Hydrogel Membranes for Wound Monitoring and Healing," in *Proceedings of the IEEE International Conference on RFID (RFID)*, April 2015, pp. 182–188.

[44] Corchia, L., G. Monti, L. Tarricone, and E. De Benedetto, "A Chipless Humidity Sensor for Wearable Applications," in *2019 IEEE International Conference on RFID Technology and Applications (RFID-TA)*, 2019, pp. 174–177.

[45] Koski, K., A. Vena, L. Sydanheimo, L. Ukkonen, and Y. Rahmat-Samii, "Design and Implementation of Electro-Textile Ground Planes for Wearable UHF RFID Patch Tag Antennas," *IEEE Antennas and Wireless Propagation Letters*, Vol. 12, No. 1, 2008, pp. 964–967.

[46] Klemm, M., and G. Troester, "Textile UWB Antennas for Wireless Body Area Networks," *IEEE Transactions on Antennas and Propagation*, Vol. 54, No. 11, November 2006, pp. 3192–3197.

[47] Lorenzo, J., A. Lázaro, R. Villarino, and D. Girbau, "Modulated Frequency Selective Surfaces for Wearable RFID and Sensor Applications," *IEEE Transactions on Antennas and Propagation*, Vol. 64, No. 10, October 2016, pp. 4447–4456.

[48] Locher, I., M. Klemm, T. Kirstein, and G. Tröster, "Design and Characterization of Purely Textile Patch Antennas," *IEEE Transactions on Advanced Packaging*, Vol. 29, No. 4, November 2006, pp. 777–788.

[49] Yuan, M., P. Chahal, E. C. Alocilja, and S. Chakrabartty, "Sensing by Growing Antennas: A Novel Approach for Designing Passive RFID Based Biosensors," in *2015 IEEE International Symposium on Circuits and Systems (ISCAS)*, 2015, pp. 2121–2124.

[50] Patron, D., W. Mongan, and T. P. Kurzweg, et al., "On the Use of Knitted Antennas and Inductively Coupled RFID Tags for Wearable Applications," *IEEE Transactions on Biomedical Circuits and Systems*, Vol. 10, No. 6, December 2016.

[51] Amin, E. M., N. C. Karmakar, and B. W. Jensen, "Polyvinayl-Alcohol (PVA)-Based RF Humidity Sensor in Microwave Frequency," *Progress in Electromagnetics Research B*, Vol. 54, November 2013, pp. 149–166.

[52] Amin, E. M., S. Bhuyian, N. C. Karmakar, and B. W. Jensen, "Development of a Low Cost Printable Chipless RFID Humidity Sensor," *IEEE Sensors Journal*, Vol. 14, No. 1, January 2014, pp. 140–149.

[53] Karmakar, N. C., R. E. Azim, S. M. Roy, R. Yerramilli, and G. Swiegers, "Printed Chipless RFID Tags for Flexible Low-Cost Substrates," in *Chipless and Conventional Radio Frequency Identification: Systems for Ubiquitous Tagging*, Hoboken, NJ: IGI Global, 2012.

[54] Nicolson, A. M., and G. F. Ross, "Measurement of the Intrinsic Properties of Materials by Time-Domain Techniques," *IEEE Transactions on Instrumentation and Measurement*, Vol. 19, No. 4, November 1970, pp. 377–382.

[55] Chenghua, X. J. S., W. Helin, X. Tianning, Y. Bo, and L. Yuling, "Optical Temperature Sensor Based on ZnO Thin Film's Temperature Dependent Optical Properties," *Review of Scientific Instruments*, Vol. 82, No. 8, 2011, pp. 084901-1–084901-3.

[56] Sahatiya, P., S. K. Puttapati, V. V. S. S. Srikanth, and S. Badhulika, "Graphene-Based Wearable Temperature Sensor and Infrared Photodetector on a Flexible Polyimide Substrate," *Journal of Flexible and Printed Electronics*, Vol. 1, No. 2, 2016, pp. 1–9.

[57] Gedela, V., S. K. Puttapati, C. Nagavolu, V. V. S. S. Srikanth, "A Unique Solar Radiation Exfoliated Reduced Graphene Oxide/Polyaniline Nanofibers Composite Electrode Material for Supercapacitors," *Materials Letters*, Vol. 152, 2015, pp. 177–180.

[58] Amin, E. M., J. K. Saha, and N. C. Karmakar, "Smart Sensing Materials for Low-Cost Chipless RFID Sensor," *IEEE Sensors Journal*, Vol. 14, No. 7, July 2014.

[59] Wu, Z., et al., "Transparent, Conductive Carbon Nanotube Films," *Science,* Vol. 305, August 2004, pp. 1273–1276.

[60] Yi, X., et al., "Thermal Effects on a Passive Wireless Antenna Sensor for Strain and Crack Sensing," *Proceedings of SPIE,* Vol. 8345, April 2012, pp. 8345F-1–8345F-11.

[61] Penza, M., and V. I. Anisimkin, "Surface Acoustic Wave Humidity Sensor Using Polyvinyl-Alcohol Film," *Sensors and Actuators A, Physical,* Vol. 76, No. 1, 1999, pp. 162–166.

[62] Karmakar, N. C., R. Koswatta, P. Kalansuriya, and E. Rubayet, *Chipless RFID Reader Architecture,* Norwood, MA: Artech House, 2013.

[63] Forouzandeh, M. M., and N. Karmakar, "Towards the Improvement of Frequency-Domain Chipless RFID Readers," in *Proceedings of the IEEE Wireless Power Transfer Conference (WPTC),* 2018, pp. 1–4.

[64] Hartmann, C. S., "A Global SAW ID Tag with Large Data Capacity," in *Proceedings of the IEEE Ultrasonics Symposium,* Vol. 1, 2002, pp. 65–69.

[65] Gandino, F., R. Ferrero, B. Montrucchio, and M. Rebaudengo, "DCNS: An Adaptable High Throughput RFID Reader-to-Reader Anticollision Protocol," *IEEE Transactions on Parallel and Distributed Systems,* Vol. 24, No. 5, May 2013, pp. 893–905.

[66] Preradovic, S., and N. C. Karmakar, "Multiresonator Based Chipless RFID Tag and Dedicated RFID Reader," *IEEE MTT-S International Microwave Symposium Digest,* 2010, pp. 1520–1523.

[67] Aliasgari, J., M. Forouzandeh, and N. Karmakar, "Chipless RFID Readers for Frequency-Coded Tags: Time-Domain or Frequency-Domain?" *IEEE Journal of Radio Frequency Identification,* Vol. 4, No. 2, June 2020, pp. 146–158.

[68] Le, T., et al., "A Novel Graphene-Based Inkjet-Printed WISP-Enabled Wireless Gas Sensor," in *2012 42nd European Microwave Conference,* 2012, pp. 412–415.

[69] Virtanen, J., L. Ukkonen, T. Bjorninen, A. Z. Elsherbeni, and L. Sydanheimo, "Inkjet-Printed Humidity Sensor for Passive UHF RFID Systems," *IEEE Transactions on Instrumentation and Measurement,* Vol. 60, No. 8, August 2011, pp. 2768–2777.

[70] Amin, E. M., N. C. Karmakar, and B. W. Jensen, "Polyvinyl-Alcohol (PVA)-Based RF Humidity Sensor in Microwave Frequency," *Progress in Electromagnetics Research B,* Vol. 54, November 2013, pp. 149–166.

[71] Gonçalves, R., et al., "RFID-Based Wireless Passive Sensors Utilizing Cork Materials," *IEEE Sensors Journal,* Vol. 15, No. 12, December 2015, pp. 7242–7251.

[72] Jain, S., and S. R. Das, "Collision Avoidance in a Dense RFID Network," in *Proceedings of the 1st International Workshop on Wireless Network Testbeds, Experimental Evaluation & Characterization, ser. WiNTECH '06,* New York, NY, ACM, 2006, pp. 49–56, http://doi.acm.org/10.1145/1160987.1160997.

[73] El-Hadidy, M., A. El-Awamry, A. Fawky, M. Khaliel, and T. Kaiser, "A Novel Collision Avoidance MAC Protocol for Multi-Tag UWB Chipless RFID Systems Based on Notch Position Modulation," in *2015 9th European Conference on Antennas and Propagation (EuCAP),* 2015, pp. 1–5.

[74] Zheng, F., Y. Chen, T. Kaiser, and A. J. H. Vinck, "On the Coding of Chipless Tags," *IEEE Journal of Radio Frequency Identification,* Vol. 2, No. 4, December 2018, pp. 170–184.

[75] Khaliel, A. M., A. El-Awamry, A. Fawky, and T. Kaiser, "A Novel Design Approach for Co/Cross-Polarizing Chipless RFID Tags of High Coding Capacity," *IEEE Journal of Radio Frequency Identification,* Vol. 1, No. 2, June 2017, pp. 135–143.

[76] Meguerditchian, C., H. Safa, and W. El-Hajj, "New Reader Anti-Collision Algorithm for dense RFID Environments," in *2011 18th IEEE International Conference on Electronics, Circuits and Systems (ICECS),* December 2011, pp. 85–88.

[77] Zhao, T., L. Kong, and X. Liu, "Prediction Method of RFID Antennas Detection Performance Based on Monte Carlo Analysis," in *2022 IEEE 9th International Symposium on Microwave, Antenna, Propagation and EMC Technologies for Wireless Communications (MAPE)*, 2022, pp. 260–263.

[78] He, Y., and M. Chen, "2.45 GHz Broadband Monopole RFID Reader Antenna Buried in the Ground of Parking Lot Near the Curb," *2016 IEEE International Conference on RFID Technology and Applications (RFID-TA),* 2016, pp. 1–5.

[79] Sarkar, S., and B. Gupta, "A Dual Frequency Circularly Polarized UHF-RFID/WLAN Circular Patch Antenna for RFID Readers," in *2019 IEEE International Conference on RFID Technology and Applications (RFID-TA),* 2019, pp. 448–452.

[80] Bajaj, C., D. K. Upadhyay, S. Kumar, and B. K. Kanaujia, "Compact Circularly Polarized 2.45/5.8-GHz Antenna for RFID Readers," in *2021 IEEE International Conference on RFID Technology and Applications (RFID-TA),* 2021, pp. 63–66.

[81] Birwal, A., V. Kaushal, and K. Patel, "Circularly Polarized Broadband Co-Planar Waveguide Fed Antenna for 2.45 GHz RFID Reader," in *2021 IEEE International Conference on RFID Technology and Applications (RFID-TA),* 2021, pp. 109–112.

[82] Okramcha, M., M. R. Tripathy, and R. K. Arya, "Ring-Shaped Tri-Band Defective Ground Reader Antenna for RFID Applications," in *2021 IEEE International Conference on RFID Technology and Applications (RFID-TA)*, 2021, pp. 105–108.

[83] Islam, M. A., N. Karmakar, and A. K. M. Azad, "Aperture Coupled UWB Microstrip Patch Antenna Array for mm-Wave Chipless RFID Tag Reader," in *2012 IEEE International Conference on RFID-Technologies and Applications (RFID-TA),* 2012, pp. 208–211.

[84] Cao, R., and S. - C. Yu, "Wideband Compact CPW-Fed Circularly Polarized Antenna for Universal UHF RFID Reader," *IEEE Transactions on Antennas and Propagation*, Vol. 63, No. 9, September 2015, pp. 4148–4151.

[85] Ebrahimi-Asl, S., M. T. Ghasr, and M. J. Zawodniok, "Design of Dual-Loaded RFID Tag for Higher Order Modulations," *IEEE Transactions on Microwave Theory and Techniques,* Vol. 66, No. 1, January 2018, pp. 410–419.

[86] Genovesi, S., F. Costa, F. A. Dicandia, M. Borgese, and G. Manara, "Orientation-Insensitive and Normalization-Free Reading Chipless RFID System Based on Circular Polarization Interrogation," *IEEE Transactions on Antennas and Propagation,* Vol. 68, No. 3, March 2020, pp. 2370–2378.

[87] Garbati, M., E. Perret, and R. Siragusa, *Chipless RFID Reader Design for Ultra-Wideband Technology: Design, Realization and Characterization,* Oxford, UK: Elsevier, 2018.

[88] Daliri, A., A. Galehdar, W. S. T. Rowe, S. John, C. H. Wang, and K. Ghorbani, "Quality Factor Effect on the Wireless Range of Microstrip Patch Antenna Strain Sensors," *Sensors,* Vol. 20, No. 1, 2014, pp. 595–605.

[89] Garbati, M., E. Perret, R. Siragusa, and C. Halopè, "Toward Chipless RFID Reading Systems Independent of Tag Orientation," *IEEE Microwave Wireless Components Letters,* Vol. 27, No. 12, December 2017, pp. 1158–1160.

[90] Forouzandeh, M., and N. Karmakar, "Application of Wideband Differential Phase Shifters with Wide Phase Range in Chipless RFID Readers," *IEEE Transactions on Microwave Theory and Techniques,* Vol. 67, No. 9, September 2019, pp. 3636–3650.

[91] Ramos, A., E. Perret, O. Rance, S. Tedjini, A. Lázaro, and D. Girbau, "Temporal Separation Detection for Chipless Depolarizing Frequency Coded RFID," *IEEE Transactions on Microwave Theory and Techniques,* Vol. 64, No. 7, July 2016, pp. 2326–2337.

[92] Behera, S. K., and N. C. Karmakar, "Chipless RFID Printing Technologies: A State of the Art," *IEEE Microwave Magazine,* Vol. 22, No. 6, June 2021, pp. 64–81.

[93] Finkenzeller, K., *RFID Handbook: Fundamentals and Applications in Contactless Smart Cards and Identification,* Second Edition, John Wiley & Sons, 2003.

[94] Sihvola, A., "Metamaterials in Electromagnetics," *Metamaterials,* Vol. 1, No. 1, March 2007, pp. 2–11.

[95] EPC Radio-Frequency Identity ProtocolsClass-1 Generation-2 UHF RFID, 2013, https://www.gs1.org/sites/default/files/docs/epc/Gen2_Protocol_Standard.pdf.

[96] Feng, L., Z. Li, C. Liu, X. Chen, X. Yin, and D. Fang, "SitR: Sitting Posture Recognition Using RF Signals," *IEEE Internet of Things Journal,* Vol. 7, No. 12, December 2020, pp. 11492–11504.

[97] Zhang, J., G. Y. Tian, and A. B. Zhao, "Passive RFID Sensor Systems for Crack Detection and Characterization," *NDT & E International,* Vol. 86, March 2017, pp. 89–99.

[98] Zhang, J., and G. Y. Tian, "UHF RFID Tag Antenna-Based Sensing for Corrosion Detection and Characterization Using Principal Component Analysis," *IEEE Transactions on Antennas and Propagation,* Vol. 64, No. 10, October 2016, pp. 4405–4414.

[99] Zhang, H., R. Yang, Y. He, G. Y. Tian, L. Xu, and R. Wu, "Identification and Characterisation of Steel Corrosion Using Passive High Frequency RFID Sensors," *Measurement,* Vol. 92, October 2016, pp. 421–427.

[100] Klair, D., K.- W. Chin, and R. Raad, "A Survey and Tutorial of RFID Anticollision Protocols," *IEEE Communications Surveys & Tutorials,* Vol. 12, No. 3, 2010, pp. 400–421.

[101] ISO Parameters for Air Interface Communications at 860 to 960 MHz, 2008, https://www.iso.org/standard/59644.html

CHAPTER 3

Smart Materials for Chipless RFID Printing

3.1 Introduction

Currently, there is an increasingly high demand for less-expensive sensors for application in a wide variety of sectors, such as biomedicine, smart packaging, plant health monitoring, transportation, farming, banking, and the IoT. Smart materials are a significant category of materials that alter their physical or chemical properties in response to external stimuli, providing sensing data of the surrounding environment. Incorporating smart materials into sensor technologies harnesses their analytical capabilities, enhancing sensitivity, expansive linear dynamic range, and specificity [1]. The properties of the material, which vary according to the change in surroundings, include relative permittivity (ε_r), relative permeability (μ_r), and conductivity (σ). For instance, phenanthrene, a temperature-sensitive material, exhibits an abrupt surge in its dielectric constant at around 800°C. Smart materials for sensing include keratin, silicon nanowires, Mitsubishi ink, Kapton, PVA, superabsorbent polymer (a subclass of hydrogel), zinc oxide (ZnO), PEDOT: PSS, phenanthrene, polyaniline (PANI), graphene, SWCNT, and nanoparticles [2]. Furthermore, integrating smart materials into textiles facilitates the development of smart fabric sensors (SFSs). These textile-based sensors can detect various physical and chemical stimuli, including alterations in temperature, pressure, mechanical force, and electrical current, among other factors. Incorporating sensing components into fabrics can be tailored to different levels contingent upon the specific structural aspects of the fabric undergoing modification or sensitization [3]. The utilization of smart materials extends far beyond the confines of sensor technology, encompassing a wide spectrum of applications that include actuators, robotics, artificial muscles, controlled drug delivery systems, and tissue engineering. In the context of their diverse utility, the advancement of smart materials finds its nexus within a multidisciplinary realm, spanning materials science, chemistry, physics, engineering, and nanotechnology [4]. Among the various active, semiactive, or chip-based RFID sensors, chipless RFID sensors are known for their flexible, printable, noninvasive, non-line-of-sight, less-expensive, multisensing, and easily adaptable (harsh environments) nature [2]. This chapter explores the smart materials that are crucial for proper sensing in a chipless RFID sensor.

3.2 Smart Materials

Smart materials with multifunctional properties are substantial enough to meet the ever-increasing demand for fast-advancing technology. Smart materials or nanostructured functional materials are a specific category of materials that produce a strong variation in their property (physical or chemical) according to a physical or chemical stimulus. The external stimuli comprise electric or magnetic field, humidity, hydrostatic pressure, ambient temperature, stress, strain, certain chemical concentrations, radiation, light intensity, and so forth. In response to these stimuli, smart or intelligent materials may change their color or physical structure, attain more strength, generate voltage, and vary permittivity, permeability, or conductivity. Smart materials are often referred to as intelligent materials due to their remarkable ability to adapt and perform in response to variations in their surroundings. Their smartness becomes evident through their capacity to undergo self-deformation, self-diagnosis, self-response, self-adaptation, and self-repair when subjected to environmental changes. This capability mimics the intricate adaptability observed in biological systems, making them a testament to the emulation of natural intelligence [5–10]. A wide range of smart materials include piezoelectric (generate a voltage in response to mechanical stress), electrostrictive/magnetostrictive (change size in response to electric/magnetic stimulus), rheological (change in liquid state in response to electric/magnetic stimulus), electrochromic (change in optical properties with respect to electric current), shape memory alloys (change in shape in response to temperature), thermoresponsive (response to temperature variations), and pH-sensitive (respond to pH variations) materials. Therefore, smart material is a suitable candidate for sensing applications. Apart from sensing, there are numerous areas where these intelligent materials have a significant role, such as in solar cells, fuel cells, actuators, designs for dams, buildings, or bridges, and ultrahigh-fidelity stereo speakers utilizing piezoelectric actuators [11, 12]. Among these applications, the focus of this chapter is on chipless RFID sensing using smart materials.

3.2.1 Smart Materials Integrated with Chipless RFID Tag

Stimuli-dependent materials integrated into the chipless tag convert the tag into a sensor. They are employed in the tag in either of the following ways [13–16]:

1. The substrate is designed using smart materials.
2. The smart materials are printed on the tag (or the tag resonators are a combination of identification and sensing elements where sensing elements are made of smart materials). The sensing material can be coated on the existing RFID tag as a superstrate.

In [14], a chipless tag design incorporating multiple resonators and an integrated temperature sensor is introduced. This chipless tag antenna is equipped with a series of N-cascaded spiral electromagnetic band gap (EBG) structures precisely tuned to resonate at N distinct frequencies, with each resonator corresponding to a specific data bit. Remarkably, sensor information is encoded at a frequency beyond the N resonant frequencies, denoted as N + 1 frequency, which operates independently from the tag's unique identification. This encoding is achieved through the variation of the sensor's impedance, which is responsive to temperature fluctuations. In [17], researchers utilized a multiresonator structure for both identification and humidity sensing. Their approach involved the addition of polyvinyl alcohol to the resonator's surface for humidity detection. Additionally, in [18], a humidity sensor was devised based on time-domain and group delay responses. This ingenious humidity-sensing mechanism entailed the deposition of silicon nanowires onto a chipless RFID tag, expanding the utility and versatility of such innovative technologies.

In general, a chipless RFID sensor can be considered as a transponder, which has a variable load impedance ($Z_L = R_L + jX_L$), where R_L is the load resistance and X_L is the load reactance. With a variation in the nearby physical condition, the load impedance alters. The dielectric constant or relative permittivity of the material is associated with polarization or capacitance, while the relative permeability is related to inductance [19]. Thus, the imaginary part of the load impedance (i.e., the reactance) acts as a variable when there is a change in ε_r or μ_r. Similarly, the real part of the load impedance (i.e., the resistance) varies when σ changes. It can be further observed experimentally that changing permittivity or permeability reflects as a shift in the resonant frequency of the received signal. Moreover, the variation in the target response level designates an alteration in the material conductivity. Both the abovementioned variations are illustrated in Figure 3.1 [20]. Therefore, it is clear that the backscattered RCS is an indication of the variation in the surrounding environment.

3.3 Classification of Smart Materials for Physical Parameter Sensing

The significant property of smart materials, the basis of chipless RFID sensors, is that their mobility is relatively low. But they are well-suited for sensing scenarios.

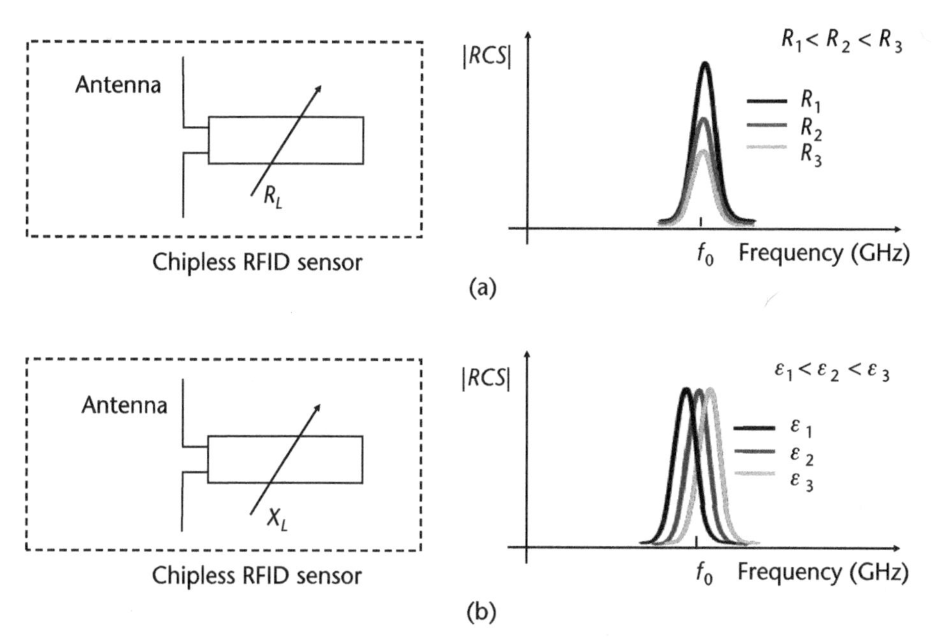

Figure 3.1 Chipless RFID sensing method: (a) change in conductivity, and (b) change in dielectric constant.

For example, materials such as silver (Ag), copper (Cu), and aluminum (Al) are good conductors, and thus utilized in microwave and millimeter-wave applications, whereas semiconductive materials such as indium tin oxide (ITO) and silver flakes serve a notable role in sensing applications. Semiconductors exhibit a remarkable ability to modulate their electrical conductivity through processes such as doping or precise adjustments using electric or magnetic fields, encompassing an extensive conductivity range from 10^{-7} Scm^{-1} to 10^{2} Scm^{-1}. These materials undergo significant alterations in their conductivity when exposed to external factors, including fluctuations in temperature, illumination, thermal influences, mechanical stress, or chemical interactions. Consequently, semiconductors emerge as highly suitable substances for sensor applications, with the design and arrangement of semiconductor devices playing a pivotal role in determining their sensing performance. Commonly employed semiconductor materials encompass silicon (Si), germanium (Ge), III-V compounds, and metal oxide semiconductors [21]. The diverse range of smart materials can be grouped into various categories based on distinct sensing parameters, which are shown in Figure 3.2 [1]. This section throws light on the materials sensitive to temperature, humidity, pH, strain, crack, gas, and light and is described in detail in the subsequent sections. The intelligent materials may be conductive polymers, semiconductive materials, metal oxides, or nanoparticles. There are a few materials that exhibit multiparameter sensing capability, such as Rogers RT/duroid® 6010.2LM, that respond to variations in both temperature and crack [22]. Several advanced characterization and manufacturing technologies are available for the synthesis of these functional materials. Therefore, the application scenarios of these materials solely rely on the availability of synthesis methods so that the processing and manipulation can be performed in an accurate way.

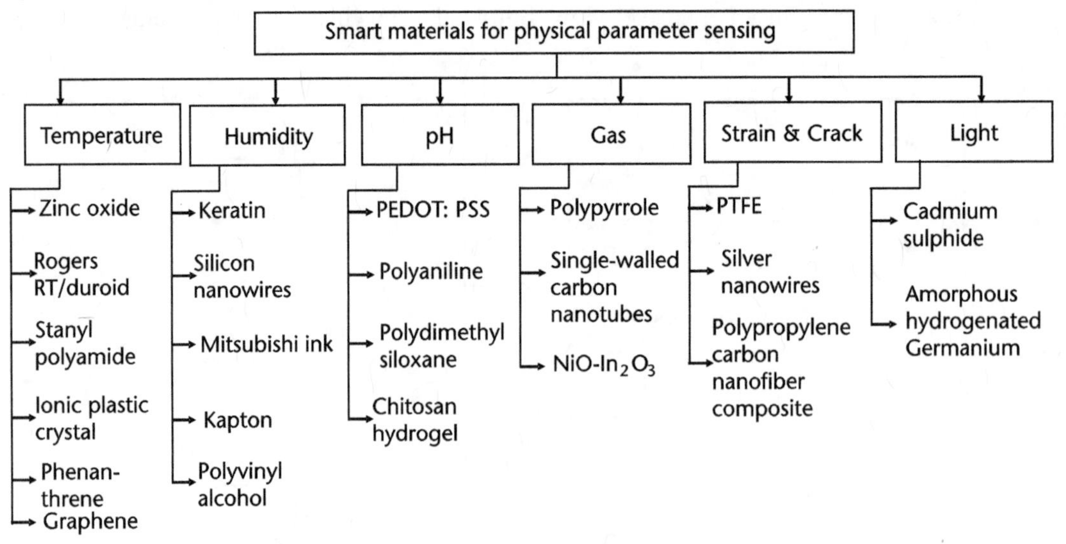

Figure 3.2 Categorization of sensing materials.

3.4 Temperature-Sensing Materials

Temperature sensing finds potential applications in pharmaceuticals, the food industry, smart homes, and structural health monitoring. In some cases, real-time monitoring of temperature is necessary, but in some other scenarios, it is required to maintain a particular threshold temperature. For example, in the food industry, milk needs to be kept at a temperature below 7°C to prevent early deterioration. For this application, a memory sensor utilizing functional materials with irreversible characteristics is necessary. Phenanthrene causes a resonant frequency shift above 80°C and will not shift back to the previous frequency if the temperature is further reduced. In the drug storage scenario, certain chemicals need to be maintained at a particular temperature. Temperature monitoring is very important as a slight variation can damage a medicine's efficacy. Another significant aspect is cold storage management in the food industry, where concurrent monitoring is required to ensure the quality and freshness of the food items. In the smart home concept, temperature is continuously monitored and machinery such as fans or air conditioners are operated automatically. Temperature sensors can also be used to prevent bushfires in the forest, thus saving flora and fauna. The smart materials used for temperature sensing are discussed in this section.

3.4.1 Zinc Oxide

Zinc oxide is a semiconductive functional material that responds to external temperature, pressure, or electric field stimuli by varying its optical and electrical properties. This wide bandgap semiconductor is thermally stable and has a high melting point of 2248K. The ZnO band gap energy has a linear relation with temperature (i.e., the rise in temperature leads to a regular red shift in the optical absorption edge of ZnO). Therefore, ZnO is well-suited for temperature sensing. An optical fiber temperature sensor based on ZnO thin film is presented in [23]. The substrate, in this case, is the sapphire material with ZnO thin film, the sensing element, deposited on top of it. The relation between the bandgap energy of ZnO and the temperature between 10K and 1,000K is expressed as [23]

$$E_g(T) = E_g(0) + \frac{rT^2}{T+\beta'} \tag{3.1}$$

where $E_g(T)$ and $E_g(0)$ are the bandgap energy of ZnO at temperatures T and zero in [K], respectively. The constant parameters of ZnO include r and β.

The optical bandgap is given by (3.2) and it is observed that there is a linear decrease in bandgap energy with ambient temperature [23]:

$$\alpha h = A\left(h - E_g\right)^{\frac{1}{2}} \tag{3.2}$$

where σ, $h\nu$, A, and E_g are the optical absorption coefficient, photon energy (h = 6.626×10^{-34} Js, Planck's constant and ν is the working frequency), constant parameter (depends on electron-hole mobility), and optical bandgap, respectively.

Another semiconducting metal oxide, ITO, which exhibits similar behavior to ZnO, can also be utilized for temperature sensing.

3.4.2 Rogers RT/Duroid

Rogers RT/duroid 6010.2LM material varies its dielectric constant with alteration in the surrounding temperature. In [22], Rogers RT/duroid 6010.2LM-based chipless RFID sensor is designed for structural health monitoring. The material is used as a substrate, and the resonators are fabricated on top of it using 0.035-mm thick copper (generally). It was observed that the resonant frequency shifted to a lower frequency of approximately 26.2 MHz for every 10°C increase in the background temperature. The thermal coefficient of the material is –425 ppm/°C in the range of –50°C to +170°C. The linear relation between the effective dielectric constant and temperature is given by [22]

$$\varepsilon_{eff}(T) = \varepsilon_0 + kT \tag{3.3}$$

where $\varepsilon_{eff}(T)$, T, ε_0, and k are effective dielectric constant at temperature T(°C), surrounding temperature (°C), absolute permittivity (8.824×10^{-12} F/m), and Boltzmann constant (1.38×10^{-23} J/K), respectively.

3.4.3 Stanyl Polyamide

The dielectric constant of Stanyl® TE200F6 polyamide has a linear relationship with temperature. The dielectric material is utilized as a temperature-sensing material for the spiral resonator-based chipless RFID sensor as in [24]. For incorporating sensing along with identification, the resonator with the highest physical size is developed with this material. This gives rise to a shift in the lowest resonant frequency while maintaining the other resonant frequencies constant. Therefore, the lowest resonant frequency indicates the surrounding temperature. The equivalent circuit of a spiral resonator can be expressed as a lumped combination of resistor, inductor, and capacitor, as shown in Figure 3.3. The resonant frequency, f_r is given by [24],

$$f_r = \frac{1}{2\pi\sqrt{L_s C_s}} \tag{3.4}$$

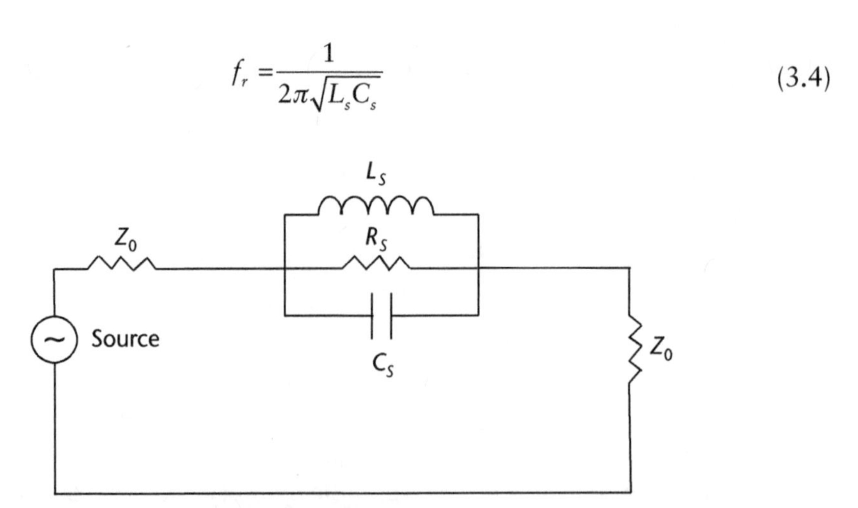

Figure 3.3 Equivalent circuit of a spiral resonator.

where f_r, L_s, and C_s are resonant frequency, equivalent inductance, and equivalent capacitance, respectively.

For the chipless sensor with Stanyl polyamide, the capacitance of the spiral is dependent on temperature as there is a variation in the relative permittivity of the Stanyl polyamide. Thus, (3.4) changes to [25]:

$$f_r = \frac{1}{2\pi\sqrt{L_s C_T}} \tag{3.5}$$

where C_T is the temperature-dependent capacitance.

3.4.4 Ionic Plastic Crystal

Plastic crystals are soft materials that are neither solid nor liquid but a stage in between these two states. The thermal behavior of the salts based on N-methyl-N-alkylpyrrolidinium cation and PF_6^- anion is explored in [25]. They possess more than one thermal transition before the melting point, which is lower (around 70°C). The varying temperature changed the conductivity of $P_{14}PF_6$ salts, which resulted in various phase transitions at –15°C, –14°C, and 42°C. The conductivity is observed to be rising as a function of temperature. This material can be utilized for concurrent temperature monitoring from –10°C to +80°C as it has a reversible behavior.

3.4.5 Phenanthrene

Phenanthrene ($C_{14}H_{10}$) responds to the surrounding temperature change by varying its phase from solid to gas (sublimation) at a transition temperature of 72°C. A study on the dielectric and pyroelectric characteristics of phenanthrene (20°C to 90°C) is carried out in [26]. The entropy value changes at the transition temperature. The material property that has an abrupt surge is the dielectric constant, ε_r after 72°C. This sharp peak in the dielectric constant is due to the phase transition. This behavior is irreversible until the vapor is desublimated. Therefore, phenanthrene can be utilized for low-cost threshold temperature sensors rather than real-time monitoring where a reversible characteristic is preferred.

3.5 Humidity-Sensing Materials

There are a wide variety of materials reacting to humidity variations in the surroundings. One of the most well known and demanding sensors is the humidity sensor. The measurement and monitoring of humidity are inevitable in our day-to-day lives, and thus the demand for accurate and less expensive humidity sensors is ever-increasing. Various humidity-sensing RFID sensors are developed in high-frequency (HF) [27] and UHF ranges [28–31]. They are manufactured either using conventional methods (photolithography) or modern techniques (screen printing, inkjet printing, etc.). The application scenarios of humidity sensing are numerous, such as in the advancement of smart grids and smart cities; in the food packaging industry to monitor the quality of food items (fresh foods such as fruits, vegetables, fish, meat, and dried grains) and to provide proper information regarding any change in the previously set threshold whose alteration can seriously affect the food quality; in structural health monitoring to identify any damage in walls or buildings due to water content; in monitoring the quality of air in a controlled room; and in self-health monitoring for diaper moisture sensing. The focus of the current research is on developing a humidity sensor with high sensitivity, a wide region of operation, stability through the dynamic range, and availability of low-cost mass fabrication. The best method for sensing is the utilization of humidity-sensitive functional materials and the integration of these with chipless RFID tags. The wide range of materials includes ceramic-based sensing materials, oxide-based sensing materials (Al_2O_3, TiO_2, SiO_2) that do not vary with temperature but respond to small variations in atmospheric humidity, and polymer-based sensing materials for a wide range of moisture conditions [32, 33]. The polymers show structural modifications in the presence of moisture, as most are hydrophilic or hydrophobic. Thus, these intelligent materials change their conductivity or permittivity according to the neighboring relative humidity. The prime objective of this section is to elaborate on these materials and their applications in terms of industrial aspects.

3.5.1 Keratin

There are several functional materials that change their properties with the surrounding relative humidity, but not all are biocompatible in nature. The biodegradable characteristic of smart material is especially relevant in the food processing industries in the packaging sector. Keratin is a group of fibrous proteins and is a biopolymer. This protein is present in the epithelial cells of hair, feathers, horns, nails, and so on. These proteins have huge contents of sulfur (S) and disulfide bonds; thus, they do not react quickly with chemical agents and are insoluble in water. In the presence of keratin, by an applied electric field, the bonds in the water molecule (H_2O) break, giving rise to H^+ and OH^- ions, which make this biopolymer well-suited for humidity sensing applications.

In [34], keratin material is utilized in a dual-polarized chipless RFID tag to monitor relative humidity. The proposed Fermat spiral resonator-based chipless sensor operates in the ultrawideband (UWB) frequency range. Moreover, the absence of a ground plane makes the compact tag (physical dimension = 10 mm^2). The biopolymer is deposited on top of the resonator with 0.03 mm thickness. The alteration in relative humidity of the surroundings is reflected in the copolar and cross-polar frequency responses, quality factor, and amplitude level. This keratin material is hydrophilic and nontoxic. Keratin is extracted from natural wool. There is a huge quantity of wool that is not suitable for the fabric industry and is thrown out as waste that can be utilized for the extraction of keratin. The numerous techniques of extraction include sulfitolysis, reduction, oxidation, alkaline hydrolysis, superheated water hydrolysis, where the disulfide bonds of the biodegradable keratin undergo breakdown. In [26], keratin is extracted using alkaline hydrolysis at 50°C (atmospheric pressure) from Merino wool, and this extraction process is less expensive when compared to other techniques where there is a high temperature and pressure requirement. The material changes its dielectric constant and loss tangent with alteration in the relative humidity. This variation can be measured by observations in the resonant frequency, which shifts accordingly. Monitoring amplitude levels and quality factor also gives an idea of the relative humidity. As keratin is not conductive, there is no change in conductivity (i.e., it remains zero throughout).

The resonator can be modeled with an RLC circuit, where R, L, and C are the equivalent resistance, inductance, and capacitance, respectively. The quality factor (Q) and resonant frequency (f_0) can be described by (3.6) and (3.7), respectively [26]:

$$Q = \frac{2\pi f_0 L}{R} \tag{3.6}$$

and

$$f_0 = \frac{1}{2\pi\sqrt{LC}} \tag{3.7}$$

Therefore, the detection methods in [34] are the observations of resonant frequency, amplitude level, and quality factor that vary with the relative humidity due to the change in loss tangent and permittivity of keratin biopolymer.

3.5.2 Silicon Nanowires

Silicon nanowires (SiNWs) have the capability to vary their electrical properties with the surrounding humidity. The variation in the dielectric constant of this material leads to a shift in the resonant frequency of the RCS plot. Thus, SiNWs are a suitable solution for humidity-sensing chipless RFID sensors. In [35], SiNW functional material is utilized for monitoring relative humidity. Catalytic chemical vapor deposition is used to deposit the material on the surface of the tag. The SiNWs are deposited on the conductive resonant elements as drops. The implementation of the sensor is compatible with printing technologies. The thickness of the deposition greatly relies on the particle size. Here, the silicon nanowire deposition is of thickness 17.5 μm. When the surrounding atmosphere contains moisture, the silicon nanowire absorbs the H_2O molecules, thus varying its dielectric constant. The dielectric constant is actually the effective dielectric constant that is acted upon as a result of the dielectric behavior of SiNWs, H_2O molecules, and the substrate dielectric (polycarbonate of relative permittivity 2.9). The relative permittivity rises when the humidity increases. This causes the resonant frequency to shift from the normal values in the absence of a humid atmosphere. It is experimentally observed that there is an upper shift in the frequency when relative humidity lowers. The measurements were carried out after a certain delay, and it proved good repeatability. Moreover, with inkjet printing technologies, repeatability can be improved.

The time-domain chipless humidity sensing tag in [36] also utilizes SiNWs. The SiNW superstrate shows a variation in permittivity and losses. As a result of this, there is a variation in RCS response (resonant frequency shift), phase, and group delay. The authors analyzed the sensitivity of the SiNW to a relative humidity of

60.2% to 88%. The chipless sensor has a sensitivity of 1.07 dB/%RH magnitude variations and 0.79 ns/%RH group delay variations.

3.5.3 Mitsubishi Ink

Commercial Mitsubishi ink, when printed on Mitsubishi paper substrate, acts as a humidity-sensitive material as proved in [37]. The superstrate (Mitsubishi ink) and the paper together act as a chemical interactive material (CIM). The paper presents a finite artificial impedance surface (AIS)-based chipless RFID humidity sensor. The AIS consists of a finite frequency selective surface (FSS) created by unit cells. The conductive Mitsubishi ink is printed on special Mitsubishi paper (NB-TP-3GU100) by inkjet printing technology, and the ink has the advantage of no requirement for curing. A thin coat of polyvinyl alcohol and aluminum oxide is applied prior to the printing of the ink to allow the easy deposition of the ink. The Mitsubishi paper is placed on cardboard with a ground plane. The metallic-backed cardboard gives the sensor an additional advantage in terms of enhanced quality factor and easy placement of the sensor on metallic objects. The ground plane will not cause any detuning effect on the tag's frequency response. The proposed design presents a coplanar frequency response. The three concentric loops of the AIS caused three deep and high-Q nulls. The dielectric permittivity of the CIM material is due to a variation in the surrounding humidity. With the increase in humidity, there is a downshift in the resonant frequency. The shift in the deep nulls according to the relative humidity shows the mechanism of sensing. A single frequency shift is sufficient for the monitoring of humidity, but multiple frequency analysis gives better results. This lowers the false reading as it provides better redundancy. The cardboard substrate on which the paper is placed is also sensitive to moisture, but as it is not exposed to the surroundings, only the Mitsubishi ink's dielectric properties change with variation in relative humidity. The adhesive tape that bonds together the Mitsubishi, cardboard, and ground plane separates the cardboard from the surrounding atmospheric conditions. The shift in resonant frequency relies on the air gap between the CIM and the resonators. If there is no air gap, the shift is the maximum. It is observed that the permittivity varied from 2–2.9 and the air gap from 0–0.1 mm. The air gap of 0 mm resulted in a remarkable lowering of frequency, while the 0.2-mm air gap led to a limited shift. It is shown that the relative humidity alteration of 50% to 90% led to a 270-MHz frequency shift. This structure illustrates lower error probability and high sensitivity but is bulky.

3.5.4 Kapton

The Kapton polyamide ($C_{12}H_{12}N_2O$) works as a capacitive humidity sensor as it varies its dielectric properties linearly with respect to surrounding moisture content. The films are created by heat-activated polycondensation. The polyamide undergoes hydrolysis when it absorbs surrounding H_2O molecules (converse reaction of polycondensation). Then, the internal carbon-nitrogen bonds break, leading to a variation in the electrical polarization. This alteration in electrical polarization changes the dielectric permittivity. This hydrophobic organic material's characteristics of dielectric constant are presented in [38]. It is observed that the functional material shows an $\varepsilon_r = 3.25$ at 25% humidity (T = 23°C). In addition, the dissipation factor varies from 0.0015 at 0% RH to 0.0035 at 100% RH. The relation between ε_r and RH is described as [38]

$$\varepsilon_r = 3.05 + 0.008 \times RH. \tag{3.8}$$

The Kapton HN polyamide is well-suited in sintering (inkjet printing of chipless sensors) as it has the capability to tolerate high temperatures. Moreover, this polyamide provides less loss, flexibility, and durability. In [39], Kapton HN polyamide is utilized in the humidity sensor developed for structural health monitoring. This can easily detect the presence of moisture content inside the floor and walls, thus identifying water damage quickly and lowering the additional damage that may result due to continuous exposure to water. As this sensor is passive and has a long-life span, it can be permanently embedded in walls, ceilings, or floors for continuous long-term monitoring of structural health. Also, its flexible nature and compact physical dimension allow it to fit in small structures. The Kapton 500 HN polyamide film by DuPont acts as the substrate on which the conducting silver nanoparticle ink (Harima NPS-J) of thickness 125 μm is deposited using inkjet printing, which forms the resonating structure. These conducting resonators act as the plate of the capacitors. When the ambient humidity varies, the dielectric permittivity of the Kapton changes, which is converted to variable capacitance using these parallel plate capacitors. This further varies the impedance matching between the integrated circuit on the tag and the resonant structures, thus changing the tag's gain with relative humidity. Here, a humidity level variation from 0% to 100% led to a 25% change in capacitance. The observed humidity sensitivity is $\frac{(198.88 \pm 14)kHz}{\%RH}$. The power on the tag, $P_{on\text{-}tag}(H)$, which is greatly dependent on humidity, is given by [39]

$$P_{on-tag}(H) = L_{fwd} P_{TS} = \frac{P_{IC}}{G_{tag} \times (1 - \tilde{A})} \tag{3.9}$$

where L_{fwd}, P_{TS}, G_{tag}, P_{IC}, and Γ are the path loss in the forward link (reader to tag), the power transmitted to activate the tag (threshold power), tag antenna gain, IC sensitivity, and power reflection coefficient, respectively.

3.5.5 Polyvinyl Alcohol

Polyvinyl alcohol is a hygroscopic polymer that exhibits variation in permittivity and conductivity with surrounding humidity. The glass transition temperature of PVA, $[-CH_2CHOH-]_n$ is about 70°C and has high molecular weight. As well, the hydrophilic nature of PVA makes it useful as a polyelectrolyte-based resistive sensor. The formation of H-H bonds in the presence of moisture varies the dielectric and conductive properties of PVA. The dielectric properties of PVA are explored in [40]. It is observed that for a frequency variation from 0.2 GHz–20 GHz, the real part of relative permittivity, ε_r', reduces with an increase in ambient humidity while for the frequency greater than 5 GHz, the imaginary part of relative permittivity,

ε_r'', increases with an increase in moisture content. This explains the dependence of both permittivity and conductivity on humidity.

The authors in [41] examined the PVA in microwave RF sensing directly and provided its potential to be implemented with mm- and μm-wave high-frequency RF sensing scenarios. A SIR structure is implemented on the Taconic TLX-0 substrate. The PVA 31-50000 polymer superstrate (deposited on top of the SIR resonators) acts as a humidity-sensitive chipless tag. The polymer obtained from Sigma Aldrich is dissolved in H_2O/ethanol 3/1 solution for about 3 hours utilizing magnetic stirring. Then droplets are deposited on the SIR resonators to attain a thickness of 0.1 mm after proper drying. It is observed that PVA has strong sensitivity and resulted in a frequency shift of 95 MHz for a humidity variation of 50% to 90%. Similarly, the PVA was coated on an electric field coupled inductor capacitor (ELC) resonator, which portrayed a frequency shift of 270 MHz from 35% to 85% relative humidity variation. Similar experiments were done for Kapton polymer where PVA showed better results and sensitivity when compared to Kapton. The hysteresis analysis is carried out to analyze the repeatability of the sensor. Kapton showed unique repeatability but high time for attaining stability. PVA has short-term repeatability, which makes it well-suited for reproducible real-world scenarios.

3.6 pH-Sensing Materials

Currently, smart packaging of food and medicines has evolved to the point where technology allows for longer product shelf life and maintaining product quality, freshness, and safety. A large quantity of food is spoiled and wasted due to improper monitoring, which can be reduced by the development of low-cost wireless sensors. When food deteriorates (food spoilage), microbes grow on these substances, which may alter the chemical composition of food, making them acidic or alkaline. Measuring the pH of food products helps in identifying the status of these items. pH provides a measure of acidity or alkalinity, thus indicating spoilage. For example, milk is spoiled due to the growth of bacteria on it, so timely monitoring can reduce waste and early intake by consumers. In the food industry, the freshness of food is vital and is maintained in the food packaging. Over and above, the sensor should be safe, biodegradable, and made of organic material, as it should be kept in close proximity to the food. Another application scenario where pH plays a significant part is digital agricultural applications [2]. The pH parameter can help in plant health monitoring by measuring soil salinity. Soil salinity is dependent on pH, and proper monitoring can result in good yield. There is a wide variety of materials showing sensitivity to pH, which include conducting polymers. The pH-sensitive polymers comprise poly acids carboxylic acids (COOH), sulfonic acids (SO_3H), and natural polymers (chitosan, hyaluronic acid, and dextran), conducting polymers (PEDOT: PSS, PANI), polydimethylsiloxane (PDMS), reduced graphene oxide (rGO), and so on. This section highlights various materials for pH sensing.

3.6.1 PEDOT: PSS

PEDOT: PSS is one of the conducting polymers that is robust and highly conductive. Even though conducting polymers have instability when exposed to environmental influences, they have the advantage of processability. PEDOT: PSS is known for its special properties, such as high conductivity, stability, perfect transparency in the doped situation, and proper film-forming capability. This smart material changes its conductivity according to the surrounding pH variation, and its variations are reversible over a wide pH range. It gives rise to low conductivity at high pH levels. PEDOT: PSS materials can disperse in water and spin-coated, thus forming transparent films. These films have high conductivity ($\sigma = 0.05–10$ S/cm) and low resistance. The conductivity and resistivity are highly dependent on the PEDOT to PSS ratio in the molecule. These materials can be printed on chipless RFID tags and can be utilized as a perfect pH sensor. The Material Engineering Department, Monash University, Clayton, Australia, has designed and fabricated PEDOT film on a plastic substrate that is 1 μm thick. The material has a conductivity of 5×10^4 S/m [1].

3.6.2 Polyaniline

Polyaniline is a conducting polymer that is well-suited for pH sensing applications due to its long-life span, excellent operational stability, and limited variation with singly charged cations. Moreover, these smart materials are easy to process and synthesize, have good stability in hydrated solutions, and have the capacity to integrate into polyaniline (PANI)-based polymer blends to confer unusual chemical precision. The control of the electrical properties of PANI is carried out through redox or protonation reactions and is reversible. The fully reduced, half-oxidized, and fully oxidized states of PANI are leucoemeraldine base (LB), emeraldine base (EB), and pernigraniline base (PB), respectively. PANI's oxidized state can be altered from LB to EB and then to PB. The protonation of EB results in the emeraldine salt (ES), the conducting form of PANI. This process increases the conductivity by about 10 orders of magnitude. This conducting state ES shows a good response to the surrounding pH level and can be used in pH sensors.

In [42], the two primary methods utilized for evolving PANI films are (1) doping PANI with camphor sulfonic acid (CSA), and (2) doping PANI with polystyrene sulfonate (PSS), where a stable response was observed in PANI-PSS. Here, the sensitive material is deposited on the resonant antenna of the RFID sensor for pH sensing. In the proposed work, the pH level variations in the surrounding cause a variation in the film resistance and capacitance between the antenna turns, thus affecting the antenna impedance. The multivariate analysis of the impedance spectra provides the estimation of the pH level.

3.6.3 Polydimethylsiloxane

Polydimethylsiloxane smart materials are utilized in UHF RFID sensors [43]. The PDMS material is coated on the RF Micron RFM 2100-AEr UHF RFID tag operated in 800 MHz–860 MHz to characterize. Different pH buffers are applied to the tag, which results in the formation of a flexible solid, leaving NH_2 groups unreacted. This group goes into reaction with protons forming NH_3^+ groups. The amount of

NH_3^+ is directly proportional to the pH value of the poured buffer. The concentration of the NH_3^+ group is the indication of the surrounding pH levels. The excellent flow properties of PDMS make it suitable for industrial applications. However, the sensitivity of the material in [43] ranges from pH 7.7 to 8.8. Microbial activity in food occurs over a range of 4 to 7 pH levels, which cannot be detected using this material. Therefore, PDMS is not recommended for food freshness monitoring.

3.6.4 Chitosan Hydrogel

The chitosan hydrogel comprises chitosan, polyethylene glycol (PEG), and acetic acid. The surrounding pH level is monitored by depositing this functional material as a superstrate on the chipless tag and observing its response. In [44], a chitosan-based hydrogel is used in the chipless RFID pH sensor operating in UWB frequencies. The proposed sensor can work over a pH range of 4 to 10 and is thus utilized for both acid and base pH sensing. The chitosan hydrogel changes its physical shape according to the variation in pH level. It was experimentally shown that the acidic medium caused the hydrogel to swell, whereas the base (alkaline) medium led to the deswelling from the previous shape. The change in the first 30 minutes stayed the same after 120 minutes for both acidic and base medium, showing consistency. The observation of S_{11} showed that the amplitude levels were reduced and a left shift of frequency response occurred over a time of 120 minutes. When exposed to an acidic medium, H^+ ions break the chemical bonds of chitosan, causing deformation in the shape, and electronic transferability is lowered. The conductivity led to increased dielectric behavior. When exposed to a base medium, similar behavior occurred even though there was not much variation in amplitude levels. Therefore, chitosan hydrogel varies its physical (shape) and electrical (dielectric constant) properties in acidic and alkaline mediums. However, their dielectric properties were observed to be almost similar in both mediums. The chitosan hydrogel smart material can be used for pH sensing in cold chain and agricultural applications.

3.7 Gas-Sensing Materials

Currently, there is a huge demand for gas sensors in monitoring the presence of harmful gases in home and work environments. In the food industry, monitoring specific gases like ammonia, carbon monoxide, and hydrogen is inevitable in maintaining the safety of food products. One of the hazardous gases produced by human activities, either by direct or indirect means, is ammonia (NH_3). A concentration above 25 ppm can seriously affect humans (respiratory tract, liver, eyes, skin, and kidneys), which calls out the need for identifying its presence. NH_3 gas detection in the human breath (0.8–14 ppm for a patient and 0.15–1.8 ppm for a healthy person) can help in the diagnosis of renal disorders or ulcers; thus, its sensing plays a significant role in the healthcare domain [45, 46]. Toluene and xylene are the two harmful aromatic hydrocarbons that affect human health. Toluene, an ingredient in the solvent of indoor paints, can cause neurological diseases such as building syndrome (SBS), and the concentration of toluene in the indoor air should be 70 ppm (World Health Organization), which can be detected by proper gas sensors [47].

N-propanol gas is a raw material used in the preparation of cosmetics and medicinal drugs whose exposure to humans for a long duration leads to nausea, headache, coma, and ultimately death [48]. A chipless RFID sensor has a major role to play in this scenario. Several materials can be integrated with chipless tags to detect the presence of these gases and are highlighted in this section.

3.7.1 Polypyrrole

Polypyrrole (PPy) is a biocompatible and stable material that exhibits excellent performance for gas sensing. Its significant aspect is that it can work at room temperature. Moreover, PPy with rGO combination are suitable gas sensors as they have high sensitivity, are easy to synthesize, and are low-cost. The high sensitivity of the PPy/rGO is due to the efficient electron transfer between NH_3 and PPy and the effective electron pathway in rGO. In addition, the response time is also improved because of the combining of rGO with PPy. According to the literature, the minimum concentration of ammonia that can be sensed by PPy nanoribbon is 0.5 ppm. The response time and recovery time are around 8 min and 3 min, respectively. In [45], a novel NH_3 sensor is designed utilizing a polypyrrole material layer deposited on graphene. The graphene is grown by chemical vapor deposition (CVD). The polypyrrole is synthesized on CVD-grown graphene at 20°C by electropolymerization. The amount of gas is detected by observing the variation in the resistance of two electrodes of the sensor (counter and reference electrodes) and is given by [45]

$$\frac{\Delta R}{R_0} = \frac{(R_s - R_0)}{R_0} \% \tag{3.10}$$

where ΔR, R_0, and R_s are the change in resistance, sensor resistance before exposure to NH_3 gas, and sensor resistance after exposure to NH_3 gas, respectively.

The polypyrrole acts similarly to a p-type semiconductor. When NH_3 molecules are adsorbed on the surface of PPy, an interaction is induced between the NH_3 molecule and PPy. NH_3, which is a donor, transfers electrons to the PPy surface. Thus, the hole concentration on the surface of PPy lowers, resulting in a surge of PPy resistance. Thus, the concentration of ammonia can be detected by the measurement of resistance. Graphene (or any metal) is required as an intermediate layer because electropolymerization cannot be performed on insulators. Graphene is used because metals are too conductive and have difficulty detecting small variations in resistance. The graphene (p-type semiconductor) permits a measurable modification in resistance and tuning of the PPy thickness down to the nanoscale. The sensor presented here detects NH_3 concentration from 1 to 4 ppm (room temperature). They show sensitivity, response time, and recovery time of 1.7%/ppm, 2 min, and 5 min, respectively. Moreover, the porous layer of PPy is ultrathin, which plays a significant part in improving sensitivity, selectivity, and immunity to humidity.

3.7.2 Single-Walled Carbon Nanotubes

Single-walled carbon nanotubes are conductive sensing substances having high electrical conductivity and transparency. Carbon nanotubes and their composites are very efficient in gas sensing. In the presence of much fewer quantities of harmful gases such as NH_3 and nitrogen gas (NO_x), the conductivity of SWCNT varies. As a result, there will be a modification in the received power level. Thus, the presence of gases can be detected by the integration of SWCNT in chipless RFID tags. The important aspects of nanotubes are as follows: (1) low carrier mobility, (2) high electron mobility, (3) light absorption dominance, and (4) reflection in the direction of the nanotube axis. The SWCNT is integrated with an RFID antenna in [46] for the detection of hazardous gases. This sensor is fully printable on a paper substrate with inkjet printing technology. CNT composites have an attraction toward gas molecules. The conductivity of the CNT varies with the absorption of NH_3 gas. As the number of CNT layers increases, the resistance reduces (here, a 25-layer film is utilized, which has a stable response). A 4% consistency is observed in the measurement. The film exhibits a fast monotonic impedance response curve according to the presence of gas. A stable response is followed by the sensor until 1 GHz. In the absence of ammonia, the impedance is 51.6–j6.1 Ω and the value rises to 97.1–j18.8 Ω in the presence of ammonia at 868 MHz. Equation (3.11) provides the power reflection coefficient, η, which is a function of load impedance, Z_{Load} and antenna terminal impedance, Z_{ANT} [46]:

$$\eta = \left| \frac{Z_{Load} - Z^{*}_{ANT}}{Z_{Load} + Z_{ANT}} \right|^2 \tag{3.11}$$

Z_{Load} varies with the concentration of ammonia, and thus, it functions like a tunable resistor. Therefore, the received power level is an indicator of the change in load, which is further dependent on the ammonia concentration of the surroundings.

3.7.3 NiO-In2O3

N-propanol gas, which is hazardous to human health, can be detected using various materials such as $LnFeO_3$ perovskite nanomaterials, PdO-decorated double-sheet $ZnSnO_3$ nanomaterials, and ZnO/NiO heterostructure materials. Metal oxide semiconductors (MOSs) are extensively used because of their properties such as simple structure, low-cost, long life span, and high sensitivity. However, the same limitation includes poor selectivity and small response. In [48], NiO-In_2O_3 is utilized for n-propanol sensing, which shows a higher response value and a much faster response and recovery time than pure In_2O_3. The sensing mechanism of In_2O_3 and NiO-In_2O_3 can be explained as follows. The surface of In_2O_3, which is an n-type MOS (electron carrier), adsorbs oxygen from air and form adsorbed oxygen ions. The creation of an electron depletion layer on the material surface occurs through the absorption of electrons from the In_2O_3 conduction band. This results in the lowering of the conductivity of In_2O_3. In the presence of n-propanol gas, the reaction between n-propanol gas molecules and the adsorbed oxygen releases back electrons into the conduction band, thus increasing the conductivity. Therefore, the measurement of conductivity can indicate the concentration of n-propanol gas. In the case of NiO-In_2O_3, a PN junction is formed on the surface of NiO and In_2O_3. The resistance of heterojunction materials is given by [48]

$$R \propto B exp\left(\frac{q\phi}{kT}\right) \tag{3.12}$$

where B, φ, k, and T are the constant (temperature dependant), heterojunction potential barriers, Boltzmann constant, and temperature, respectively.

In the presence of gas, the n-propanol molecules react with the surface, and the electrons reenter the conduction band, resulting in the reduction of the potential barrier. The resistance has an exponential relationship with the barrier; therefore, a small variation in the barrier can be observed in the resistance value, which aids in the measurement of gas concentration. It was observed that in the NiO-In_2O_3 sensor, the response value was enhanced by 5.2 times than pure In_2O_3.

3.8 Strain and Crack Sensing Materials

With the demand for various fields such as structural health monitoring and pressure sensing, strain and crack sensors have attained relevance in the current scenario. With the construction of a structure, a sensor is embedded into it to detect any fault or deformation. This aids in the deformation detection of any damage in the initial phase, thus preventing serious destruction later. There need not be any extra cost for diagnosis and maintenance as the sensors are already embedded. Metals and semiconductors utilized in conventional strain sensors are the reason for their low strain range, sensitivity, and toughness. The major components of a strain sensor comprise a sensitivity element, preferably made of a flexible, sensitive material, an electrode, and a wire. The currently used material for excellent strain sensing due to high electrical conductivity and mechanical characteristics are carbon-based materials, metal nanoparticles, conduction polymers, nanowires, and so on. Examples of carbon-based nanomaterials include multiwalled carbon nanotubes (MWCNTs), graphene, carbon-based nanofibers, and carbon black nanoparticles (CBNPs). The combination of different polymers and carbon-based nanomaterials results in strain sensors with high flexibility and stability. The substrates preferred for the protection and the enhancement of interfacial adhesion of these nanomaterials include thermoplastic polyurethane (TPU), PET, PDMS, polyimide (PI), natural rubber, polypropylene (PP), and epoxy resin. Generally, pure carbon-based materials have a higher sensitivity than carbon-polymer composites. However, the polymer adds advantages such as enhanced stability, strain range, and linear response. For better performance characteristics, the interaction between the carbon-based material and polymers should be stable. In the case of composite-sensing material, the volume fraction of conductive fillers such as carbon nanoparticles (0D geometry), carbon nanotubes (1D geometry), graphene nanoplatelets (2D geometry), and so on, is incorporated in an insulative matrix material. The conduction of these fillers helps in the creation of percolation networks (conductive path). If any deformation or damage occurs to the structure, the percolation networks get changed or broken. Thus, monitoring the electrical resistance of these networks gives the measure of strain, which is beneficial in structural health monitoring [49]. The healthcare application of strain sensing includes wearable strain sensors for monitoring body movements like finger flexion and leg movement. Another application where strain sensing is significant is the electronic-skin (e-skin). Multifunctional sensing has an important part to play in the smooth manipulation and natural human-machine interaction of smart robotics and prosthesis [50].

This section explores various materials for strain and crack sensing. Several materials like glass microfiber-reinforced PTFE composite, nickel-titanium (Nitinol) alloy, polyester-based stretchable fiber, and silver nanowires can be utilized for strain and crack sensing where PTFE has excellent electrical behavior and a large variation in resonant frequency with alteration in temperature that contributes to the high dielectric permittivity of the material.

3.8.1 PTFE

Most sensors, such as acoustic, optical, or piezoelectric, provide good results but are very expensive. This calls out the need for sensors utilizing smart materials for application in structural health monitoring. A substrate material, Rogers RT/duroid 6202, a ceramic-filled PTFE material, is utilized in the RFID antenna sensor in [51] for strain and crack sensing. The layers of the RFID sensor are a copper ground plane (bottom), substrate (middle), and copper cladding with an RFID chip (top). The results of two substrates such as Rogers 5880, a glass microfiber-reinforced PTFE composite, and Rogers 6202, a ceramic-filled PTFE composite, are analyzed and compared. The resonant frequency of the folded patch antenna under zero strain is given by [51]

$$f_{R_0} = \frac{c}{4(L+L')\sqrt{\varepsilon_r}} \tag{3.13}$$

where c, L, L', and ε_r are the speed of light, the physical length of the copper cladding, length compensation because of fringing effect, and permittivity, respectively. When there is a strain acting upon the sensor, the physical length, L, changes. During strain (at room temperature), the resonant frequency is given by [51]

$$f_R = \frac{c}{4(L+L')(1+\varepsilon)\sqrt{\varepsilon_r}} \approx f_{R_0}(1-\varepsilon) \tag{3.14}$$

where ε represents a small strain ($<10^{-3}$). The theoretical strain sensitivity at room temperature is given by [51]

$$S_\varepsilon = -f_{R_0} \tag{3.15}$$

The shift in resonant frequency indicates the amount of strain where a positive ε shifts the resonant frequency to the left while a negative ε does the reverse, which can be observed by (3.14). It is experimentally shown that 1 $\mu\varepsilon$ strain rise led to a 750 Hz lowering of the resonant frequency. Moreover, by crack tests, it is shown that the sensor has the capacity to detect milli-inch crack width and trace its propagation. It is observed that Rogers 6202 maintains a stable permittivity when temperature varies compared to Rogers 5880, thus enhancing the sensor reliability.

The coefficient of thermal expansion in Rogers 6202 is 8 ppm/°F, 8 ppm/°F, and 17 ppm/°F in x, y, and z directions, respectively which are reduced by half of that of Rogers 5880 in the x and y directions and 87% in the z direction. Therefore, it proves a stable behavior under thermal fluctuations. Moreover, dielectric constant change is also smaller in Rogers 6202 than in Rogers 5880 under different thermal conditions.

3.8.2 Silver Nanowires

Strain sensing with the use of smart materials such as silver nanowires (AgNWs) is widely applied in e-skin, health monitoring, and smart textiles. In [52], finger movement is monitored using smart gloves where AgNWs and MWCNTs are doped as conductive materials on an ecoflex flexible substrate. The article shows experimental results of the sensor performance with different strain values and different concentrations of AgNWs (prepared by the polyol method). The sensor with high sensitivity, stability, and repeatability by different amounts of AgNWs is found. The use of ecoflex rubber (platinum catalyzed silicone resin) is easy. Moreover, the lower value of viscosity helps in easy mixing and dehydration. Excellent characteristics are observed by the mixture of MWCNTs and ecoflex. The combination of carbon nanotubes with AgNWs improves the conductivity of AgNWs. The relative resistance of the sensor system varies with a variation in the strain. The electromechanical property of the strain sensor is given by R/R_0 where $R = (R - R_0)$, R_0 and R are the resistances before and after deformation due to the applied strain, respectively. The gauge factor (GF) of the material is given by [52]

$$GF = \frac{\Delta R / R_0}{\varepsilon} \tag{3.16}$$

where $\Delta R/R_0$ and ε are the variation of the sample electric resistance and elongation change, respectively. The factors responsible for the improvement of sensitivity are the increase in AgNW concentration and stretch rate. However, a very large amount of AgNWs can decrease sensor sensitivity. The best linearity of the sensor is achieved with a concentration of AgNW equal to 0.15 mg/cm^2.

3.8.3 Polypropylene Carbon Nanofiber Composite

Carbon nanofibers (CNFs) are very effective fillers for polymer nanocomposites due to their good thermal and mechanical properties and low-cost employment into polymers in comparison with carbon nanotubes (CNTs). With the deformation of the material, the conductive characteristics of the CNF vary, thus giving a measure of strain. In [53], the piezoresistive property of polypropylene CNF composite is analyzed for strain measurement scenarios. Piezoresistivity is the phenomenon where the resistivity of a material alters when mechanical stress is applied. It is observed that as the concentration of CNF increases, the variation of resistivity with strain decreases. That is, it led to less piezoresistive response, which means less piezoresistive behavior is obtained when the samples are more conductive. With small strain values (up to 3×10^{-3}), the gauge factor remains constant while a linear relationship between strain and GF is observed up to 9×10^{-3} (maximum obtained GF is 2.46). It is also evident that the electrical resistance varies nonlinearly with strain. The composite has a positive piezoresistive coefficient. The strain gauge based on the materials presented here can tolerate high deformation at low-cost when compared to metal or semiconductor strain gauges.

3.9 Light-Sensing Materials

The sensing and continuous monitoring of light intensity have received great attention in today's world due to its tremendous applications such as indoor lighting, aquaculture, safety and security, and agriculture. In the popular concept of smart homes, the proper monitoring of light intensity aids in power saving. In addition to power consumption, the control of light delivers a suitable environment for the consumer [54]. In aquaculture, the monitoring of fish existence and the quality of water is essential. An optical sensor utilized in underwater wireless sensor networks (WSNs) helps farmers monitor fish movement. Such a system is designed and verified in [55], which is based on light-dependant resistors (LDRs). LDRs are devices that change their resistivity according to the variation in light intensity. These devices are made of semiconductor materials such as cadmium selenide (CdSe) or cadmium sulfide (CdS), whose conductivity has a linear relationship with light intensity. Most of the devices use microcontrollers and complementary metal-oxide-semiconductors (CMOS), thus leading to a surge in the cost of the light sensor. This calls out the need for sensors made of smart materials. Thus, chipless RFID sensors with light dependence are in huge demand in this scenario. There are a number of light-sensitive materials that change their conductivity with light intensity, such as indium antimonide (InSb), lead sulfide (PbS), and lead selenide (PbSe), and materials that permanently vary their dielectric properties on exposure to ultraviolet (UV) radiations such as SU-8, and poly(methylmethacrylate) (PMMA). In [56], an azobenzene polymer-based UV light sensor is proposed, which induces strain on the optical fiber where this polymer is coated. This measurement of strain indicates the amount of UV exposure. This section highlights various light-sensitive materials.

3.9.1 Cadmium Sulfide

One of the popular light-dependent resistor materials is CdS. Light-dependent resistors are also known as photoconductors. CdS has a very highly sensitive response in the visible range (450–700 nm), thus matching with the human eye. In [57], a metal-oxide semiconductor field effect transistor (MOSFET) is combined with photoconductive material (CdS) to act as a light sensor MOSFET (LiSFET). A chipless tag integrated with CdS, thus giving the performance of a chipless light sensor, is designed in [58]. With the increase in light intensity, the CdS shows a lowering of resistance and is commonly utilized in low-frequency scenarios. The multiresonator-based chipless tag designed here consists of three SIRs and an additional resonator loaded with a photoresistive element. All the experiments are performed in a closed box under suitable light conditions. The variation of light intensity is carried out with the aid of two cross-polarized lenses in front of a bright light source. With various light intensities, a variation in the impedance (both resistance and reactance) is observed. Moreover, there is a change in the resonant frequency and received power level with a light intensity varying from 0 to 1,000 lux. The shift in resonant frequency is evident from the photoresistive material coated SIR. The maximum frequency shift obtained is 70 MHz, and the power variation is 7.5 dB. It can be concluded that the light sensitivity of CdS is less in RF when compared to DC, but it provides light-dependent scattering parameters when integrated with radio frequency circuits matching to 50 Ω. Therefore, they are well-suited for sensing in chipless RFID tags.

3.9.2 Amorphous Hydrogenated Germanium

A germanium semiconductor, having an indirect bandgap at 0.66 eV and a direct bandgap at 0.8 eV, is generally utilized in thin-film transistors, optoelectronics, photodetectors, radiation detectors, and so on. Its excellent properties, such as high charge carrier mobility, low melting point, linear response, and low process temperature, make it well-suited for wearable and flexible electronics. In [59], a recrystallized amorphous hydrogenated germanium (a-Ge: H) thin film on polyimide is utilized as a near infrared (NIR) light sensor. It acts as a photoresistor, which varies its resistivity in accordance with the light intensity. a-Ge: H is an intrinsic semiconductor, and so has low electrical conductivity. Therefore, the conductivity is improved by laser annealing. In this work, a NIR light-emitting diode (LED) at 850 nm is utilized to analyze the response of the sensor. The resultant photocurrent is measured, which is an indication of the amount of light intensity. It is observed that the sensor shows a large dynamic range, and the photocurrent follows an inverse square distance relation. Thus, a-Ge: H is a suitable candidate for NIR light sensing.

3.10 Graphene

Graphene is widely used in sensors due to its excellent electrical at high frequency (especially in THz) and thermal and mechanical properties such as high electron mobility, tensile, transparency, and large surface area, which is due to the result of its physical (two-dimensional honeycomb carbon lattice) and electronic structure. At room temperature, the electron charge mobility is higher than 100,000 cm^2/Vs and current densities of 10^8–10^9 cm^2 Ns. The thermal conductivity is 5,000 W/mK. The different methods to attain high quality graphene include mechanical exfoliation (low scale) and chemical vapor deposition (wafer scale). It can be made of layers of multilayered graphite as it can be made by the deposition of one layer of carbon onto another material. The fundamental building blocks of graphite, carbon nanotubes, and quantum dots are graphene. Even though it has low weight, it is 100 times more potent than steel [60, 61].

Two-dimensional materials such as graphene play a significant role in wearable biosensors due to their good electrical conductivity, mechanical flexibility, and stability to biological electrolytes. Several tattoo-like devices based on graphene were previously reported for vital signs monitoring such as heartbeat and body temperature. The subnanometer thickness of graphene results in easy bending and stretching with the skin. In [62], a graphene-based light sensor is proposed to monitor blood oxygen levels and human pulse where light is passed through the tissue.

The graphene is coated with nanomaterial. The light is absorbed by the blood (i.e., hemoglobin, which carries out the oxygen transparent in the blood). By monitoring the light intensity passing through the tissue (photoplethysmograms), the heart rate can be measured. With the absorption of light, arterial oxygen saturation (SpO_2) can also be measured as it will depend on the presence of oxygen in the cells.

Various physiological parameters from the skin can be measured using skinlike electronic systems such as epidermal electronic systems (EESs). In [63], a skin-conformable inkjet-printed temperature sensor is designed based on graphene/PEDOT: PSS. The composite of graphene/PEDOT: PSS is suitable for stretchable applications. Here, the composite is deposited on a polyurethane substrate, which acts like the epidermis. Figure 3.4(a) depicts the sensor multilayer structure and Figure 3.4(b) illustrates the device mounted on skin. The sensor was exposed to different temperature conditions. It is observed that the graphene/PEDOT: PSS composite acts like a negative temperature coefficient (NTC) material as the resistance lowers when the temperature rises.

Nanomaterials are known for strain sensors because of their excellent electromechanical properties. In [64], graphene woven fabrics (GWFs) are utilized for monitoring strain. With the application of deformation, the polycrystalline structure of GWFs transforms, which varies the resistivity of the structure. A gauge factor of ~10^3 is observed for 2% to 6% strain and ~10^6 for higher values of strain. A sample of the sensor is shown in Figure 3.5(a), with the optical images under 20% and 50% strain shown in Figure 3.5(b). The electrical resistance has a linearly increasing behavior with strain. The relative resistance ($\Delta R/R_0$) increases with a nonlinear behavior up to 2% strain and later, with higher strain, exponentially increases as a result of crack formation and structural changes included in GWFs. This sensor shows excellent repeatability and sensitivity.

3.11 Summary

Wireless passive sensing has gained a lot of attention in various sectors like healthcare, agriculture, retail, the textile industry, and banking. Chipless RFID sensors play a vital role in this due to their low-cost, flexible, and printable nature. The

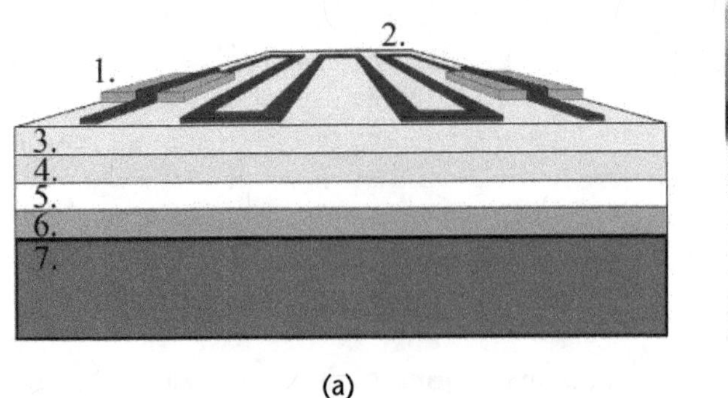

(a)

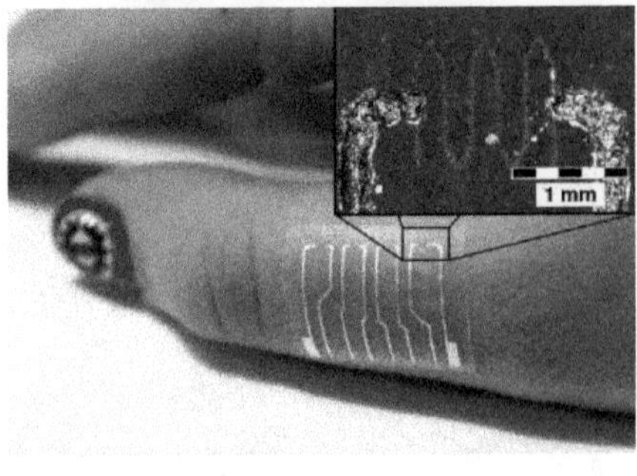

(b)

Figure 3.4 (a) Sensor multilayer structure: 1. screen-printed silver conductors, 2. wave-patterned graphene/ PEDOT: PSS temperature sensor, 3. PU surface layer, 4. adhesive layer, 5. protective paper, 6. PET film, and 7. cooling/heating element. (b) Device mounted on the skin [63].

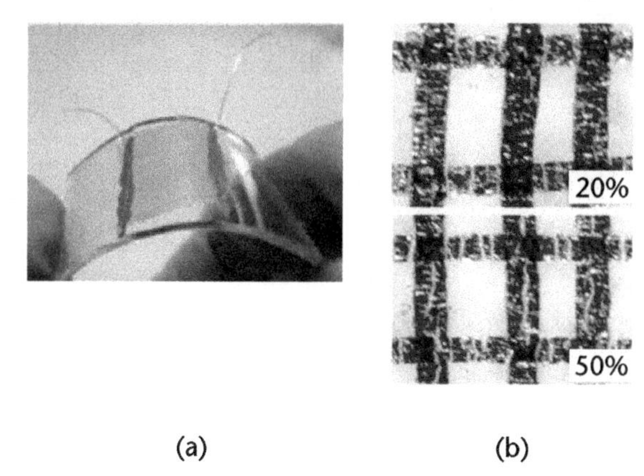

Figure 3.5 (a) A sample of the sensor, and (b) optical images under various strain conditions [64].

sensing capability of chipless sensors can be enhanced with the use of smart, functional materials. In this chapter, various smart materials were classified based on their sensitivity to different physical parameters. The smart materials printed on the chipless tag vary their properties in accordance with the surrounding conditions, thus enabling sensing. Different temperatures, humidity, pH, gas, strain and crack, and light sensors were discussed in this chapter. Graphene, which plays a significant role in various sensors, was explored. The materials used in several real-world applications were highlighted, thus showing that the materials are practically viable. It has been discovered that some materials have the capability of sensing more than one physical parameter. Thus, chipless RFID sensors are well-suited for multiparameter sensing. Fully printable chipless RFID sensors are the future of green technology and pollution-free disposable sensor nodes. These unique, less-expensive systems have the potential to monitor any object and connect them across the world through the advent of the Internet of Things.

References

[1] Amin, E., M. J. K. Saha, and N. C. Karmakar, "Smart Sensing Materials For Low-Cost Chipless RFID Sensor," *IEEE Sensors Journal,* Vol. 14, No. 7, July 2014, pp. 2198–2207.

[2] Karmakar, N. C., E. M. Amin, and J. K. Saha, *Chipless RFID Sensors,* Hoboken, NJ: John Wiley & Sons, 2015.

[3] Castano, L. M., and A. B. Flatau, "Smart Fabric Sensors and E-Textile Technologies: A Review," *Smart Materials and Structures*, Vol. 23, No. 5, 2014, 053001.

[4] Erdem, Ö., et al., "Smart Materials-Integrated Sensor Technologies for COVID-19 Diagnosis," *Emergent Materials*, Vol. 4, 2021, 169–185.

[5] Basheer, A. A., "Advances in the Smart Materials Applications in the Aerospace Industries," *Aircraft Engineering and Aerospace Technology,* Vol. 92, No. 7, 2020, 1027–1035.

[6] Su, M., and Y. Song, "Printable Smart Materials and Devices: Strategies and Applications, *Chemical Reviews*, Vol. 122, No. 5, 2021, pp. 5144–5164.

[7] Hassabis, D., D. Kumaran, C. Summerfield, and M. Botvinick, "Neuroscience-Inspired Artificial Intelligence," *Neuron,* 2017, Vol. 95, pp. 245–258.

[8] Koo, J. H., D. C. Kim, H. J. Shim, T.- H. Kim, and D.- H. Kim, "Flexible and Stretchable Smart Display: Materials, Fabrication, Device Design, and System Integration," *Advanced Functional Materials,* Vol. 28, 2018, 1801834.

[9] Fischer, P., B. J. Nelson, and G.- Z. Yang, "New Materials for Next Generation Robots," *Science Robotics,* Vol. 3, No. 18, 2018, eaau0448.

[10] Karunakaran, R., S. Ortgies, A. Tamayol, F. Bobaru, and M. P. Sealy, "Additive Manufacturing of Magnesium Alloys," *Bioactive Materials,* Vol. 2020, pp. 44–54.

[11] Schwartz, M. (ed.), *Smart Materials,* Boca Raton, FL: CRC Press, 2009.

[12] Fathi, P., N. C. Karmakar, M. Bhattacharya, and S. Bhattacharya, "Potential Chipless RFID Sensors for Food Packaging Applications: A Review," *IEEE Sensors Journal,* Vol. 20, No. 17, September 2020, pp. 9618–9636.

[13] McGee, K., P. Anandarajah, and D. Collins, "A Review of Chipless Remote Sensing Solutions Based on RFID Technology," *Sensors*, Vol. 19, No. 22, 2019, p. 4829.

[14] Preradovic, S., and N. Karmakar, "Chipless RFID Tag with Integrated Sensor," *Sensors,* 2010, pp. 1277–1281.

[15] Cui, L., Z. Zhang, N. Gao, Z. Meng, and Z. Li, "Radio Frequency Identification and Sensing Techniques and Their Applications—A Review of the State-of-the-Art," *Sensors,* 2019.

[16] Yeo, J., J. I. Lee, and Y. Kwon, "Humidity-Sensing Chipless RFID Tag with enhanced Sensitivity Using an Interdigital Capacitor Structure," *Sensors*, Vol. 2, No. 19, 2021, p. 6550.

[17] Amin, E. M., and N. C. Karmakar, "Development of a Low Cost Printable Humidity Sensor for Chipless RFID Technology," in *2012 IEEE International Conference on RFID-Technologies and Applications (RFID-TA),* 2012, pp. 165–170.

[18] Nair, R. S., Perret, E., Tedjini, S., and Baron, T., "A group-delay-based chipless RFID humidity tag sensor using silicon nanowires," *IEEE Antennas and Wireless Propagation Letters, 12*, 2013, pp. 729–732.

[19] Sebastian, M. T., R. Ubic, and H. Jantunen, *Microwave Materials and Applications,* John Wiley & Sons, 2017.

[20] Vena, A., E. Perret, D. Kaddour, and T. Baron, "Toward a Reliable Chipless RFID Humidity Sensor Tag Based on Silicon Nanowires," *IEEE Transactions on Microwave Theory and Techniques,* Vol. 64, No. 9, September 2016, pp. 2977–2985.

[21] Raju, P., and Li, Q., "Semiconductor Materials and Devices for Gas Sensors," *Journal of the Electrochemical Society*, Vol. 169, No. 5, 2022, p. 057518.

[22] Javed, N., M. A. Azam, and Y. Amin, "Chipless RFID Multisensor for Temperature Sensing and Crack Monitoring in an IoT Environment," *IEEE Sensors Letters,* Vol. 5, No. 6, June 2021, pp. 1–4.

[23] Chenghua, X. J. S., W. Helin, X. Tianning, Y. Bo, and L. Yuling, "Optical Temperature Sensor Based on ZnO Thin Film's Temperature Dependent Optical Properties," *Review of Scientific Instruments,* Vol. 82, No. 8, 2011, pp. 084901-1–084901-3.

[24] Amin, E. M., and N. Karmakar, "Development of a Chipless RFID Temperature Sensor Using Cascaded Spiral Resonators," in *Proceedings of Sensors,* October 2011, pp. 554–557.

[25] Golding, J., et al., "N-Methyl-N-Alkylpyrrolidinium Hexafluorophosphate Salts: Novel Molten Salts and Plastic Crystal Phases," *Chemical Materials,* Vol. 13, No. 2, February 2001, pp. 558–564.

[26] Kroupa, J., J. Fousek, N. R. Ivanov, B. Brezina, and V. Lhotska, "Dielectric Study of the Phase Transition in Phenanthrene," *Ferroelectrics,* Vol. 79, No. 1, March 1988, pp. 189–192.

[27] Voutilainen, J., "Methods and Instrumentation for Measuring Moisture in Building Structures," PhD dissertation, Helsinki University of Technoogy, Espoo, Finland, 2005.

[28] Chang, K., Y. H. Kim, Y. J. Kim, and Y. J. Yoon, "Functional Antenna Integrated with Relative Humidity Sensor Using Synthesised Polyimide for Passive RFID Sensing," *Electronics Letters,* Vol. 43, No. 5, March 2007, pp. 7–8.

[29] Yi, J., M. Heiss, F. Qiuyun, and N. A. Gay, "A Prototype RFID Humidity Sensor for Built Environment Monitoring," in *Proceedings of the 2008 International Workshop on Education Technology and Training & 2008 International Workshop on Geoscience and Remote Sensing,* Shanghai, China, 2008, pp. 496–499.

[30] Oprea, A., N. Barsan, U. Weimar, M. Bauersfeld, D. Ebling, and J. Wollenstein, "Capacitive Humidity Sensors on Flexible RFID Labels," *Sensors and Actuators B, Chemical,* Vol. 132, No. 2, June 2008, pp. 404–410.

[31] Oprea, A., N. Barsan, U. Weimar, M. Bauersfeld, D. Ebling, and J. Wollenstein, "Capacitive Humidity Sensors on Flexible RFID Labels," in *Transducers/Eurosensors XXI, Proceedings of the 14th International Conference on Solid-State Sensors, Actuators, and Microsystems and 21st European Conference on Solid-State Transducers,* 2007, pp. 2039–2042.

[32] Ansbacher, F., and A. C. Jason, "Effects of Water Vapour on the Electrical Properties of Anodized Aluminium," *Nature,* Vol. 171, 1953, pp. 177–178.

[33] Chen, Z., et al., "Humidity Sensors with Reactively Evaporated Al2O3 Films as Porous Dielectrics," *Sensors and Actuators B: Chemical,* Vol. 2, 1990, pp. 167–171.

[34] Fathi, P., S. Bhattacharya, and N. C. Karmakar, "Dual-Polarized Keratin-Based UWB Chipless RFID Relative Humidity Sensor," *IEEE Sensors Journal,* Vol. 22, No. 3, February 1, 2022, pp. 1924–1932.

[35] Vena, A., E. Perret, D. Kaddour, and T. Baron, "Toward a Reliable Chipless RFID Humidity Sensor Tag Based on Silicon Nanowires," *IEEE Transactions on Microwave Theory and Techniques,* Vol. 64, No. 9, September 2016, pp. 2977–2985.

[36] Nair, R. S., E. Perret, S. Tedjini, and T. Baron, "A Group-Delay-Based Chipless RFID Humidity Tag Sensor Using Silicon Nanowires," *IEEE Antennas Wireless Propagation Letters,* Vol. 12, 2013, pp. 729–732.

[37] Borgese, M., F. A. Dicandia, F. Costa, S. Genovesi, and G. Manara, "An Inkjet Printed Chipless RFID Sensor for Wireless Humidity Monitoring," *IEEE Sensors Journal,* Vol. 17, No. 15, August 1, 2017, pp. 4699–4707.

[38] Chen, Z., and C. Lu, "Humidity Sensors: A Review of Materials and Mechanisms," *Sensor Letters,* Vol. 3, 2005, pp. 274–295.

[39] Virtanen, J., L. Ukkonen, T. Bjorninen, A. Z. Elsherbeni, and L. Sydänheimo, "Inkjet-Printed Humidity Sensor for Passive UHF RFID Systems," *IEEE Transactions on Instrumentation and Measurement,* Vol. 60, No. 8, August 2011, pp. 2768–2777.

[40] Sengwa, R. J., and K. Kaur, "Dielectric Dispersion Studies of Poly(Vinyl Alcohol) in Aqueous Solutions," *Polymer International,* Vol. 49, 2000, pp. 1314–1320.

[41] Amin, E. M., N. Karmakar, and B. Winther-Jensen, "Polyvinyl-alcohol (PVA)-Based RF Humidity Sensor in Microwave Frequency," *Progress in Electromagnetics Research B,* Vol. 54, 2013, pp. 149–166.

[42] Potyrailo, R. A., C. Surman, T. Sivavec, and T. Wortley, "Passive Multivariable RFID pH Sensors," in *2011 IEEE International Conference on RFID-Technologies and Applications,* 2011, pp. 533–536.

[43] Hillier, A., V. Makarovaite, S. J. Holder, C. W. Gourlay, and J. C. Batchelor, "A Passive UHF RFID pH Sensor (Smart Polymers for Wireless Medical Sensing Devices)," in *Loughborough Antennas & Propagation Conference (LAPC 2017),* 2017, pp. 1–2.

[44] Athauda, T., P. C. Banerjee, and N. C. Karmakar, "Microwave Characterization of Chitosan Hydrogel and Its Use as a Wireless pH Sensor in Smart Packaging Applications," *IEEE Sensors Journal,* Vol. 20, No. 16, August 15, 2020, pp. 8990–8996.

[45] Tang, X., et al., "A Fast and Room-Temperature Operation Ammonia Sensor Based on Compound of Graphene with Polypyrrole," *IEEE Sensors Journal,* Vol. 18, No. 22, November 15, 2018, pp. 9088–9096.

[46] Yang, L., R. Zhang, D. Staiculescu, C. P. Wong, and M. M. Tentzeris, "A Novel Conformal RFID-Enabled Module Utilizing Inkjet-Printed Antennas and Carbon Nanotubes for Gas-Detection Applications," *IEEE Antennas and Wireless Propagation Letters,* Vol. 8, 2009, pp. 653–656.

[47] Deng, L., X. Ding, D. Zeng, S. Zhang, and C. Xie, "High Sensitivity and Selectivity of C-Doped WO_3 Gas Sensors Toward Toluene and Xylene," *IEEE Sensors Journal,* Vol. 12, No. 6, June 2012, pp. 2209–2214.

[48] Yuan, Z., J. Li, S. Luo, H. Zhang, and F. Meng, "Preparation of NiO-In_2O_3 Ordered Porous Thin Film Materials with Enhanced n-Propanol Gas Sensing Properties," *IEEE Sensors Journal,* Vol. 22, No. 16, August 15, 2022, pp. 15716–15723.

[49] Asfar, Z., S. Nauman, G. ur Rehman, F. Mumtaz Malik, Y. Ayaz, and N. Muhammad, "Development of Flexible Cotton-Polystyrene Sensor for Application as Strain Gauge," *IEEE Sensors Journal,* Vol. 16, No. 24, December 15, 2016, pp. 8944–8952.

[50] Zhang, W., Q. Guo, Y. Duan, C. Xing, and Z. Peng, "A Textile Proximity/Pressure Dual-Mode Sensor Based on Magneto-Straining and Piezoresistive Effects," *IEEE Sensors Journal,* Vol. 22, No. 11, June 1, 2022, pp. 10420–10427.

[51] Yi, X., et al. "Thermal Effects on a Passive Wireless Antenna Sensor for Strain and Crack Sensing," *Sensors and Smart Structures Technologies for Civil, Mechanical, and Aerospace Systems,* Vol. 8345, 2012.

[52] Zhang, Y., et al., "Highly Sensitive, Low Hysteretic and Flexible Strain Sensor Based on Ecoflex-AgNWs- MWCNTs Flexible Composite Materials," *IEEE Sensors Journal,* Vol. 20, No. 23, December 1, 2020, pp. 14118–14125.

[53] Rocha, J. G., A. J. Paleo, F. W. J. van Hattum, and S. Lanceros-Mendez, "Polypropylene-Carbon Nanofiber Composites as Strain-Gauge Sensor," *IEEE Sensors Journal,* Vol. 13, July 2013, No. 7, pp. 2603–2609.

[54] Elgailani, M. A., A. H. H. Al-Masoodi, N. B. Sariff, and N. Abdulrahman, "Light Dependent Resistor Sensor Used for Optimal Power Consumption for Indoor Lighting System," in *2021 2nd International Conference on Smart Computing and Electronic Enterprise (ICSCEE),* 2021, pp. 237–242.

[55] Basterrechea, D. A., M. Botella-Campos, L. Parra, and J. Lloret, "Implementation of an Optical Sensor to Detect Fish in Aquiculture Tanks," in *2020 Global Congress on Electrical Engineering (GC-ElecEng),* 2020, pp. 71–76.

[56] Ahn, T. -J., G. -S. Seo, and O. -R. Lim, "Distributed UV Light Sensor Probe Based on Photo-Responsive Polymer Coated Optical Fiber," *IEEE Photonics Technology Letters,* Vol. 31, No. 12, June 15, 2019, pp. 987–989.

[57] Hashim, U., K. A. Rahman, and A. R. A. J. Abdullah, "Mask Design and Fabrication of LiSFET for Light Sensor Application," in *2008 International Conference on Electronic Design,* 2008, pp. 1–5.

[58] Amin, E. M., R. Bhattacharyya, S. Sarma, and N. C. Karmakar, "Chipless RFID Tag for Light Sensing," in *2014 IEEE Antennas and Propagation Society International Symposium (APSURSI),* 2014, pp. 1308–1309.

[59] Ferrone, A., et al., "Flexible Near Infrared Photoresistors Based on Recrystallized Amorphous Germanium Thin Films," *IEEE Sensors*, 2016, pp. 1–3.

[60] Liu, J., S. Bao, and X. Wang, "Applications of Graphene-Based Materials in Sensors: A Review," *Micromachines*, Vol. 13, No. 2, 2022, p. 184.

[61] Zhang, H., D. Wang, and J. Liu, et al., "Review on Thermal Conductivity of the Graphene Reinforced Resin Matrix Composites," *IOP Conference Series: Materials Science and Engineering,* Vol. 562, No. 1, 2019, p. 012018.

[62] Akinwande, D., and D. Kireev, "Wearable Graphene Sensors Use Ambient Light to Monitor Health," *Nature,* Vol. 576, 2019, pp. 220–221.

[63] Vuorinen, T., et al., "Inkjet-Printed Graphene/PEDOT: PSS temperature Sensors on a Skin-Conformable Polyurethane Substrate," *Scientific Reports*, Vol. 6, No. 1, 2016, pp. 1–8.

[64] Li, X., et al., "Stretchable and Highly Sensitive Graphene-on-Polymer Strain Sensors," *Scientific Reports,* Vol. 2, No. 1, 2012, pp. 1–6.

CHAPTER 4

Characterization of Smart Materials for Printing

4.1 Introduction

A wise choice of smart sensing materials has the potential to meet the high requirement of cost-effective sensor technologies such as printable chipless RFID sensors. Materials with low conductivity (e.g., indium tin oxide, silver flakes, and silver nanoparticles) are well-suited for sensing. These materials can be treated as smart sensing materials. They have good mechanical and electrical properties [1, 2]. In contrast, those with high conductivity (e.g., silver, copper, and aluminum) are suitable for identification. However, a breakthrough in material engineering has led to the emergence of certain materials convenient for identification and sensing (e.g., graphene). Graphene, a remarkable two-dimensional carbon material, has sparked extensive and diverse research endeavors due to its exceptional electromagnetic, mechanical, electrical, and thermal characteristics. Graphene enables the propagation of surface plasmon polaritons (SPPs), which exhibit strong wave confinement, moderate energy loss, and a unique tunability via electrical, magnetic, or chemical means. Notably, this plasmonic response occurs in the terahertz and infrared frequency ranges, establishing graphene as a highly promising foundation for terahertz transceivers and optoelectronic systems [3–10]. Nevertheless, graphene-based sensors have versatile utility extending to lower frequency domains, encompassing HF [11] and UHF [12] applications. This underscores the adaptability and broad spectrum of applications for graphene-based sensing technologies. With the futuristic demand for consumer electronics subservient to microwave and millimeter-wave frequencies, it is significant to explore the characteristics of these functional materials prior to their utilization in both high-frequency and traditional low-frequency regimes [13, 14]. In the context of material science, characterization deals with the procedure of probing and measuring material structure and trails, which aids in the scientific perception of engineering materials. The characterization may be microscopic (imaging atoms on chemical bonds in the angstroms range) or macroscopic (imaging coarse grain structure in the centimeters range). Material characterization encompasses a systematic and meticulous procedure involving analysis, measurement, testing, modeling, and simulation to generate comprehensive qualitative and quantitative information about materials tailored to diverse application contexts [15]. For instance, when considering chipless RFID sensors at microwave and millimeter-wave frequencies, particular attention is directed toward assessing key properties such as dielectric constant, loss tangent, conductivity, and other relevant factors. The variation in these parameters with the surrounding environmental changes of ambient temperature, humidity, light, pressure, and so on, is monitored for efficient sensing. In addition to the existing characterization techniques like optical microscopy, which has been explored for centuries, several novel technologies such as electron microscopes and secondary ion mass spectrometry have transformed this domain. This has led to a deep understanding of the behavioral changes of certain materials due to the imaging on very small scales. While microscopy utilizes techniques to map the surface and subsurface structure of a material, spectroscopy deals with the methodologies to expose the chemical composition and crystal lattice of material [16]. This chapter highlights the characterization techniques for smart sensing materials to be chosen for the printing of RFID tag sensors.

4.2 Characterization of Sensing Materials

The central aim of this chapter is to delve into the attributes of smart materials having low conductivity that are well-suited for sensing purposes. Various properties such as optical, chemical (chemical composition), physical (structural), thermal, electrical (conductivity, stability), and scattering parameters (attenuation, complex permittivity, reflection loss, dielectric loss) in microwave and millimeter-wave regimes are investigated. In order to examine the structure of materials, various techniques such as optical microscopy, transmission electron microscopy, and scanning electron microscopy are employed, while X-ray diffraction and transmission electron microscopy play a crucial role in analyzing crystal structure and defects. The methods used to characterize functional materials are categorized into distinct classes based on their application in investigating different material properties. These classes include (1) structural analysis (microstructural examination, surface morphology evaluation, Raman spectroscopy, secondary ion mass spectroscopy, spectroscopic ellipsometry, Fourier transform infrared spectroscopy, and transmission electron microscopy), (2) surface analysis (spectroscopic ellipsometry, scanning electron microscopy, and atomic force microscopy), (3) optical analysis (ultraviolet-visible spectroscopy and spectroscopic ellipsometry), (4) electrical analysis (conductivity measurement), and (5) microwave analysis (assessment of complex dielectric permittivity and reflection and dielectric loss). The following sections provide detailed explanations of these various diagnostic techniques.

4.3 X-Ray Diffraction

X-ray diffraction has emerged as one of the prominent technologies in recent times, finding applications in both qualitative and quantitative evaluations, as well as in central research concerning the characteristics and structures of polymers. The methodology's nondestructive nature makes it incredibly convenient as it is frequently essential to acquire more information with just a limited amount of material

[17]. Theories of the structure of matter are primarily concerned with the distribution of electrons within atoms, the placement of atoms within crystal structures, the separation between the atoms in the molecule, the strength of thermal vibrations, and the bond angles. Examples of X-ray diffraction problems that are of great practical significance include the validation of crystalline materials, the identification of desired alignment, the assessment of particle dimensions, the analysis of strain, the research of the absence of a discernible pattern in structure, and the charting of phase equilibrium diagrams [18].

In accordance with Bragg's law of diffraction, when the difference in path length between two beams reflected by adjacent atomic planes with a specific interplanar spacing, denoted as d, is a whole-number multiple of the incoming monochromatic X-ray beam's wavelength, represented as λ, the phenomenon of diffraction takes place. The X-ray diffraction principle is portrayed in Figure 4.1 [19].

X-ray diffraction characterization aids in identifying the appropriate smart material for printing, which is evident as follows. The crystallographic structure and crystalline quality of the sensing material sample can be well-analyzed using this technique [20]. Small conjugated organic molecules have the potential to employ high-performance organic field-effect transistors (OFETs) that could serve as inexpensive foundations for the development of RFID tag sensors. The distinctive challenge in designing functional organic thin films arises from the extremely anisotropic characteristics of organic molecules, which consequently influence the anisotropic traits of these films on a larger scale. Therefore, a decisive aspect for accurately modeling the electronic band structure and the inherent charge-transport characteristics is figuring out the alignment of molecules and arrangement of organic semiconductors in thin-film configurations, which can vary noticeably from their overall structure. However, the controlled growth of organic thin films and practical investigation using surface-specific X-ray diffraction techniques have made it feasible to determine the three-dimensional arrangement of these thin films. A full molecular 3D packing of thin organic films of PTCDI-C_8 is determined in [21] with the advent of surface-ray diffraction methods.

In a referenced study [22], researchers have successfully created gas sensors based on Ru-doped ZnO that exhibit high sensitivity to gaseous NH_3 at room temperature during the course of exposure to low gas concentrations. The produced

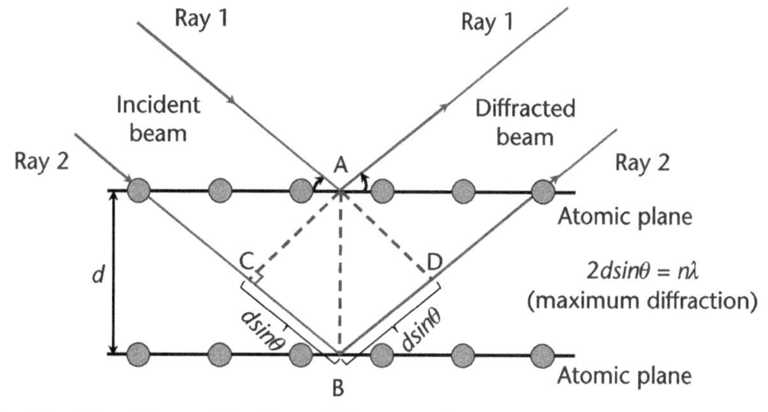

Figure 4.1 Principle of X-ray diffraction within a crystal.

nano powders are analyzed using X-ray diffraction to determine their physical, chemical, and crystalline structures. The observed diffraction patterns identified the hexagonal wurtzite structure. The results further demonstrated that the effect of the doping process on the crystal structure of synthetic ZnO/Ru nanostructures varies with concentration. There is a peak at 33° in the ZnO pure pattern that could be diethanolamine.

4.4 Raman Scattering Spectroscopy

Raman scattering (RS) is extensively utilized in fundamental spectroscopic investigations of excitations in various states of matter, including solids, liquids, and gases, as well as in the characterization of materials. Raman spectra are obtained through the process of inelastic scattering of quantized electromagnetic energy. When the interaction of photons takes place with a medium, they can either transfer energy to the sample (Stokes scattering) or acquire energy from it (anti-Stokes scattering), as depicted in Figure 4.2. This phenomenon occurs because the incident photon stimulates the molecule, causing it to transition into a virtual energy state. Consequently, in the majority of Raman measurements, the focus is on the detection of Stokes-shifted light [23].

Raman spectroscopy is a powerful technique for characterizing advanced materials and devices since it is noncontact and noninvasive, allowing for safe operation in ambient or controlled environment settings without affecting the material's intrinsic properties. Furthermore, RS gives sufficient electric and vibrational information with excellent energy resolution, with energy resolution down to μeV. The energy resolution achieved through Raman spectroscopy surpasses that of other nanotechnology techniques, such as electron microscopy, scanning probe

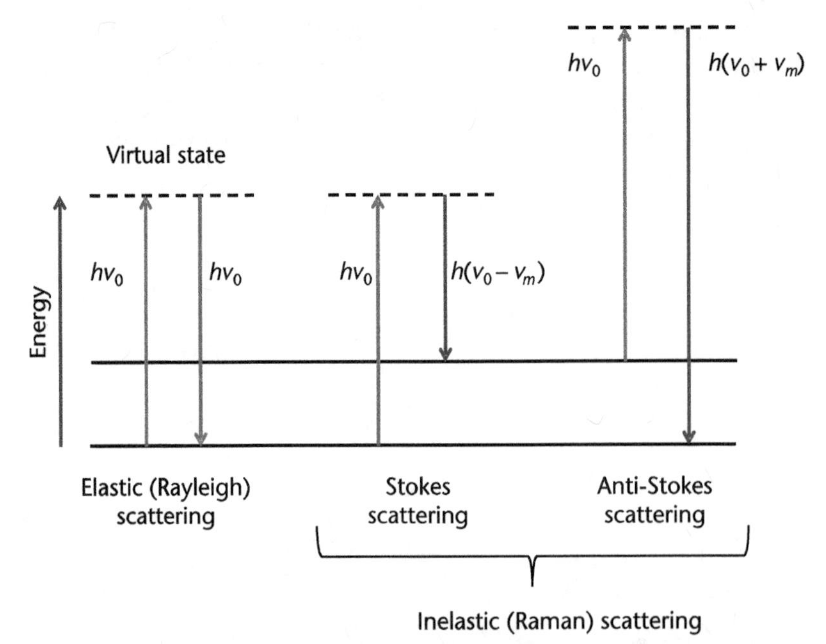

Figure 4.2 Energy level diagrams illustrating Stokes and anti-Stokes inelastic scattering.

spectroscopy, and certain optical techniques that rely on less precise energy levels. Additionally, Raman spectroscopy provides a distinctive spectral fingerprint of a material by measuring the inelastically scattered light emitted from it. Due to its exceptional sensitivity, Raman spectroscopy holds substantial value in device characterization and serves as a significant metric in the context of CNT-based sensors. In a referenced study [24], the authors explore the application of Raman spectroscopy in various carbon nanotube applications, including RFID gas sensors. The advanced materials are characterized by Raman spectroscopy to check the chemical properties necessary for the printing of tag sensors.

In a study conducted by the authors [25], laser-induced graphene (LIG) was investigated using Raman spectroscopy to verify its composition as a carbon allotrope. The relative intensity of the G band to the D band (I_D/I_G) served as an indicator of the disordered carbon structure. The I_D/I_G ratios observed in the LIG samples ranged from 0.668 to 0.942, signifying fluctuating carbon defects within the lattice structure of different LIG specimens. The Raman measurements of all five samples exhibited consistent D and G band wavelengths, specifically 1330.039 cm^{-1} and 1557.68 cm^{-1}, respectively. This suggests that the samples were composed of graphene oxide. Notably, variations in the count of graphene layers, as reflected by the I_D/I_G ratio, could account for the discrepancies in electrical conductivity observed across different LIG samples.

4.5 Secondary Ion Mass Spectrometer

Secondary ion mass spectrometry (SIMS) is a well-established surface and thin-film analysis method that originated in the early 1960s and has a robust instrumentation foundation. It is a widely adopted and commercially available technology employed in both industrial and academic settings. SIMS is utilized to determine the isotopic, elemental, or molecular composition of small specific regions at a microscopic level on the surface and/or near-surface region of solid materials. It can also be utilized in the analysis of frozen liquids in certain instances. Ion spectroscopy is a type of spectrometry that records ions rather than electrons or photons. When a surface is bombarded by heavy ions, such as oxygen or heavier elements, at energies of a few kilo electron volts (keV), the surface material undergoes a process called sputtering, resulting in the ejection of atoms from the surface. Generally, the typical production is in the range of one to several atoms per ion impact. During the sputtering process, a portion of the ejected material becomes ionized, with the degree of ionization dependent on the chemistry of both the surface and the sputtered species. This means that individual atoms, clusters of atoms, and entire chemical species are emitted and ionized to altering degrees of efficiency. By analyzing the secondary ions generated during sputtering, the chemical determination of the sputtered species is achieved using mass spectrometry. This technique allows for the identification of all elements involving hydrogen and helium, and current advancements in sampling technologies have resulted in its relevance in the field of molecular mass spectrometry [26].

Illustrated in Figure 4.3, the pertinent information in SIMS is contained inside the ions that are expelled from the outermost layer of a material succeeding the bombardment by energetic ions. Typically, a concentrated beam of primary ions with energies ranging from 0.1 to 50 keV is employed. The primary ions are

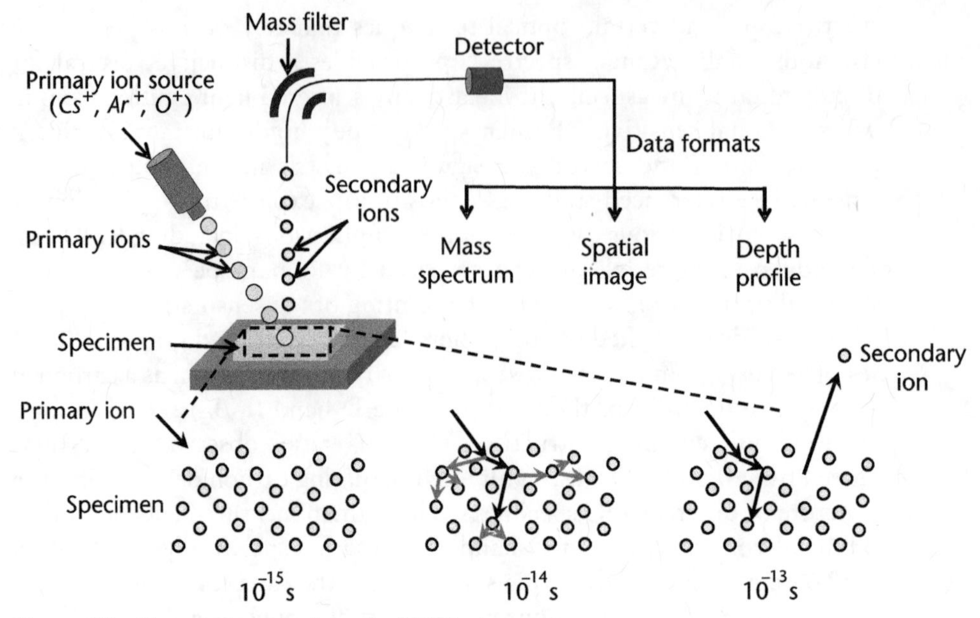

Figure 4.3 The working principle of a SIMS instrument.

responsible for the impact, while the ions released from the surface are referred to as secondary ions [27].

The two different modes in which SIMS analysis can be performed are static SIMS and dynamic SIMS. In static SIMS, the analysis focuses on arrangements of elements and/or molecules within the unaffected outermost layer of material. In contrast, dynamic SIMS offers information about the arrangements of elements and/or molecules across multiple successive layers of atoms at different depths. While the underlying mechanisms involved in signal generation are similar in both static and dynamic modes, there are differences in analytical requirements and data acquisition processes. Since static SIMS may be employed to probe the chemical composition and offer molecular signatures specific to the surface of interest's outermost monolayer, it has sparked substantial curiosity in disciplines ranging from materials sciences to biosciences, whether in academic or industrial settings. For precise elemental analysis in restricted spaces, dynamic SIMS utilizing chemically active atomic or small molecular primary ion beams (C_s^+, O^-, O_2^+) offers the most superior detection thresholds. As a result, this method is favored in the semiconductor industry to accurately monitor dopant distributions at various depths and in the field of earth sciences for mapping isotope distributions.

SiNWs have gained significant interest for their favorable electrical properties and substantial surface-to-volume ratio, rendering them highly suitable for applications involving RFID sensing. Nevertheless, enhancing their performance is accomplished by the alteration of their surface characteristics using organic or inorganic molecules, underscoring the importance of appropriate characterization techniques, which is beneficial for RFID tag printing. In a study presented in [28], the authors examined the efficacy of phosphine (PH_3) utilized like a gas dopant throughout the growth of Au-catalyzed SiNWs. They explored the impact of modulating the ratio of phosphorus to silicon in the gas stream on the resulting SiNWs. The doped SiNWs demonstrated a gradual rise in surface coarseness compared to the approximately undoped specimens. To quantify the amount of phosphorus present in the SiNWs, SIMS was employed.

4.6 Transmission Electron Microscopy

Transmission electron microscopy (TEM) has evolved into a sophisticated imaging technology with subnanometer resolution. One of the primary purposes of TEM is to observe and determine the crystallographic nature of a wide range of material properties, from the actual crystal structure to the exact atomic configurations and chemistry surrounding defects for chipless RFID printing techniques [29, 30].

In traditional TEM, a thin sample is exposed to an electron beam with uniform current density. Electrons used in this context usually have energies ranging from 60 to 150 keV, commonly around 100 keV, or from 200 keV to 3 MeV in the situation of a high-voltage electron microscope. Electrons are released from either cathodes made of tungsten hairpins or rods composed of LaB_6 via thermionic emission or field emission, respectively, from sharp tungsten filaments. Field emission is utilized in the case of strong brightness requirements. The illumination aperture and the area of the specimen that is illuminated can be adjusted using a two-stage condenser-lens setup. The electron intensity distribution behind the specimen is then projected onto a fluorescent screen using a lens with a configuration of either three or four stages. The resulting image can be captured by directly exposing a photographic emulsion within the vacuum environment. Due to significant objective lens aberrations, achieving a resolution of around 0.2–0.5 nm requires working with extremely minuscule objective apertures (approximately 10 to 25 milliradians). Bright-field contrast is obtained by either absorbing electrons scattered at angles beyond the objective aperture or by the intrusion among the scattered and incident waves at the image point where the former is known as scattering contrast and the latter is called phase contrast. The objective lens' wave aberration introduces a phase shift in the electron waves behind the specimen. Additionally, the energy dispersion of the electron gun restricts the contrast transmission at higher spatial frequencies. Electron interactions with atoms involve both types of scattering (elastic and inelastic). Therefore, there is a requirement for extremely thin samples, explicitly ranging from 5 nm to 0.5 μm for 100-keV electrons, based on the density, elemental composition, and desired resolution of the sample. This necessitates specialized procedures such as electropolishing metal foils and ultramicrotomy of dyed and implanted biomatter. A high-voltage electron microscope can examine thicker specimens [31].

The TEM (Figure 4.4) can achieve exceptional resolution due to the fact that elastic scattering, which occurs within the localized region surrounding an atomic nucleus's shielded Coulomb potential, is an interactive process. Still, inelastic scattering is more dispersed, diffusing around a nanometer. Modern TEM offers the additional feature of generating small electron probes with diameters ranging from 2 to 5 nm. This is made possible through a condenser lens setup with three phases, where the final lens field serves as the objective prefield positioned in front of the sample. This configuration promotes the TEM to function in a scanning transmission mode, providing a resolution resolved by the diameter of the electron probe. This capability is handy for visualizing samples that are either thick or crystalline

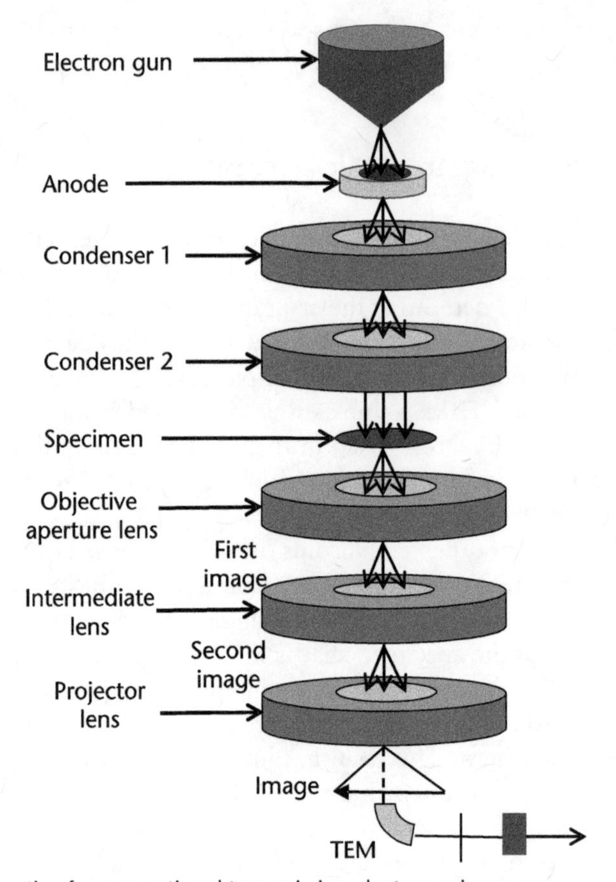

Figure 4.4 Schematic of a conventional transmission electron microscope.

and allows the capture of secondary and backscattered electrons, cathodoluminescence, and electron-beam-induced currents. The primary benefit of outfitting a TEM with a scanning TEM (STEM) connection is the production of a very small electron probe, which may be used for the purpose of conducting elemental analysis and microdiffraction on highly localized regions [31].

With the rapid growth of industrial technology, the emission of exhaust gases, including nitric oxide (NO), has led to a significant increase in air pollution. To address this issue, chipless RFID gas sensors have emerged as a promising solution, and ZnO stands out as a suitable material for this application. Zinc oxide exhibits numerous advantageous characteristics, including a broad direct bandgap, elevated exciton binding energy, excellent electron mobility, nontoxic nature, customizable morphology, and excellent chemical and thermal stability. In a study conducted by the authors in [32], they focused on fabricating a thin film of (002)-oriented ZnO on a glass substrate using diethanolamine. The gas-sensing characteristics of randomly distributed ZnO nanorods (sample A) and (002)-oriented ZnO nanorods (sample B) were investigated to assess the influence of growth alignment on the features of the NO gas sensor. To analyze the structure, orientation, and morphology of the ZnO film samples, TEM is applied. The TEM images depicted in Figure 4.5(a) and the selected area electron diffraction (SAED) patterns shown in Figure 4.5(b) revealed important insights. The SAED analysis approved that sample A

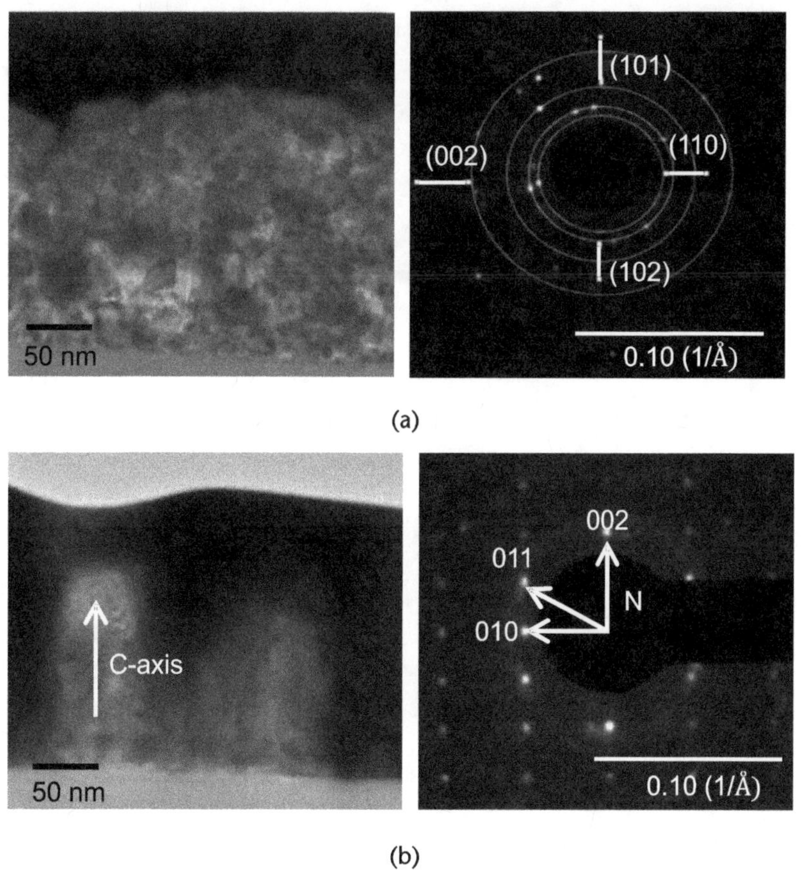

Figure 4.5 TEM figures (left) and SAED patterns (right) of ZnO films produced (a) without diethanolamine in (sample A), and (b) with diethanolamine (sample B) in the seeding procedure [32].

exhibited arbitrary positioning, as evidenced via a diffraction pattern in the form of a ring. Conversely, sample B exhibited a hexagon-shaped crystal-like assembly with a diffraction pattern aligned along the substrate surface's normal direction, indicating that the ZnO film was primarily formed along the (002) orientation.

4.7 Scanning Electron Microscope

A standard optical microscope's utmost usable magnification is around one thousand times. To achieve enhanced resolution in electron microscopy, it is necessary to decrease the imaging energy wavelength. In the context of electron microscopy, electrons are commonly ramped up to high energies, typically ranging from 2 to 1,000 keV, which corresponds to wavelengths ranging from 0.027 to 0.0009 nm. As illustrated in Figure 4.6, the high-energy electron beam can undergo various interactions with the atoms in the object. When the sample is relatively thin, electrons can pass through it without being absorbed and can be utilized to generate the image in TEM. However, a thicker sample does not allow complete transmission. Therefore, only particles escaping from the surface provide information. This data is utilized in a conventional scanning electron microscope (SEM) [33].

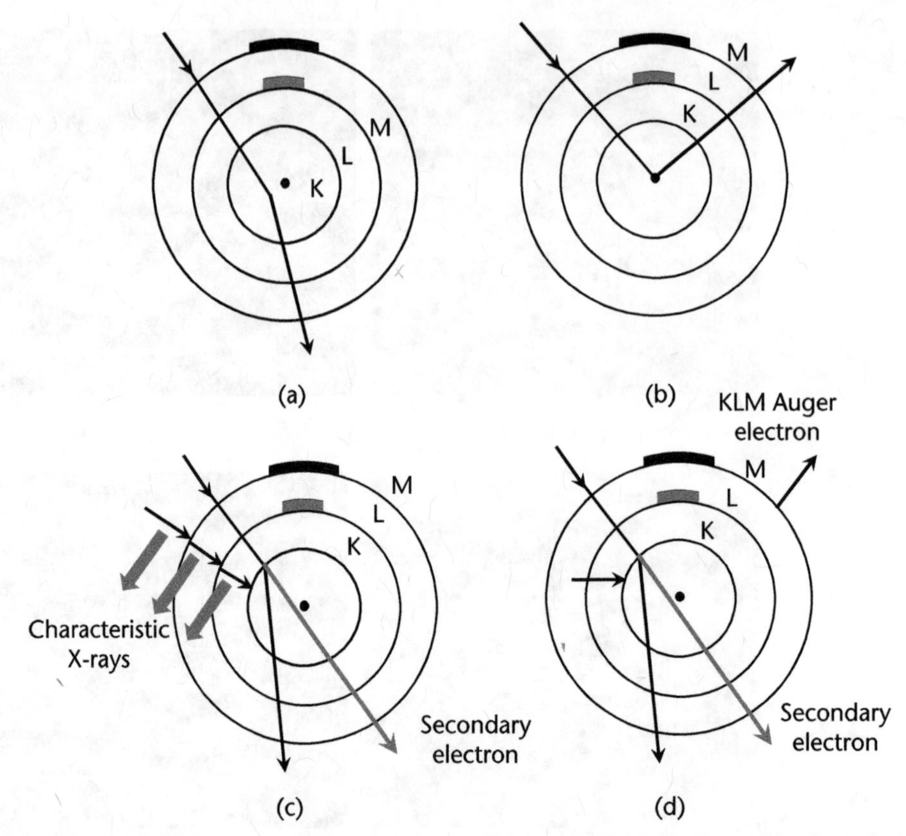

Figure 4.6 Atom high energy electron interaction. (a) Low-angle scattering: minimal energy loss, electrons scatter to next atom layer, (b) high-angle scattering: significant energy loss, electrons scatter back, (c) X-ray emission: atoms emit X-rays and secondary electrons, and (d) Auger electron emission: atoms emit Auger electron and a secondary electron.

The SEM is an efficient tool for studying and analyzing microstructure morphology and chemical composition characterizations. SEM provides valuable insights into the surface topography, crystalline structure, chemical composition, and electrical properties of a specimen, specifically targeting the top 1 micrometer of its surface. SEM setups can be equipped with specific stages designed for different purposes, such as in situ mechanical testing or observing the behavior of the specimen under different temperature conditions (hot or cold stages). These capabilities enable the study of the specimen's characteristics and responses in various experimental situations. For instance, cathodoluminescence (light emission) is substantially stronger at temperatures approaching absolute zero than at normal temperatures. Hence, images created from light released by a chilled sample exhibit significantly reduced noise level. Furthermore, there are several advantages of SEM over TEM. First, SEM can characterize large samples (more than 200-mm diameter wafers) compared to TEM (less than 3-mm diameter wafers). Second, SEM aids in the nondestructive analysis of samples, whereas specimen preparation required in TEM reveals its destructive testing approach. Finally, specimen preparation is less time-consuming and straightforward than TEM [33].

4.7.1 Principle of Operation of SEM

The schematic of an SEM is displayed in Figure 4.7. The incident electrons from an electron gun in an SEM usually have energy ranging from 2 to 40 keV. Using two or three electromagnetic condenser lenses, the electron beam is carefully reduced in size to form a fine probe. This probe is then glided over a designated zone of the sample's surface using scan coils. When the electrons interact with the sample under test, they enter in a teardrop-shaped volume, the specific sizes of which are resolved via factors such as the electron beam energy, the elemental atomic mass of the test material, and the incidence angle (or the angle at which the electron beam impinges the surface) of the test material. It should be noted that the depth at which the electrons penetrate the specimen, also known as the penetration depth, surges with enhanced electron beam energy, greater angular incidence, and reduced atomic masses. Secondary, backscattered, Auger electrons, X-rays, and possibly light are produced by the electron beam's interaction with the material and captured through several detectors in the specimen chamber. Each detector's signal is sent into a monitor, which is restored in time with the electron beam. The enlargement of the image is defined by the fraction of the monitor display's side length to the side length of the raster on the sample [33, 34].

To generate images in the SEM, it is necessary to record the signals resulting from the contact among the electron beam and the specimen, including both elastic and inelastic scattering events. Elastic scattering takes place once the impinging electron is rebounded by either the atomic nucleus of the specimen or the electrons

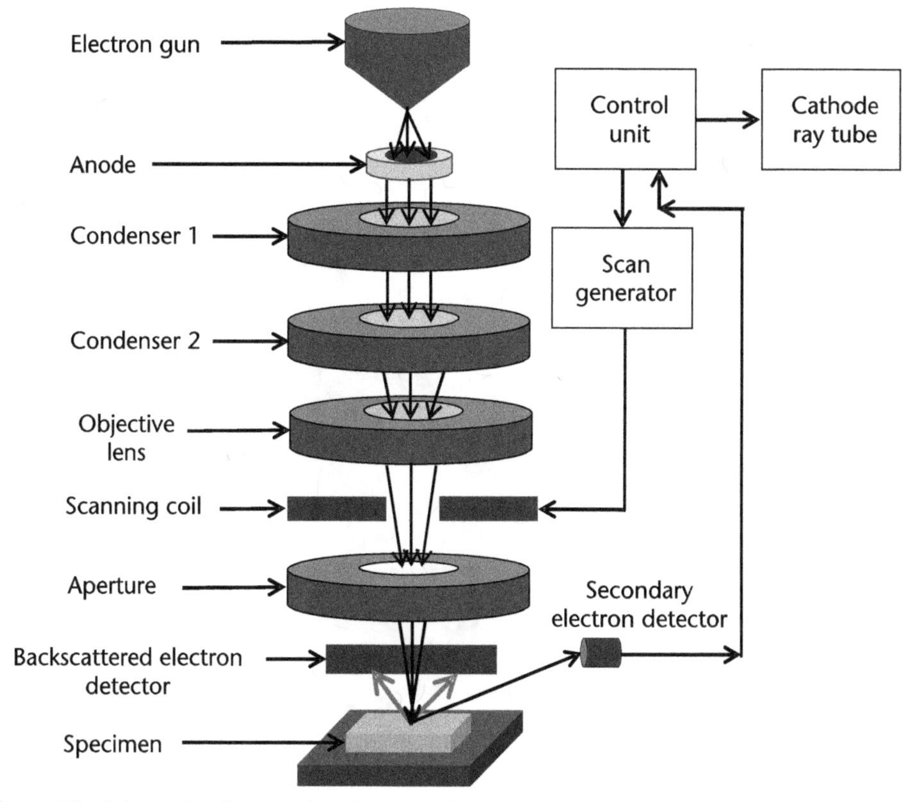

Figure 4.7 Schematic of a scanning electron microscope.

located in the outermost shells with alike levels of energy. Backscattered electrons (BSE) are a type of incident electron that undergoes elastic scattering at an angle greater than 90°. These backscattered electrons provide valuable information for imaging the material. On the other hand, inelastic scattering occurs due to various contacts among the incident electrons and the electrons and atoms within the specimen. In the above process, the primary beam electron transfers a significant amount of energy to the atoms or electrons of the specimen. The stimulation of specimen electrons during atom ionization generates secondary electrons (SE), typically described as having energies less than 50 eV and can be utilized to scan or examine the specimen under test. Figure 4.8 depicts the areas from which various signals are perceived [35].

4.7.2 Characterization of Sensing Materials with SEM

As industries develop, several air pollutants are discharged into the environment from diverse sources. High CO_2 gas concentrations have a negative effect on the earth's environment, causing harm to human and animal life. Due to the increasing concerns about global warming and the growing consciousness of environmental issues, the need for CO_2 measurement has skyrocketed. The study described in [36] focuses on the production of a carbon dioxide (CO_2) gas sensor using a metal-oxide material consisting of La_2O_3 and SnO_2. The structural and morphological features of sensing material are explored with the field-emission scanning electron microscopy (FESEM) characterization method. FESEM at 16.5 KV, as displayed in Figure 4.9, examines the surface properties of three pastes with a thick film com-

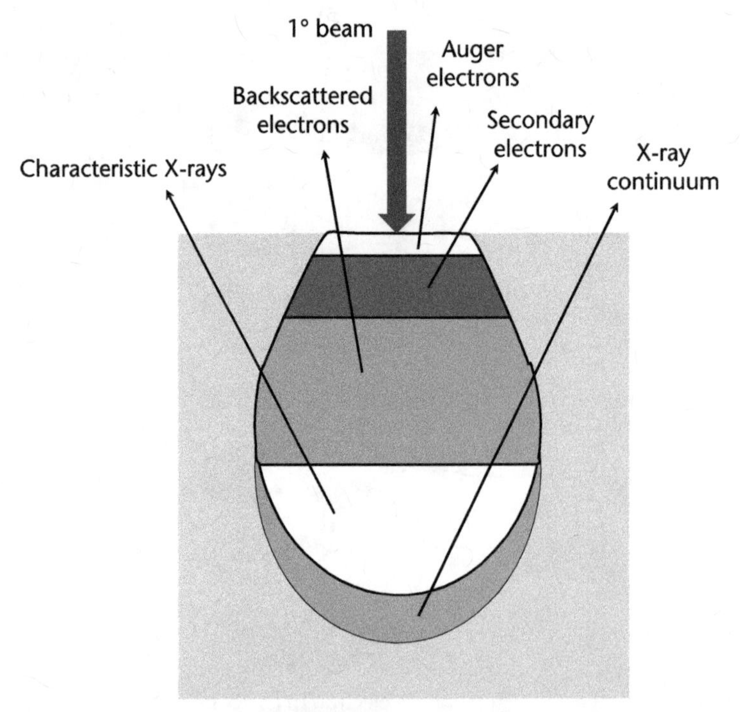

Figure 4.8 The signals created by the electron beam-specimen interaction in the SEM.

Figure 4.9 FESEM images of (a) a thick film composed of La_2O_3/SnO_2, which was screen printed onto an alumina substrate, (b) La_2O_3/SnO_2 doped with 1wt. % Pt, and (c) La_2O_3/SnO_2 doped with 3wt. % Pt [36].

position that are sensitive to specific conditions that have undergone sintering at a temperature of 750°C.

On the surface of the manufactured sensor, in Figure 4.9(a) the La_2O_3/SnO_2 thick layer exhibits a granular structure with a distinct hexagonal pattern. The findings indicate that the sensitive paste applied on the alumina substrate (via printing) has a nearly uniform dimensional distribution. Additionally, the composition of the sensitive powder appears to be homogeneous. Minimizing uncertainties and unwanted errors during the sensitivity calculation can help to produce a more accurate data evaluation. Figures 4.9(b) and 4.9(c) illustrate the surface structures of thick films of La_2O_3/SnO_2 doped with platinum (Pt) of 1 wt.% and 3 wt.%, respectively. A uniform apportionment of platinum dopant is found within both samples. With a high amount of Pt, sample forms tend to be elongated and hexagonal.

Another toxic air pollutant whose detection becomes challenging due to its colorless and odorless nature is CO. The authors in [37] detail the production and analysis of gas sensors that can be worn as wearable devices. These sensors are fabricated using wet solution-based methods to create graphene/ZnO composites upon cotton fabrics, and their properties are thoroughly examined. The morphological structures of the resultant graphene/ZnO nanostructure layers are explored using SEM (see Figure 4.10). Because of the intense surface contact, graphene has agglomerated flake shapes (Figure 4.10(a)).

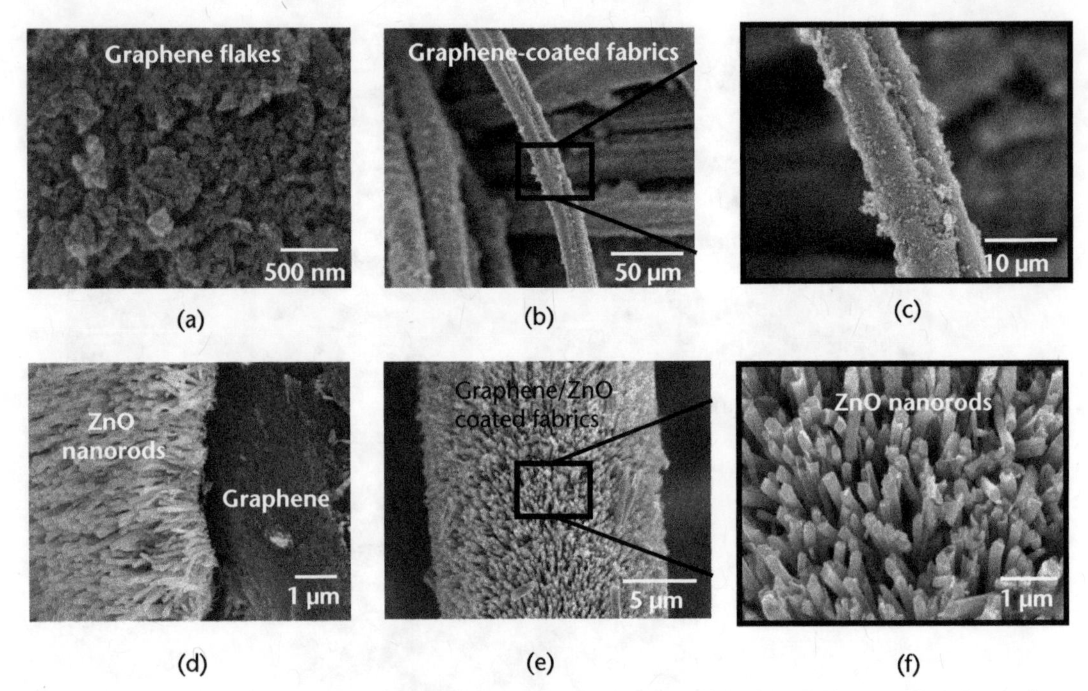

Figure 4.10 SEM images of (a) graphene flakes, (b) graphene-coated fabric, and (c) magnified view indicating uniform material coating. (d) The boundary between ZnO nanorods and graphene. (e, f) Vertical ZnO nanorods developed on graphene-coated fabrics [37].

4.8 Atomic Force Microscopy

Atomic force microscopy (AFM) is a highly precise microscopic technique that offers exceptional lateral resolution. Unlike scanning electron microscopy, which is typically performed under vacuum conditions, AFM can be used in air or even liquids, expanding its applicability to a wider range of samples. However, imaging irregular samples with AFM presents challenges as it requires accounting for interactions and accurately resulting in the surface of the sample depending on the forces among the scanning tip and the specimen. Nonetheless, AFM provides the advantage of measuring various physical properties beyond surface topography. It enables the generation of images that reveal the positioning of individual atoms within a sample or the molecular structure of each molecule. AFM has the significant benefit of being able to scan practically any sample, whether it is very soft, like highly flexible polymers, human cells, or individual DNA molecules, or very hard, like the surface of a ceramic substance or a dispersion of metallic nanoparticles [38].

An AFM differs from other microscopes in that it does not, unlike an optical or electron microscope, create an image of a surface by focusing light or electrons. An AFM creates a map of the sample's surface height by physically feeling the sample's surface using a pointed probe. A conventional microscope captures a two-dimensional projection of a sample's surface, which lacks height information and requires additional analysis or sample manipulation to deduce such details. In contrast, AFM provides a means to obtain height information along with surface topography. However, to generate an image that resembles what would be observed under a traditional microscope, the data acquired from AFM needs to undergo processing [38].

4.8.1 AFM Technique

The scanning probe microscope that has been employed the most recently is AFM. The following can be used to describe its fundamental operation (see Figure 4.11). The cantilever tip used in AFM is commonly constructed from materials such as Si or silicon nitride (Si_3N_4) and typically possesses a tip radius ranging from 1 to 20 nm. However, it can occasionally be bigger. A specified spring constant allows the cantilever to oscillate at frequencies between 10 and a few hundred kilohertz. The cantilever holder's z-position is often adjusted by detecting the signal on a photodiode to maintain consistent force exerted between the tip and the surface. The shifting z-position has an impact on the surface topography, as mentioned in [39].

In the absence of additional tip functionalization, the forces acting between the tip and the surface primarily consist of repulsive electrostatic Coulomb forces and attractive van der Waals forces. These forces are feeble, ranging from approximately 1011 to 107N. In ideal conditions, achieving a resolution of around 0.1 nm is feasible, particularly when dealing with exceptionally smooth surfaces due to the short separation from the surface to the tip, specifically ranging from 0.1 to 10 nm. However, other forces may superimpose the ones mentioned earlier, which must be considered while interpreting AFM images. There are various ways the cantilever may reach the sample surface, including the contact mode (see Figure 4.12(a)) and the dynamic (tapping) mode (see Figure 4.12(b)). While the contact mode of AFM may offer higher resolution, it can pose challenges for soft or uneven surfaces due to the lateral forces generated by friction between the tip and the sample. These

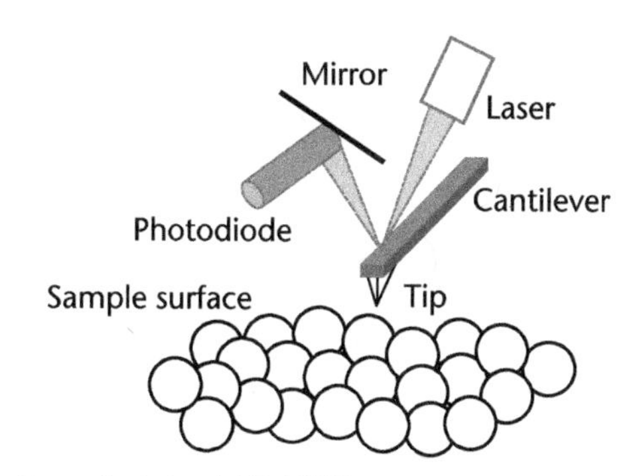

Figure 4.11 Basic operating principle of AFM [39].

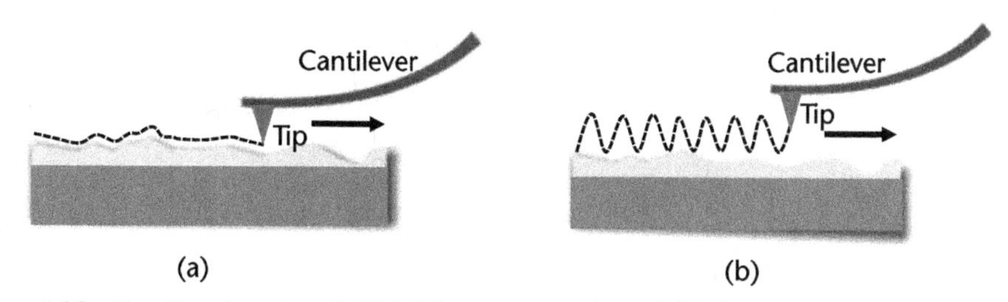

Figure 4.12 Functional modes of AFM: (a) contact mode, and (b) dynamic mode [39].

lateral forces can adversely affect the imaging quality and potentially cause damage to the surface being examined. Therefore, the dynamic mode is frequently preferred for such samples [39].

4.8.2 AFM Characterization

In various fields, such as robotics, human-machine interaction, and in biomedical equipment, there is a need for intelligent pressure or force sensors capable of accurately detecting and responding to dynamic contact situations. In [40], tactile sensor devices based on ground on pliable MOSFET and aluminum nitride (AlN) piezocapacitors are highlighted. As the transducer layer is merged with a MOSFET, the AlN shows a piezoelectric trait without the usual high voltage for poling, making it a prime entrant for sensors. The roughness of the film is characterized with the AFM technique. To harness the strong transversal piezoelectric effect along the *c*-axis, it is crucial to carefully augment the manufacturing course of AIN films. This involves delivering sufficient energy during film growth to promote oriented growth with high-quality texture, resulting in excellent piezoelectric properties. The use of AFM scanning over a $2\,\mu m \times 2\,\mu m$ area reveals compact microstructures with atypical surface unevenness of 5.18 nm (as depicted in Figure 4.13). Analysis conducted with NanoScope software estimates a grain magnitude of 16 nm.

Paper is found nearly everywhere in a wide range of typical applications because of its industrial development, affordable price, and accessibility in numerous forms, according to specific requirements and conditions. As the requirement for cost-effective, pliable, and ecoconscious technologies grows, paper substrates have emerged as attractive options for electronic applications. The electrical traits of paper substrates, as well as the performance of thin films created using silver nanoparticles through printing methods, dielectric layers, and metallization processes utilizing catalysts, are explored in [41] using AFM characterization techniques. Silver nanoparticle inks are commonly used for inkjet-printed electronics on various substrates, including paper substrates, due to their low thermal sintering

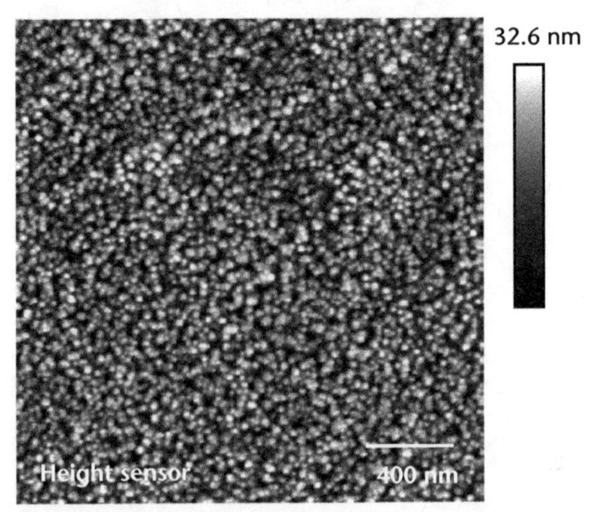

Figure 4.13 AFM scan of the top surface of AIN, revealing a densely packed nanostructure having a grain dimension of ~16 nm [40].

temperature and higher conductivity compared to other nanoparticle-based inks like Cu or gold (Au) nanoparticle inks. Figure 4.14 illustrates a visual of a silver nanoparticle (printed using inkjet technology) thin film on a glass substrate, which is scanned utilizing the AFM technique. The film is printed using a 10-pL (picoliter) ink cartridge with a droplet spacing of 20 μm (1024 dpi), followed by thermal sintering at 150°C for 2 hours under air pressure. The experimentally determined arithmetic average and root mean squared coarseness results are 11.4 and 14.4 nm, respectively.

4.9 UV-Visible Spectrophotometers

Ultraviolet-visible (UV-vis) spectroscopy is an approach used to evaluate the contact of molecules that possess electromagnetic energy within the ultraviolet and visible portions of the electromagnetic spectrum. The absorption of light energy (ranging from 150 to 400 kJ mol–1) causes electrons to transition from the ground state to an excited state. By measuring the light absorption as a function of frequency or wavelength, a spectrum can be obtained. Molecules containing electrons present in aromatic systems that are spread out or shared over multiple atoms frequently exhibit light absorption in the near-UV range (150–400 nm) or the visible range (400–800 nm) [42].

Ultraviolet and visible absorption spectrophotometry is a method used to measure the reduction in the intensity of electromagnetic radiation when it passes through an absorbing substance. This reduction can be attributed to various factors such as reflection, scattering, absorption, or interference. The degree of absorption, indicated by the absorbance, is directly related to the concentration of the substance being analyzed and the path length traveled by the light during the measurement. Beer's law is generally expressed as $A = \varepsilon bc$, where A, ε, b, and c are the absorbance, the molar absorbance coefficient (mol^{-1} L cm^{-1}), the route length (cm), and the absorber concentration (mol L^{-1}), respectively. This linear relationship can be influenced by various factors, including spectrophotometer characteristics, photodegradation of molecules, scattering or absorbing interferences in the

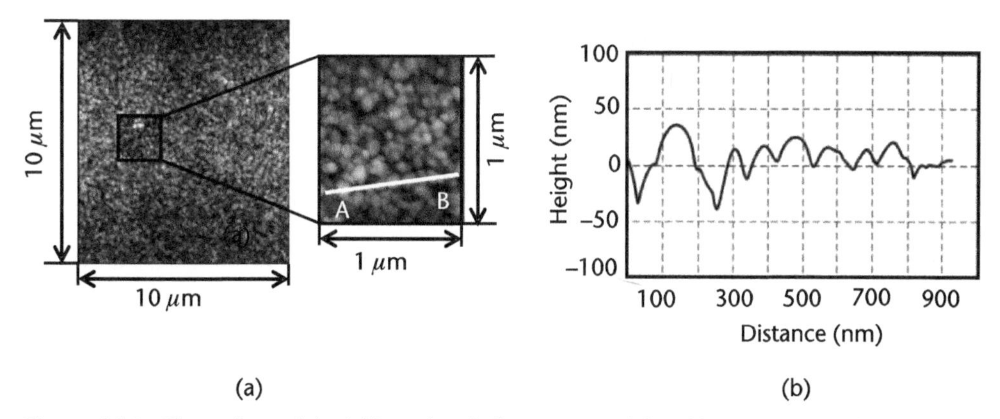

Figure 4.14 The surface of the inkjet-printed silver nanoparticles: (a) AFM images of 10 μm × 10 μm and 1 μm × 1 μm areas, and (b) a cross-sectional view along the line AB [41].

sample, fluorescent compounds in the sample, interactions between the analyte and the solvent, and pH [43].

4.9.1 UV-Visible Spectrophotometry Instrumentation

A spectrophotometer typically has two light sources: a deuterium lamp for the UV region and a tungsten-halogen lamp for the visible range. In the process of ultraviolet and visible absorption spectrophotometry, the light is directed through a monochromator or optical filter and then focused into a cuvette. The intensity of the transmitted light is monitored by a photomultiplier or a photodiode. In double-beam instruments, a reference cuvette containing a buffer is placed in the reference beam, and its absorbance is subtracted from the absorbance measured for the specimen. By continuously adjusting the wavelength of the incident light, a spectrum of absorbance values is obtained. In diode-array spectrophotometers, the entire spectrum of light emitted by the lamp illuminates the sample. The transmitted light is then split into its constituent components by a prism and detected by an array of diodes, typically spaced at 2-nm intervals. Unlike conventional devices, diode-array spectrophotometers record the complete spectrum simultaneously rather than through a time-dependent scan. Thus, spectral changes in a wide range of wavelengths can be tracked simultaneously [42].

A typical single-beam spectrophotometer is shown in Figure 4.15. The experimental setup involves directing the light beam, originating from a UV or visible source, into the monochromator through an entrance slit. A concave holographic grating, optimized for dispersion at 220 nm, is utilized to separate the different wavelengths of light. The dispersed beam exits the monochromator through an exit slit and is further filtered to eliminate stray light before reaching the sample compartment cell and ultimately reaching the detector. It should be noted that the sample chamber of the Beckman DU-60 series is designed to accommodate only one cell at a time. Initially, the reference solution is introduced into the sample compartment, and the corresponding detector signal is recorded and stored in the instrument's computer. The sample solution is then added to the same cell, and the

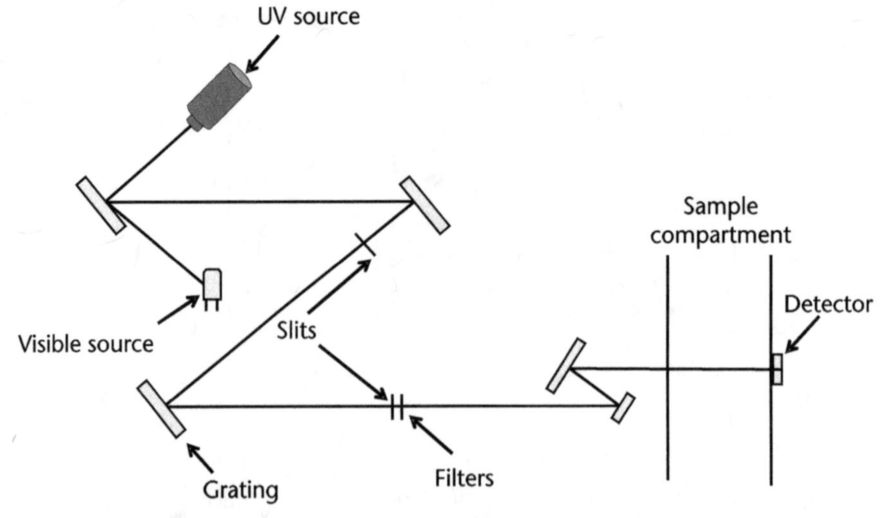

Figure 4.15 Schematic of a typical single-beam spectrophotometer.

computer determines the absorbance value using the two stored detector signals. This type of absorbance measurement necessitates the light source's energy output being stabilized (stable-beam technology). Although the primary application field of single-beam instruments is absorbance measurement for quantitative analysis, stable-beam technology allows single-beam devices to be used for scanning spectra: the entire spectrum of the reference can be stored in the computer, and after scanning the sample solution, the former is subtracted from the latter to obtain the spectrum [44].

4.9.2 Material Characterization with UV-Vis Spectrophotometry

In recent decades, heavy metal pollution has significantly impacted the environment, with mercury being a major contributor to contamination and bioaccumulation in various organisms and food chains. To address this issue, a cost-effective, portable spectrophotometer has been designed [45] to detect mercuric ions (Hg^{2+}) in an aqueous medium using a chemo sensor coumarin derivative. The compound 7-(diethylamino)-2H-chromene-2-thione was synthesized and tested with different metals, only displaying a colorimetric, fluorometric, and UV-Vis spectrometric effect with Hg^{2+} (as shown in Figure 4.16). Upon introducing mercuric ions to the compound solution, a significant reduction in the absorption band at 483 nm is observed, accompanied by a shift toward longer wavelengths to 513 nm, known as bathochromic displacement. Over time, a decrease in overall absorption intensity, referred to as a hypochromic shift, is observed, along with the emergence of a new absorption band at 388 nm in the UV-Vis spectrum. The proposed approach involves utilizing a chemo sensor as a device and employing spectrophotometric techniques and ^{13}C nuclear magnetic resonance (NMR) spectroscopy for characterization. NMR spectroscopy utilization is a distinctive aspect of this research, enabling the identification and analysis of the compound's structural features.

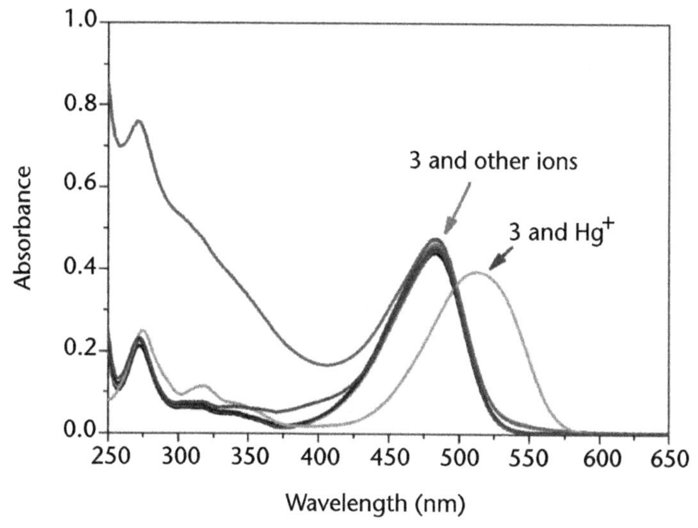

Figure 4.16 Absorption spectra of the synthesized compound (5×10^{-6} M) with the inclusion of different ions (1×10^{-4} M) [45].

4.10 Electrical Conductivity Measurement

The electrical conductivity characterization plays a vital role in the development and performance evaluation of chipless RFID sensors, which are widely used in various applications such as wireless communication, inventory management, and security systems. Electrical conductivity measurements are particularly important for materials used in electronic devices, as they directly impact the performance of these devices and also provide critical information about a material's electrical properties. In addition, the electrical conductivity of materials can also be an indicator of other properties, such as thermal conductivity and mechanical strength. Conductivity measurements can be used to determine the effectiveness of electrical insulation, the presence of impurities, and the degree of material degradation over time. The electrical conductivity of a material can also be influenced by external factors such as temperature, humidity, and pressure, making it a valuable tool for monitoring changes in material properties under different conditions [46].

The demand for polymer nanocomposites (PNCs) stems from their ability to exhibit excellent electrical conductivity, making them essential for real-world scenarios in electronic devices and sensors. Among the various nanofillers with rod-like structures, such as carbon nanotubes and nanowires, those having great aspect ratios are the focus of considerable scrutiny for fabricating PNCs with enhanced conductivity. However, silver nanowires (AgNWs) have emerged as highly desirable materials due to their exceptional conductivity (6.3×10^7 S/m) and suitability for applications in the fields of sensing, electronics, and protection against electromagnetic interference (EMI). Moreover, silver nanowires possess superior biocompatibility compared to various nonmetallic conductive nanomaterials, such as carbon nanotubes, as they exhibit microbicidal traits. These attributes contribute to the attractiveness and versatility of AgNWs in various fields.

A theoretical model has been developed in [46] to investigate the electrical conductivity of PNCs comprising AgNWs. The study investigates the impact of several factors, including interphase thickness, tunneling separation, nanowire waviness, aspect ratio, and a fraction of the filler material that contributes effectively to the percolation performance and overall electrical conductivity of samples strengthened with AgNWs. The findings reveal that AgNWs with higher aspect ratios form extensive networks with a lower percolation threshold, leading to enhanced conductivity in the nanocomposite. Additionally, increasing the interphase thickness promotes conductivity by expanding the network, while nonwavy AgNWs possess higher conductivity in comparison to their wavy counterparts. Surprisingly, the surface energies of the polymer channel and AgNWs participate in minimal influence on the specimen's conductive nature. Conversely, the volume fraction and aspect ratio of the AgNWs, as well as the interphase depth and tunnelling separation, have a significant impact on the nanocomposite's conductive nature. These results provide valuable insights into the design and optimization of AgNW-reinforced systems for improved electrical conductivity. Overall, these findings provide valuable insights for designing and optimizing conductive polymer/AgNWs nanocomposites for various practical applications. The experimental results for the conductivity, as presented in Figure 4.17, exhibit a high degree of consistency with the proposed theoretical model.

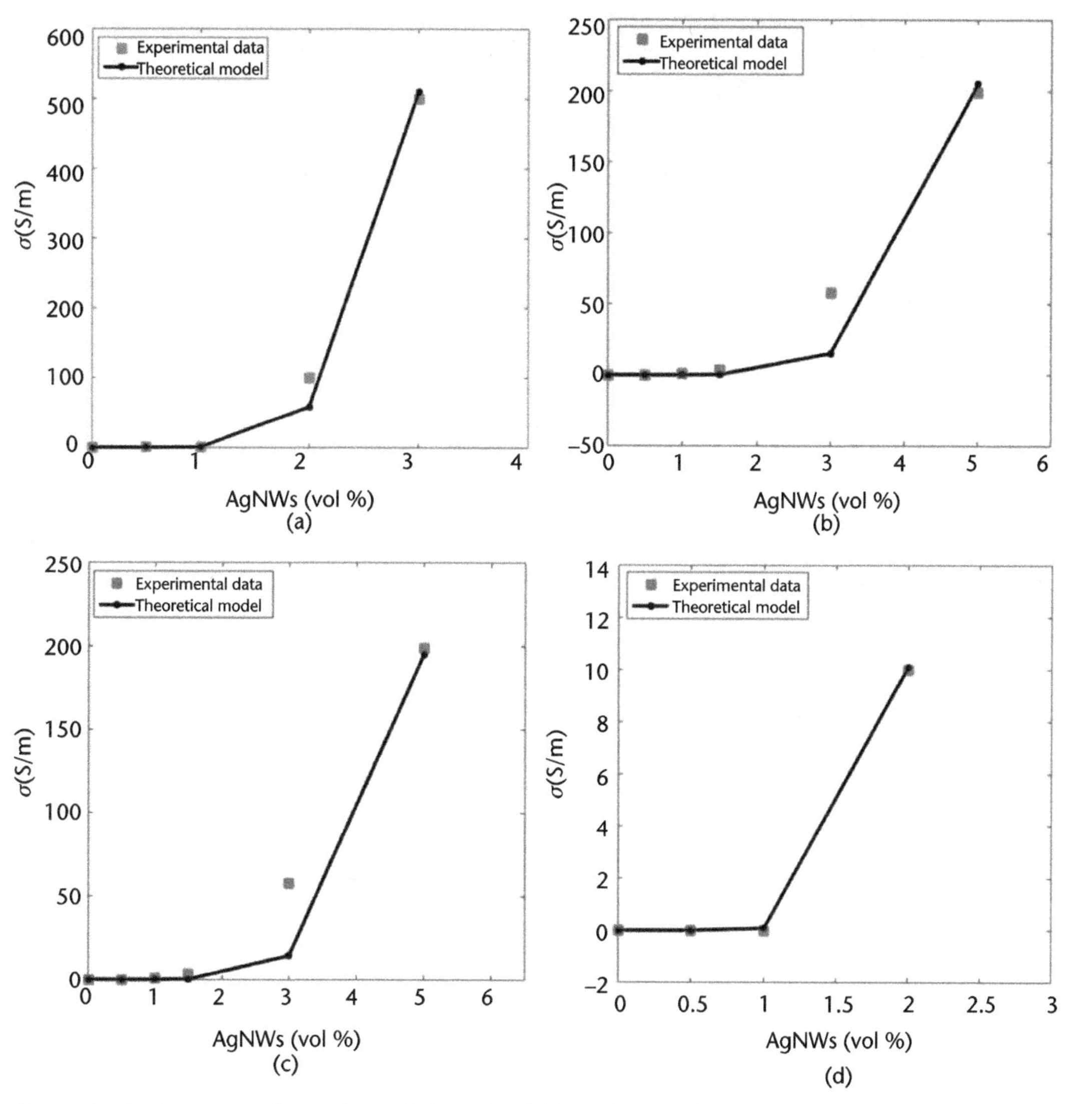

Figure 4.17 Experimental and theoretical conductivities of various PNCs containing AgNWs at different nanofiller concentrations for (a) PI/AgNWs, (b) PEKK/AgNWs, (c) PLA/AgNWs, and (d) PMMA/AgNWs [46].

PANI is a type of electrically conductive polymer that has received significant attention due to its low-cost synthesis methods and stability under ambient conditions, along with its controllable electrical conductivity. In [47], composite fibers made of polyacrylonitrile-polyaniline (PAN/PANI) were analyzed to determine the effect of temperature on their electrical conductivity. The study employed a two-electrode system to measure the current-voltage plot of the fibers. The results indicated that the fiber conductivity at room temperature was approximately 150 μS (6.7 kΩ), whereas pure PAN fibers exhibited a resistance of approximately 100 GΩ. Notably, the conductance of the fibers showed a decrease both below and above ambient temperature, with the most pronounced changes occurring at 50°C. Intriguingly, the conductance of fibers subjected to low temperatures demonstrated fully reversible changes, whereas fibers exposed to temperatures exceeding 60°C exhibited irreversible alterations in conductance. The conductance value exhibited the most noteworthy variations over time during the measurement carried out at a temperature of 50°C, as illustrated in Figure 4.18. These findings provide valuable insights into the behavior of composite fibers made of PAN/PANI under different temperature conditions.

4.11 Basic Microwave-Material Interaction Aspects

Microwave characterization of sensing materials involves the use of electromagnetic waves in the microwave frequency range to study the properties of the materials used in sensors. This technique is useful in understanding the behavior of materials under different environmental conditions, such as changes in temperature, pressure, and humidity. Microwave characterization can be performed using various methods, such as time-domain reflectometry, microwave spectroscopy, and resonant cavity techniques. These methods involve measuring the microwave signal that interacts with the sensing material and analyzing the changes in the signal that occur as a result of the material's properties. One of the advantages of microwave characterization is that it allows for nondestructive testing of the sensing material, which is important for ensuring the longevity and reliability of sensors. Additionally, microwave characterization can provide information on the dielectric properties of the material, such as its permittivity and loss tangent, which are important for the design and optimization of sensors. Overall, microwave characterization is a powerful tool for studying the properties of sensing materials and can provide valuable insights into the behavior of sensors in various environments [48].

When microwaves are directed toward a substance, their energy engages with the material through three distinct mechanisms: reflection, transmission, and absorption. The dielectric characteristics of the substance determine the proportions of energy involved in these interactions. In the context of microwave electronics,

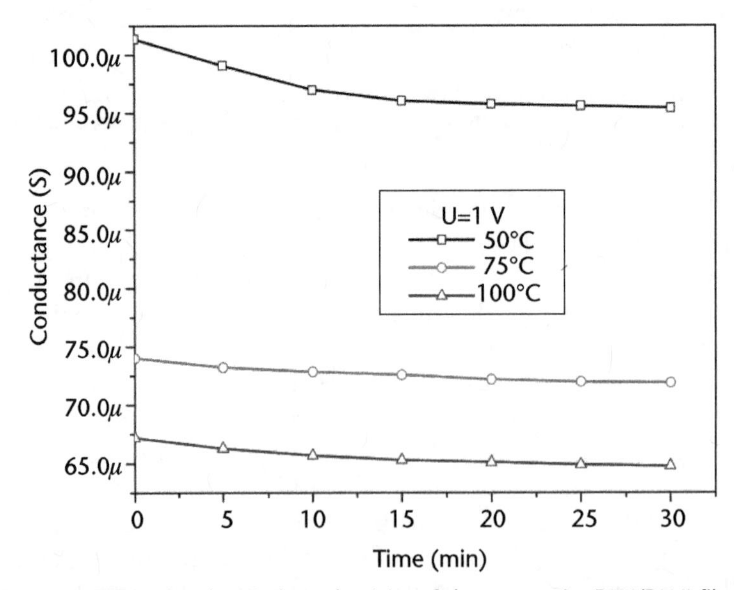

Figure 4.18 Alteration in the electrical conductivity of the composite PAN/PANI fibers in time at 50°C, 75°C, and 100°C [47].

the complex relative permittivity, denoted by ε, is the crucial electrical property that characterizes the nature of the interactions between the microwaves and the material and is expressed as [48]

$$\varepsilon = \varepsilon_0 \varepsilon_r = \varepsilon_0 \left(\varepsilon_r' - j\varepsilon_r''\right) \tag{4.1}$$

Where $\varepsilon_0(8.854 \times 10^{-12}\ F/m)$, ε_r, ε_r', and ε_r'' are the permittivity of free space, relative permittivity, real part of relative permittivity, and imaginary part of relative permittivity, respectively. The real part of the permittivity describes the amount of energy that can be stored by the material when subjected to an external electric field. On the other hand, the imaginary part of the permittivity measures the degree of dissipation or loss experienced by the material in the presence of the electric field. While the real part is sometimes referred to as the dielectric constant, it is not a constant and can vary with factors such as frequency, temperature, and pressure. The imaginary part, also known as the loss factor, can be used to calculate the loss tangent or tan δ, which is the ratio of energy lost to energy stored by the material, which is expressed as [48]

$$\tan\delta = \frac{\varepsilon_r''}{\varepsilon_r'} \tag{4.2}$$

Microwave techniques for material characterization can be broadly classified into two types: resonant and nonresonant. Resonant methods provide highly accurate information about the dielectric traits of substances at a specific frequency or a limited region of frequencies. Nonresonant methods, on the other hand, offer a more general understanding of the electromagnetic properties of materials across a range of frequencies. The choice of method depends on the specific application and the desired level of accuracy. In the following discussion, we will explore the various methods used for the microwave characterization of materials [49].

4.11.1 Nicolson-Ross-Weir Method

The Nicolson-Ross-Weir (NRW) method is a widely used approach for the electromagnetic characterization of materials based on transmission and reflection measurements of the test specimen. This method determines the properties of the material by analyzing its impedance and wave velocities. However, the primary limitation of the NRW technique is associated with the electrical thickness of the sample under analysis. Ambiguous results arise when the specimen's thickness corresponds to a whole-number multiple of half the wavelength within the material, resulting from the periodic phase of the electromagnetic wave. Unambiguous outcomes require prior knowledge of the specimen. Additionally, the NRW method is not suitable for thin samples or materials with low reflectivity due to the inversion procedure employed for parameter extraction. Figure 4.19 presents two examples of transmission lines utilized for material characterization, where the test sample of d units size is positioned within a coaxial cable or a waveguide [50].

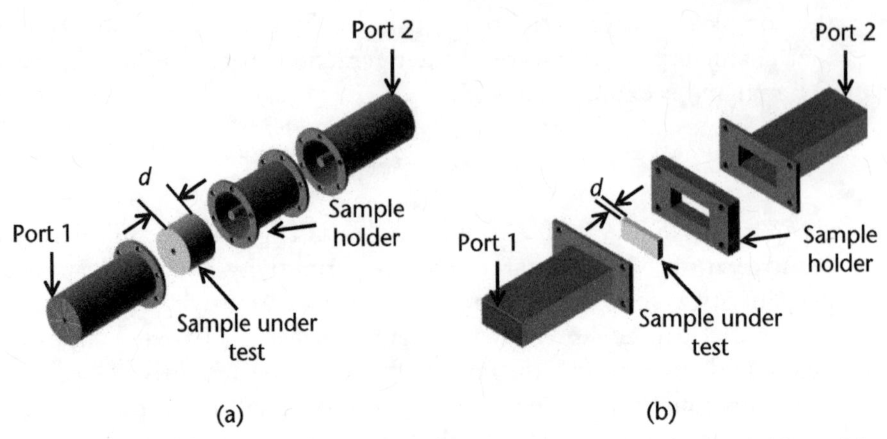

Figure 4.19 Transmitting/reflecting devices with specimen under investigation: (a) coaxial cable, and (b) waveguide [50].

4.11.2 Coaxial Line Reflection Method

The open-circuited reflection method entails establishing direct contact between the material being tested and the open end of a coaxial line. This arrangement causes the reflection of electromagnetic waves at the interface due to impedance mismatches. By measuring the reflectivity at the interface, valuable information about the material's properties, such as permittivity and permeability, can be derived. However, it is essential to conduct an adequate number of independent reflection measurements to ensure accurate characterization. Although various transmission lines can be used, the use of coaxial probes is common due to their broadband nature, simplicity, and nondestructive nature. However, the method has limited accuracy and is most suitable for liquids or semisolids. To use the open-ended coaxial probe, the material must be nonmagnetic, isotropic, and homogenous, with a flat surface if it is solid and thick enough to be considered semi-infinite relative to the diameter of the probe's aperture. Moreover, it is imperative to ensure that there are no discontinuities or voids between the probe and the sample, eliminating any possibility of air gaps. While the coaxial line reflection method is convenient, its accuracy is limited under certain conditions when compared to other techniques, such as the transmission line and resonator methods [49].

Reference [51] highlights the significance of precise characterization of the dielectric features of living tissues to ensure the effective development and assessment of diagnostic and therapeutic methods utilizing microwaves. The open-ended coaxial probe (OECP) strategy is a commonly employed technique that offers numerous advantages, including minimal sample preparation, a broad measurement range, and ease of use. However, it is important to acknowledge that the OECP method is susceptible to significant measurement inaccuracies that can arise from errors associated with both the tissue being analyzed and the equipment used for measurement. While system calibration can help reduce equipment-related errors, accurate DP characterization relies heavily on proper equipment selection, especially in terms of selecting an appropriate probe with a proper sensing depth (SD). While previous studies have investigated the SD of OECPs with varying aperture sizes and tissue types, it is required to establish a clear definition of the SDs of OECPs having small aperture sizes for accurate measurement of the dielectric properties of thin layered tissues, like skin tissue, composed of multiple layers. The experimental arrangement is depicted in Figure 4.20.

4.11.3 Resonator Perturbation Method

Resonant perturbation techniques are used to determine the dielectric properties of materials by introducing the material under test (MUT) within a resonating structure to analyze the change in resonant characteristics. The cavity perturbation process, which involves the use of hollow metallic cavities, is commonly employed in these techniques. The theoretical foundations and equations for resonant cavities were extensively explored by Waldron, as documented in the referenced publication [52]. The researchers in [53] scrutinized the cavity perturbation technique in the waveguide designed for the X-band frequency range (typically between 8 and 12 GHz) and found it to be a convenient and accurate method for measuring dielectric characteristics. The streamlined perturbation equation is said in the assessment of the dielectric constant and loss tangent through the placement of the specimen at the location of the extreme electric field within the rectangular waveguide cavity. Furthermore, Waldron provided a more comprehensive discussion of the perturbation formulas in the case of resonators and waveguides, as described in [54]. The paper highlighted that these formulas have limitations when the frequency shift is minimal.

The Damaskos model 600T open resonator, which comprises two spherical mirrors placed coaxially, is utilized to measure the DPs by positioning the material under test in the cavity center, as illustrated in Figure 4.21. The resonator of this kind was first familiarized by Treacy in 1966 [55], and subsequently, in 1970,

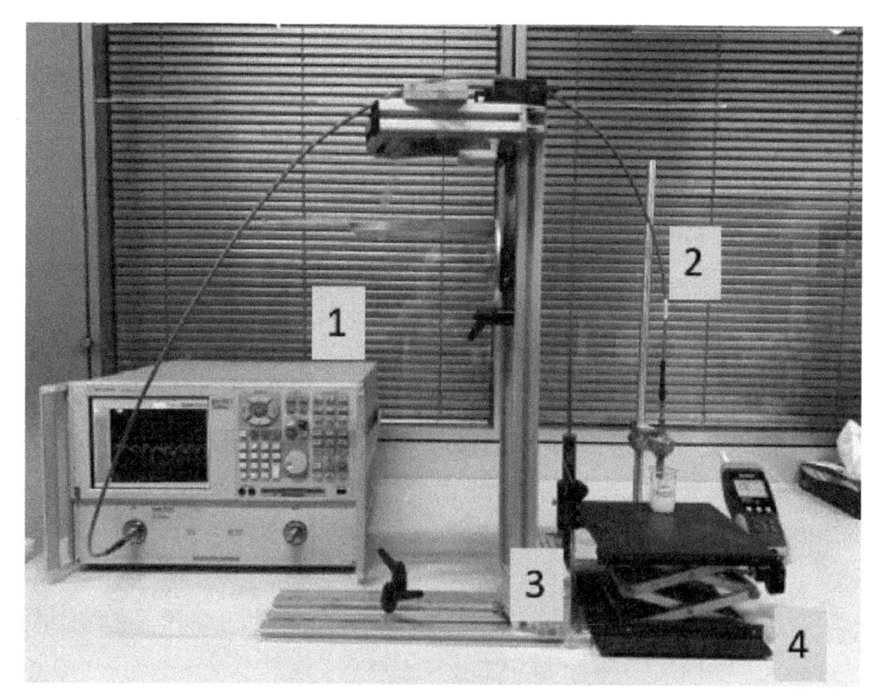

Figure 4.20 Experimental apparatus: (1) Agilent N5230A PNA-L Series Network Analyzer, (2) Agilent Slim Form Probe, (3) digital calliper, and (4) variable stand [51].

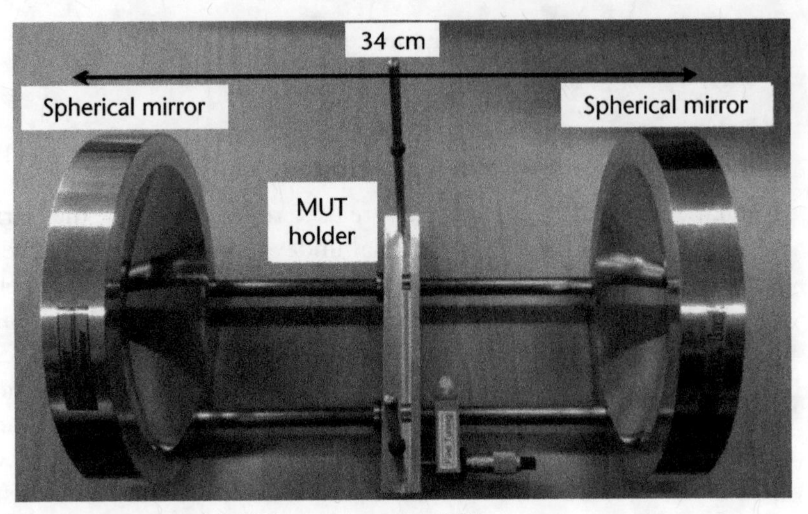

Figure 4.21 Damaskos model 600T open resonator [48].

Yu utilized it as an innovative approach for figuring out the permittivity and loss tangent [56]. Detailed descriptions regarding the measurement process, apparatus, experimental setups, and outcomes pertaining to this technique were provided. The authors in [57] demonstrated that the ambiguity of the dielectric constant for polystyrene and Perspex was ±0.25%. Moreover, they established a solid theoretical basis for the open resonator technique used to measure the dielectric constant and loss tangent.

4.12 Analysis on Characterization of Smart Materials

Characterization of materials is a crucial aspect of materials engineering analysis that aims to understand the physical, chemical, and electrical properties of materials to select the appropriate material for a particular application. A wide variety of characterization techniques are employed to identify materials, including scanning electron microscopy, transmission electron microscopy, X-ray diffraction, X-ray fluorescence (XRF), NMR, and synchrotron techniques. However, it is not feasible to rely on a single technique to fully analyze a material. Practical research work necessitates the use of two or three characterization techniques to draw conclusions. The characterization techniques for a broad range of materials can be classified into microscopy and spectroscopy. Microscopy is a fundamental technique for material characterization and scientific research involving the study of materials and surfaces using microscopes. Different types of microscopes use various methods to produce magnified images of materials and surfaces. The microstructural analysis is performed using microscopy techniques such as optical microscopy, scanning electron microscopy, and transmission electron microscopy. Spectroscopy involves the study of the interaction between radiation and matter, and the analysis and measurement of the resulting absorption or emission spectra provide valuable information regarding the material being tested. Different spectroscopy methods, like NMR, Raman, Fourier transform infrared spectroscopy (FTIR), XRD, XRF, and SIMS, are generally used for chemical composition analysis. Moreover, there are other techniques such as microwave- and electrical conductivity-based characterization [58].

Table 4.1 provides a comprehensive overview of the characterization methods employed for sensing materials, accompanied by the essential material properties relevant to each technique. The variation of microwave properties is used for calibrating measurable sensing data, and the development of microwave sensing proficiencies in functional materials is a new and exciting domain for the development of passive RFID sensors. This interdisciplinary research area has the potential to lead to the development of highly sensitive, passive sensor nodes.

4.13 Summary

The characterization of sensing materials is critical to understand their behavior and performance in detecting a particular parameter. Accurate microwave characterization of smart materials is essential in developing appropriate materials for chipless RFID sensors that are tailored for specific sensing applications. The characterization procedure plays a vital role in determining the fundamental properties of materials, like their relative permittivity and conductive dielectric losses at microwave frequencies. These parameters are essential for the accurate modeling of chipless RFID sensors operating in the microwave range. Furthermore, the characterization process enables the evaluation of various other significant material properties, including extraction, surface roughness, stability, and structural characteristics, specifically within the microwave regime.

This chapter introduced a range of innovative analysis and characterization techniques for intelligent materials, covering microstructural and surface morphology, optical, electrical, and thermal properties (such as DC conductivity and stability), as well as microwave scattering parameters such as complex permittivity, dielectric loss, and reflection loss in the gigahertz range for sensing materials. Based on this information, the subsequent conclusions can be presented:

Table 4.1 Assessment Methodologies for Functional Sensing Materials

Methods	*Category*	*Material Property*
XRD	Structural	Crystallinity
Raman	Structural	Crystallinity
SIMS	Structural	Isotopic, elemental, or molecular composition
TEM	Structural	Crystallographic nature with subnanometric resolution
SEM	Structural, surface, optical, electronic	Surface topography, crystalline structure, chemical composition, and electrical behavior
AFM	Surface	Surface roughness
UV-visible spectrophotometer	Optical, electronic	Transmission, reflection, and absorption
Electrical conductivity	Electrical	Conductivity
Microwave	Optical, electrical, electronic	Permittivity, permeability, loss tangent

- The structural portrayal of intelligent materials intended for microwave sensing necessitates the application of diverse methodologies, such as XRD, Raman spectroscopy, SIMS, SEM, TEM, and FTIR. These methods are essential in identifying and verifying materials, as well as obtaining definitive structural information that can be used to design numerous intelligent materials for the purpose of microwave sensing.
- Intelligent materials for microwave sensing require surface characterization, which can be achieved through the use of SEM and AFM techniques. These techniques provide information on the topography of the material surface, which is essential in the development of various smart materials for microwave sensing.
- SE and UV-visible spectroscopy are optical characterization methods that are employed to analyze the optical characteristics of intelligent materials utilized in microwave sensing applications. SE, in combination utilizing an optical model, allows for the extraction of information such as complex refractive index, dielectric constant, layer depth, and surface unevenness for each layer of the material. On the other hand, UV-vis spectroscopy offers evidence on the transmittance, reflection, absorption coefficient, band gap, and other properties in thin films. These optical characterization techniques are crucial in understanding and developing functional materials for sensing applications in microwave regimes.
- To effectively use smart materials for microwave sensing, it is essential to characterize their electrical and microwave properties. Electrical conductivity measurements are used to evaluate dielectric/semiconductor materials for microwave sensing. Additionally, microwave characterization techniques such as measuring scattering parameters, complex dielectric permittivity, dielectric, and reflection loss at microwave and mm-wave regimes are used. These measurement outcomes lead to variations in microwave properties and can help develop highly sensitive passive sensor nodes.

In conclusion, materials characterization is an important process in material science and engineering aimed at understanding the physical, electrical, and chemical properties of materials in the direction of chipless RFID printing. Although different characterization techniques for smart materials were discussed in this chapter, only some are suitable for printing tag sensors.

References

[1] Amin, E. M., J. K. Saha, and N. C. Karmakar, "Smart Sensing Materials for Low-Cost Chipless RFID Sensor," *IEEE Sensors Journal,* Vol. 14, No. 7, July 2014, pp. 2198–2207.

[2] Dey, S., J. K. Saha, and N. C. Karmakar, "Smart Sensing: Chipless RFID Solutions for the Internet of Everything," *IEEE Microwave Magazine,* Vol. 16, No. 10, November 2015, pp. 26–39, doi: 10.1109/MMM.2015.2465711.

[3] Correas-Serrano, D., and J. S. Gomez-Diaz, "Graphene-Based Antennas for Terahertz Systems: A Review," 2017, *arXiv preprint arXiv:1704.00371.*

[4] Zhu, Y., et al., "Graphene and Graphene Oxide: Synthesis, Properties, and Applications," *Advanced Materials,* Vol. 22, No. 35, 2010, pp. 3906–3924.

[5] Castro Neto, A. H., F. Guinea, N. M. R. Peres, K. S. Novoselov, and A. K. Geim, "The Electronic Properties of Graphene," *Reviews of Modern Physics,* Vol. 81, No. 1, 2009, pp. 109–162.

[6] Huang, W., "Special Issue on Graphene," *Chinese Science Bulletin,* Vol. 57, No. 23, November 2012, p. 2947.

[7] Lee, C., X. Wei, J. W. Kysar, and J. Hone, "Measurement of the Elastic Properties and Intrinsic Strength of Monolayer Graphene," *Science,* Vol. 321, No. 5887, July 2008, pp. 385–388.

[8] Elias, D. C., et al., "Control of Graphene's Properties by Reversible Hydrogenation: Evidence for Graphene," *Science,* Vol. 323, No. 5914, 2009, pp. 610–613.

[9] Berger, C., et al., "Ultrathin Epitaxial Graphite: 2D Electron Gas Properties and a Route Toward Graphene-Based Nanoelectronics," *The Journal of Physical Chemistry B,* Vol. 108, No. 52, 2004, pp. 19912–19916.

[10] Balandin, A. A., "Thermal Properties of Graphene and Nanostructured Carbon Materials," *Nature Materials,* Vol. 10, No. 8, 2011, p. 569.

[11] Ozek, E. A., S. Tanyeli, and M. K. Yapici, "Flexible Graphene Textile Temperature Sensing RFID Coils Based on Spray Printing," *IEEE Sensors Journal,* Vol. 21, No. 23, December 1, 2021, pp. 26382–26388.

[12] Akbari, M., J. Virkki, L. Sydänheimo, and L. Ukkonen, "Toward Graphene-Based Passive UHF RFID Textile Tags: A Reliability Study," *IEEE Transactions on Device and Materials Reliability,* Vol. 16, No. 3, September 2016, pp. 429–431.

[13] Brinker, K., et al., "Review of Advances in Microwave and Millimetre-Wave NDT&E: Principles and Applications," *Philosophical Transactions of the Royal Society A,* Vol. 378, No. 218, 2020, 20190585.

[14] Randazzo, A., K. M. Donnell, and Y. S. (B.) Lee, "Microwave and Millimeter-Wave Sensors, Systems, and Techniques for Electromagnetic Imaging and Materials Characterization," *International Journal of Microwave Science and Technology,* 2012.

[15] Zhang, S., L. Li, and A. Kumar, *Materials Characterization Techniques,* Bocan Raton, FL: CRC Press, 2008.

[16] Leng, Y., *Materials Characterization: Introduction to Microscopic and Spectroscopic Methods,* John Wiley & Sons, 2009.

[17] Ryland, A. L., "X-ray diffraction," *Journal of Chemcial Education,* Vol. 80, 1958.

[18] Warren, B. E., "X-ray Diffraction Methods," *Journal of Applied Physics,* Vol. 12, No. 5, 1941, pp. 375–384.

[19] Karmakar, N. C., E. M. Amin, and J. K. Saha, *Chipless RFID Sensors,* Hoboken, NJ: John Wiley & Sons, 2015.

[20] Roggero, U. F. S., et al., "Graphene-Biopolymer-Based RFID Tags: A Performance Comparison," *Materials Today Communications*, Vol. 31, 2022, 103726.

[21] Krauss, T. N., et al., "Three-Dimensional Molecular Packing of Thin Organic Films of PTCDI-C8 Determined by Surface X-Ray Diffraction," *Langmuir,* Vol. 24, No. 22, 2008, pp. 12742–12744.

[22] Ali, I., et al., "Ruthenium (Ru) Doped Zinc Oxide Nanostructure-Based Radio Frequency Identification (RFID) Gas Sensors for NH_3 Detection," *Journal of Materials Research and Technology,* Vol. 9, No. 6, 2020, pp. 15693–15704.

[23] Smith, E., and G. Dent, *Modern Raman Spectroscopy: A Practical Approach*, Hoboken, NJ: John Wiley & Sons, 2019.

[24] Jorio, A., and R. Saito. "Raman Spectroscopy for Carbon Nanotube Applications," *Journal of Applied Physics,* Vol. 129, No. 2, 2021, 021102.

[25] Kothuru, A., C. Hanumanth Rao, S. B. Puneeth, M. Salve, K. Amreen, and S. Goel, "Laser-Induced Flexible Electronics (LIFE) for Resistive, Capacitive and Electrochemical Sensing Applications," *IEEE Sensors Journal,* Vol. 20, No. 13, July 1, 2020, pp. 7392–7399.

[26] Williams, P., "Secondary Ion Mass Spectrometry," *Annual Review of Materials Science,* Vol. 15, No. 1, 1985, pp. 517–548.

[27] Van der Heide, P., *Secondary Ion Mass Spectrometry: An Introduction to Principles and Practices*, Hoboken, NJ: John Wiley & Sons, 2014.

[28] Wang, Y., et. al., "Use of Phosphine as an n-Type Dopant Source for Vapor-Liquid-Solid Growth of Silicon Nanowires, " *Nano Letters,* Vol. 5, 2005, pp. 2139–2143.

[29] De Graef, M., *Introduction to Conventional Transmission Electron Microscopy*, Cambridge, UK: Cambridge University Press, 2003.

[30] Yue, L., et al., "Analytical Transmission Electron Microscopy for Emerging Advanced Materials," *Matter,* Vol. 4, No. 7, 2021, pp. 2309–2339.

[31] Reimer, L., *Transmission Electron Microscopy: Physics of Image Formation and Microanalysis*, Vol. 36, New York: Springer, 2013.

[32] Septiani, N. L. W., D. R. Adhika, A. G. Saputro, Nugraha, and B. Yuliarto, "Enhanced NO Gas Performance of (002)-Oriented Zinc Oxide Nanostructure Thin Films," *IEEE Access,* Vol. 7, 2019, pp. 155446–155454.

[33] Vernon-Parry, K. D., "Scanning Electron Microscopy: An Introduction," *II-Vs Review,* Vol. 13, No. 4, 2000, pp. 40–44.

[34] Reimer, L., "Scanning Electron Microscopy: Physics of Image Formation and Microanalysis," *Measurement Science and Technology,* Vol. 11, No. 12, 2000, pp. 1826–1826.

[35] Zhou, W., et al., "Fundamentals of Scanning Electron Microscopy (SEM)," *Scanning Microscopy for Nanotechnology: Techniques and Applications*, 2007, pp. 1–40.

[36] Ehsani, M., M. N. Hamidon, A. Toudeshki, M. H. S. Abadi, and S. Rezaeian, "CO_2 Gas Sensing Properties of Screen-Printed La_2O_3/SnO_2 Thick Film," *IEEE Sensors Journal,* Vol. 16, No. 18, September 15, 2016, pp. 6839–6845.

[37] Utari, L., et al., "Wearable Carbon Monoxide Sensors Based on Hybrid Graphene/ZnO Nanocomposites," *IEEE Access,* Vol. 8, 2020, pp. 49169–49179.

[38] Eaton, P., and P. West, *Atomic Force Microscopy*, Oxford, UK: Oxford University Press, 2010.

[39] Joshi, J., S. V. Homburg, and A. Ehrmann, "Atomic Force Microscopy (AFM) on Biopolymers and Hydrogels for Biotechnological Applications—Possibilities and Limits," *Polymers,* Vol. 14, No. 6, 2022, pp. 1267.

[40] Gupta, S., N. Yogeswaran, F. Giacomozzi, L. Lorenzelli, and R. Dahiya, "Touch Sensor Based on Flexible AlN Piezocapacitor Coupled With MOSFET," *IEEE Sensors Journal,* Vol. 20, No. 13, July 1, 2020, pp. 6810–6817.

[41] Kim, S., "Inkjet-Printed Electronics on Paper for RF Identification (RFID) and Sensing," *Electronics,* Vol. 9, No. 10, 2020, pp. 1636.

[42] Schmid, F. X., "Biological Macromolecules: UV-Visible Spectrophotometry," *e LS,* 2001.

[43] Passos, M. L. C., and M. L. M. F. S. Saraiva, "Detection in UV-Visible Spectrophotometry: Detectors, Detection Systems, and Detection Strategies," *Measurement,* Vol. 135, 2019, pp. 896–904.

[44] Görög, S., *Ultraviolet-Visible Spectrophotometry in Pharmaceutical Analysis*, Boca Raton, FL: CRC Press, 2018.

[45] González-Morales, D., et al., "Development of a Low-Cost UV-Vis Spectrophotometer and Its Application for the Detection of Mercuric Ions Assisted by Chemosensors," *Sensors,* Vol. 20, No. 3, 2020, p. 906.

[46] Mohammadpour-Haratbar, A., Y. Zare, and K. Y. Rhee, "Simulation of Electrical Conductivity for Polymer Silver Nanowires Systems," *Scientific Reports,* Vol. 13, No. 1, 2023, p. 5.

[47] Karbownik, I., et al., "The Effect of Temperature on Electric Conductivity of Polyacrylonitrile-Polyaniline Fibers," *IEEE Access,* Vol. 9, 2021, pp. 74017–74027.

[48] Lee, C.- K., et al., "Evaluation of Microwave Characterization Methods for Additively Manufactured Materials," *Designs,* Vol. 3, No. 4, 2019, p. 47.

[49] Tsipogiannis, C., "Microwave Materials Characterization Using Waveguides and Coaxial Probe," Master's thesis, Department of Electrical and Information Technology Faculty of Engineering, LTH, Lund University, Lund, Sweden, 2012.

[50] Costa, F., et al., "Electromagnetic Characterisation of Materials by Using Transmission/Reflection (T/R) Devices," *Electronics,* Vol. 6, No. 4, 2017, p. 95.

[51] Aydinalp, C., et al., "Characterization of Open-Ended Coaxial Probe Sensing Depth with Respect to Aperture Size for Dielectric Property Measurement of Heterogeneous Tissues," *Sensors,* Vol. 22, No. 3, 2022, p. 760.

[52] Waldron, R. A., "Perturbation Theory of Resonant Cavities," *Proceedings of the IEE–Part C, Monographs,* Vol. 107, 1960, pp. 272–274.

[53] Dube, D.C., M. T. Lanagan, J. H. Kim, S. J. Jang, "Dielectric Measurements on Substrate Materials at Microwave Frequencies Using a Cavity Perturbation Technique," *Journal of Applied Physics,* Vol. 63, 1988, pp. 2466–2468.

[54] Waldron, R. A., "Perturbation Formulas for Elastic Resonators and Waveguides," *IEEE Transactions on Sonics and Ultrasonics,* Vol. 18, 1971, pp. 16–20.

[55] Treacy, E. B, "The Two-Cone Open Resonator," *Proceedings of the IEEE,* 1966, Vol. 54, pp. 555–560.

[56] Yu, P. K., "Measurements Using an Open Resonator," Doctoral thesis, London, UK: University College London, 1970.

[57] Cullen, A. L., and P. K. Yu, "The Accurate Measurement of Permittivity by Means of an Open Resonator," *Proceedings of the Royal Society London,* 1971, Vol. 325, pp. 493–509.

[58] Ananthapadmanaban, D., "Summary of Some Selected Characterization Methods of Geopolymers," in *Geopolymers and Other Geosynthetics*, M. Alshaaer and H. Y. Jeon (eds), IntechOpen, 2018.

CHAPTER 5

Passive Printable Chipless RFID Tag for Biomedical Applications

5.1 Introduction

The current digital era anticipates replacing manual systems with automation. A well-known wireless technology—RFID—enters the market for auto-ID technology by removing the barrier of human participation. Although optical barcodes have long been the standard due to their low cost, their short-read distance, poor information storage capacity, and heavy reliance on manual labor make an alternative solution necessary. All of the aforementioned flaws of an optical barcode are fixed by standard RFID. However, one significant issue with RFID is its cost. The advantages of chipless RFID over current auto-ID technology are discussed in this chapter. The fundamentals of the RFID system, types of RFID tags, and contrast between chipless and chipped RFID tags are described. The classifications of chipless RFID tags, the difficulties in developing a chipless RFID, the operation of a backscattering chipless tag, and design criteria like the RCS are covered in this chapter to understand the future relevance of chipless RFID. The chapter proceeds with a thorough exploration of the world of RFID technology with a clear emphasis on its passive, printable, and chipless varieties specifically created to meet the unique and changing requirements of the biomedical industry. RFID has emerged as a transformative force within healthcare, demonstrating immense potential in patient care, pharmaceutical traceability, laboratory management, and overall operational efficiency within medical facilities. This chapter begins by providing a foundational understanding of RFID, elucidating its core principles, and tracing its evolutionary path from active to passive RFID, where tags no longer necessitate an internal power source. Additionally, the chapter comprehensively investigates the disruptive capabilities of chipless RFID, an innovative solution that offers cost-effective, highly adaptable features, making it particularly compelling for the biomedical sector. Real-time applications such as patient monitoring and management, drug authentication, blood and tissue tracking, inventory control of medical supplies, and equipment tracking are highlighted in the chapter. By delving into these varied applications, the chapter offers a vivid glimpse into the transformative potential of chipless RFID technology in reshaping healthcare delivery and management, signifying a promising and exciting future for the healthcare industry. This chapter also discusses RFID tag printing, chipless RFID data capture methods, specific tag production materials, chipless RFID-based sensors, potential applications, present obstacles, and chipless RFID tag directions. There is currently a considerable increase in the demand for less expensive sensors to measure a variety of physical factors, including humidity, pressure, strain, temperature, and biological parameters, including blood pressure, heart rate, glucose level, pH of body fluids (blood, saliva, urine, gastric juice, etc.), and body temperature. Planar technology-compatible passive chipless RFID sensors have lower production and maintenance costs and are therefore essential for a variety of real-world applications, including biomedicine, defense, industrial, smart packaging, healthcare, and food safety. The unique characteristics of chipless sensors include immaculate assimilation into fabric materials, contemporaneous monitoring, fewer negative environmental effects, longer lifespan, and low cost. Furthermore, chipless sensors can deliver precise results under challenging circumstances where chip-based RFID sensors encounter restrictions as a result of the existence of ASIC. The idea of chipless sensors is new for the Internet of Things [1–7]. In Figure 5.1, a straightforward chipless RFID sensing device is shown. The reader emits an UWB signal received by the sensor. Both the sensing data and the individual identification (ID) replies are backscattered by the sensor. There are two different kinds of resonators on the tag, one of which is only used for ID answers and the other for gathering data [8]. Transforming physical data into EM properties is done by chipless sensors. A frequency signature made up of the environment's sensory data is produced by the electromagnetic waves that are reflected [2].

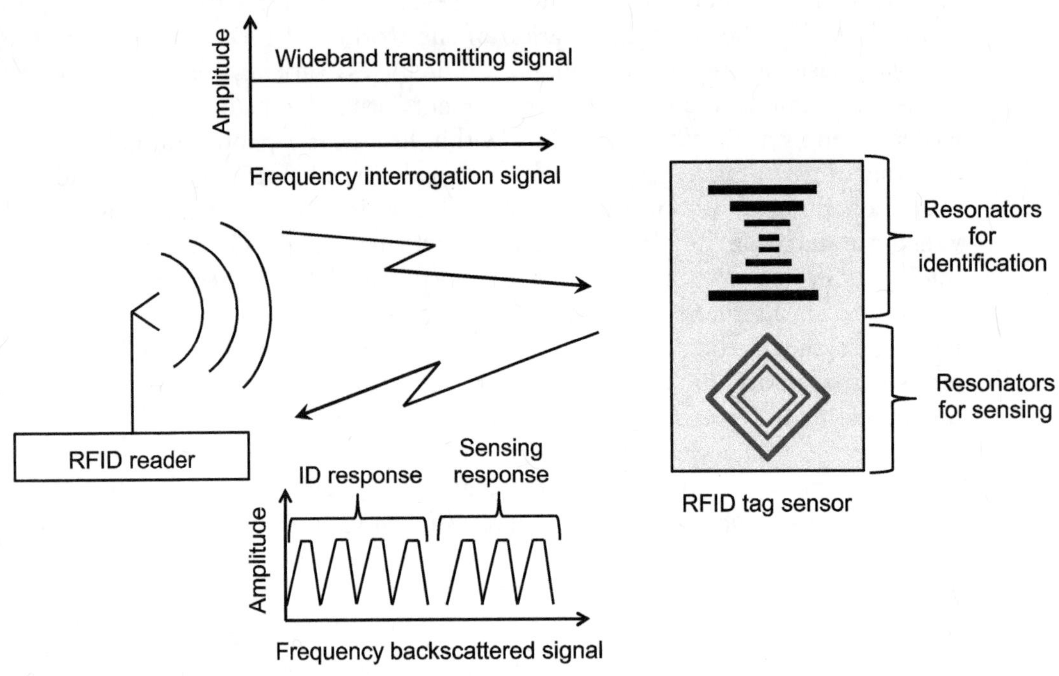

Figure 5.1 Basic operating principle of a chipless RFID sensing system.

5.1.1 Comparison of Chipped and Chipless Sensors

The overall amount of a conventional RFID system turns away its possibility to be an advancement that substitutes barcodes. This sky-scraping price (about 10 cents) of the traditional RFID sensor emerges from the ASIC in the chip-based RFID tag sensor. The comparable price of chipless RFID sensors (about 1 cent) to barcodes is an alternative to the chipped sensors. However, their progress is a difficult accomplishment [9]. In a typical RFID system, the tracking distance is primarily influenced by factors such as frequency of operation, the power of the interrogator, the antenna design, and the sensitivity of the RFID tags. In a chip-based RFID system, the tracking distance can be influenced by the power transmitted by the interrogator, but it's not directly proportional. Increasing the power doesn't guarantee a linear increase in the tracking distance due to other limiting factors like the tag's sensitivity and the environment's RF interference. Chipless RFID systems, on the other hand, often have a shorter tracking distance compared to chip-based RFID systems. This is because chipless RFID tags rely on backscattering waves, often resulting in a weaker signal compared to chip-based tags that can actively respond and communicate with the interrogator. Therefore, chipless RFID systems are more appropriate for applications where shorter tracking distances are acceptable. The shorter tracking distance in chipless systems is a trade-off for their potential cost-effectiveness and flexibility, making them ideal for applications where close proximity tracking suffices. In chipless RFID, the term "chipless" implies that the tags do not have a traditional integrated circuit (IC) chip that stores data or information. Instead, chipless RFID tags rely on unique patterns or features in the frequency domain, typically in response to electromagnetic waves, to convey information. These patterns represent data without the need for a physical memory storage component. The storage capabilities in chipless RFID are fundamentally different from chip-based RFID systems. Instead of storing data in a memory element like a traditional IC chip, chipless RFID tags encode information using various techniques such as frequency modulation, phase modulation, or spectral signature. The amount of information that can be encoded into a chipless RFID tag is not tied to frequency bands but rather to the design, structure, and encoding methods used for that specific chipless tag. The storage capabilities of a chipless RFID tag are not dependent on the considered frequency bands but rather on the encoding mechanism and the tag's physical design. Different encoding techniques and designs can result in different storage capacities and data representation, making chipless RFID a unique and innovative approach for certain applications. Passive chipless and chip-based RFID tag sensors are available where the design of the complete system in the former is complex, and the latter is simple because of the reader side complexity. Chip-based sensors and chipless sensors each possess distinct characteristics and advantages, making it difficult to assert one as universally more flexible or inflexible than the other. Chip-based sensors typically involve the integration of a microchip (IC) within the sensor, providing precise and tailored functionality for specific applications. These sensors can be highly flexible in terms of their ability to incorporate various functionalities within the chip, offering features like signal processing, memory, and communication capabilities. On the other hand, chipless sensors, which do not rely on a traditional IC chip, are often praised for their flexibility in design and fabrication. Their structure and design can be adjusted for specific applications without the constraints of integrating an IC chip. However, the perception of flexibility also depends on the context and requirements of a particular application. In some cases, chipless sensors may offer more design freedom, while in others, the tailored functionality and precision of chip-based sensors might be more beneficial. Ultimately, the flexibility of a sensor technology should be evaluated based on the specific needs and goals of the intended application. The multitag detection capacity varies in both chipped and chipless sensors, with respect to certain experiments in [10], 1,000 chip-based tags, whereas only one chipless tag can be detected (three tags theoretically) simultaneously. A comparison of the chip-based and chipless sensors is illustrated in Table 5.1.

5.1.2 Categories of Chipless Sensors

Chipless sensors can be characterized into the following categories based on how they are encoded. As shown in Figure 5.2, chipless sensors are based on (1) TDR, (2) frequency modulation, and (3) phase modulation. The delay line-based tags are more energy-efficient and smaller. Data is encoded into the frequency domain when using sensors that use frequency modulation. This family of chipless sensors that encrypt data using resonant structures has the largest coding capacity. Also, they are less costly and smaller. However, ambient noise has a noteworthy effect on detection. Phase-encoded tags, which store data in the phase, comprise the third category of chipless sensors. This type of sensor benefits from consecutive operation within a restricted bandwidth (BW), allowing for efficient data storage and transmission [2].

This is especially true when comparing the power requirements of chipless RFID systems to standard chipped RFID (from Table 5.1). In chipless RFID systems, the interrogating power can often be lower compared to standard chipped RFID due to fundamental differences in their operating principles. Standard chipped RFID tags contain an IC chip that requires a certain amount of power to operate and respond to the interrogating signal. The IC chip needs power to process the incoming signal, modulate its response, and transmit data back to the reader. In contrast, chipless RFID tags do not have a traditional IC chip and usually rely on passive mechanisms to respond to the interrogating signal. These passive mechanisms, such as changes

Table 5.1 Comparison of Chipped and Chipless Sensors

Aspects	*Chip-Based RFID Sensor*	*Chipless RFID Sensor*
ASIC	Exists	Does not exist
Power radiated from the interrogator	High	Low
Sensor design	Simple	Complicated
Tracking distance [2, 3, 4, 8]	Long (10–20m)	Short (1–3m only)
Mode of activation of tag sensor	Inbuilt power sources/interrogation signal	Interrogation signal
Physical property	Fragile and nonprintable	Pliable and printable
Cost	Exorbitant cost (10 cents)	Less cost (1 cent)
Sensing environment	Does not operate in harsh environments	Operates in extreme temperatures
Sensor lifespan	Less	More

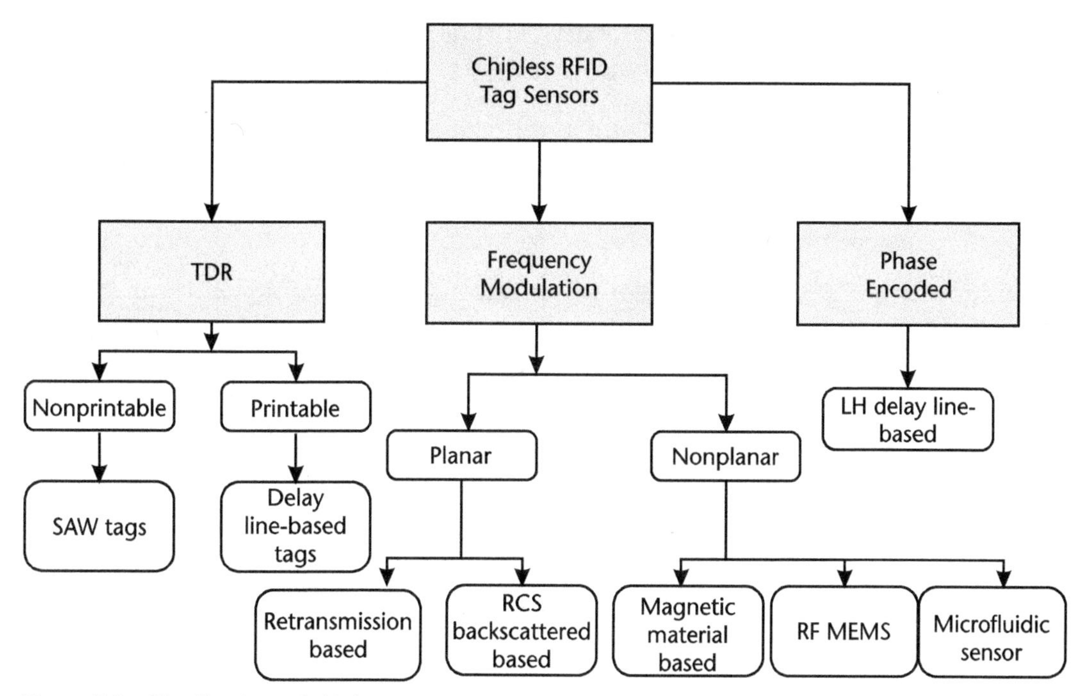

Figure 5.2 Classifications of chipless sensors.

in resonance or reflection, can be induced with lower power levels. Chipless tags often derive their uniqueness from alterations in material properties or geometric patterns, allowing them to reflect a distinctive response without the need for an energy-demanding active circuit. The absence of an energy-consuming IC chip and the passive nature of chipless RFID mechanisms make it possible to achieve reliable communication and data retrieval with lower power levels during interrogation. This is especially beneficial in applications where minimizing power consumption is a priority, such as in battery-free or energy-harvesting scenarios or in environments where power availability is limited or expensive. Therefore, chipless RFID systems generally exhibit a lower interrogating power requirement compared to their chipped RFID counterparts.

5.1.3 Design Challenges of a Chipless RFID System

The reader radiator structure is the foremost difficulty in a chipless RFID system. Chipless RFID works in the ultrawideband, and the utmost power emitted in this band is less than that transmitted by the chip-based interrogator. Therefore, the tracking distance in the case of the chipless system is limited, which further lowers the backscattered signal processing capacity. Moreover, the received signal is low, which is further infested by surrounding noise [8].

The next difficult task is the tag antenna design. The tag's physical dimension is compact in a higher frequency band of operation, which reduces the reflected signal toward the reader. However, the utilization of antenna arrays at the cost of the tag's physical footprint can act as a solution to this issue [11].

An additional difficulty arises in the proper capturing of data from the tag sensor as the tag doesn't have signal processing ability. Several tags can be detected simultaneously by the implementation of anticollision algorithms [9]. Another crucial factor in the tag sensor design is the right choice of sensing material.

5.2 Design Principle: Passive Resonator-Based Tags

A transponder's capability to backscatter electromagnetic energy in the pathway of the interrogator's receiving radiator is known as its radar cross section. The target is easily spotted once the RCS of a tagged entity, RCS_{tag}, is improved because this enhances the strength of the backscattered signal, S_{tag}. The usual radar equation for determining the received power is (5.1), where λ is the operating wavelength, d is the separation between the tag and reader antenna, G_R is the gain of the reader radiator, and P_T is the transmitted power. The RCS and the operational wavelength are the only two parameters that, if the measurement equipment is left unaltered, define the obtained power [12]:

$$S_{tag} = \frac{G_R^2 P_T RCS_{tag} \lambda^2}{(4\pi)^3 d^4} \tag{5.1}$$

An approximation of the monostatic RCS of a scattering tag can be calculated as in (5.2), where the RCS plate is the RCS of a metallic square plate with a physical size identical to the tag, Γ is the tag's reflection coefficient (Γ), and A is the tag's area of cross section [8]:

$$RCS_{tag} = RCS_{plate} (\Gamma)^2 = \frac{4\pi A^2}{\lambda^2} (\Gamma)^2 \tag{5.2}$$

The following variables serve as the primary means by which the chipless RFID tag RCS is expressed: the cross-sectional part of the tag (A in m^2), and the directivity (D), as shown in (5.3) [13]:

$$\sigma = A|\Gamma|^2 D \tag{5.3}$$

5.2.1 Fabrication Techniques of Chipless Sensors

Nanofabrication, printing, and micromachining (surface/bulk) technologies are the three main categories of production processes for chipless RFID sensors. Comparatively speaking, a roll-to-roll mass production method used for printable sensor tags offers a less expensive option. Nevertheless, precision high-bit-capacity tag manufacture is made possible by nanofabrication and micromachining [14].

Chipless sensors are made using nanofabrication processes such as electrodeposition, physical and chemical vapor deposition, and lithography. In order to obtain the original pattern, the etching procedure requires removing extraneous materials from the laminate. Fine line pattern transfer techniques are used in plasma processing, which is suitable for substrates with limited heat stability. Electron beam lithography is also capable of producing intricate patterns as small as 100 nm. Due to its high energy concentration and directionality, laser processing offers a spatial resolution of greater than 10 nm [15].

Microelectromechanical systems (MEMS) or nanoelectromechanical systems (NEMS) are used in the manufacture of miniature sensors using bulk or surface micromachining processes. Surface micromachining has a higher construction concentration than bulk micromachining and is appropriate for producing freestanding objects [14].

The low-cost printing approach uses an additive approach to deposit ink directly on either flexible or rigid substrate material; as a result, labor-intensive procedures like etching, stripping, or cleaning are not necessary. In [13], a number of printing technologies are discussed. Inkjet printing, as demonstrated in Figure 5.3, operates without the need for a main printing plate that has already been embedded, making it ideal for mass production, batch sizes that are small, and straightforward configuration [16]. A simple, quick, and useful approach for evenly dispersing ink over substrates is doctor-blading. Chipless tags are created using a thermal transfer printer [17]. When using a roller to apply pressure to the substrate during the screen-printing process, a mask is needed to contain the areas where the ink must be printed. Moreover, by using this method, substantial layers of a solid can be deposited, which enhances conductivity and dependability [18]. One method that helps to achieve great accuracy and speed is gravure printing technology. In [19], a textile-based chipless tag is created using a 3D printer. Especially for wireless systems, the technology delivers excellent precision because it operates in a microprocessor-based environment.

5.2.2 Smart Sensing Materials for Chipless RFID Sensors

Sensors, solar cells, and fuel cells frequently make use of nanostructured functional substances or smart materials. Smart materials are excellent for detection but have low carrier mobility, making them undesirable candidates for RF applications. Before utilizing the materials for sensing applications, it is essential to consider that

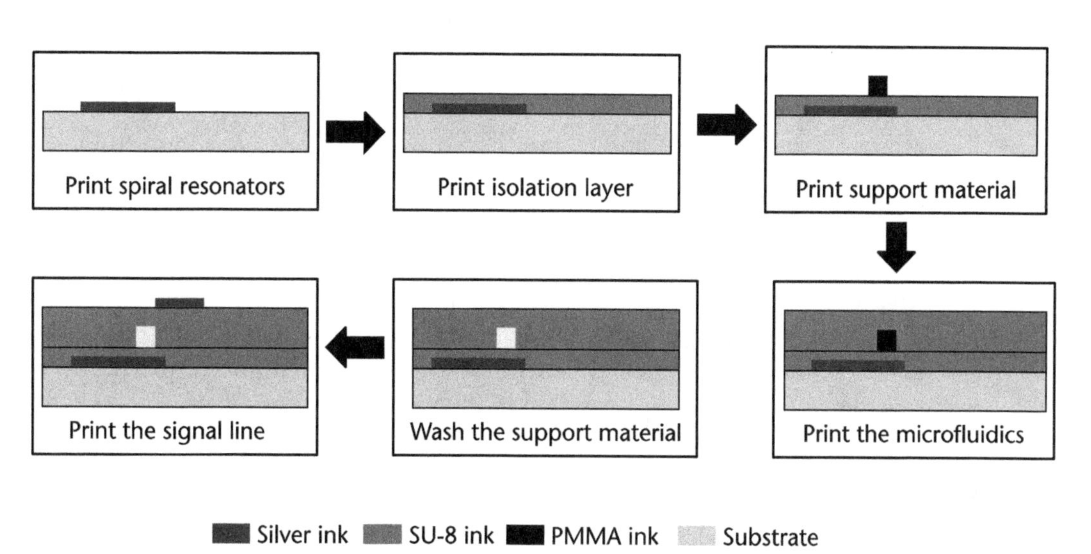

Figure 5.3 Inkjet printing steps.

variations in the physical parameters of the immediate surroundings can alter the EM responses of the tag sensor. It is simple to determine the source of emitted electrons and photons from tested materials from the electronic structure of the sensing material. Ag, Cu, and Al are suitable materials to use in the construction of passive microwave and millimeter-wave circuits due to their increased conductivity. For sensing applications, materials with lower conductivity, such as ITO, silver flakes, and silver nanoparticles, can be employed. The following methods can be used to characterize the structural characteristics of a microfabricated device, such as thickness and surface roughness. Techniques for material characterization include (1) optical and scanning electron microscopy, (2) transmission electron microscopy, (3) X-ray diffraction, and (4) microstructural and surface morphology (XRD, AFM, SEM, TEM, etc.) [20].

5.2.3 Applications of Chipless Sensors

The use of RFID technology is currently expanding quickly in the healthcare, biomedical, and IoT sectors [21, 22]. The chipless RFID system solves the drawbacks of a chipped RFID system and may be an alternative in the IoT. These techniques provide the benefit of chipless RFID, which is becoming more and more common in IoT and mm-wave 5G cellular network systems.

Chipless RFID represents a promising advancement in healthcare and wearable biomedical applications, offering a new dimension of flexibility and functionality. Unlike chip-based sensors, chipless RFID sensors provide improved comfort and wearability, addressing issues related to the fragility of chips and connectivity problems. For instance, [24] introduces a chipless wearable sensor capable of human identification, highlighting the potential of chipless technology in wearables. One of the notable advantages of chipless sensors is their seamless integration with fabrics, enabling their use in monitoring critical biological parameters like blood glucose, pressure, temperature, and heart rate. This integration is a significant leap forward in creating nonintrusive and comfortable biomedical monitoring solutions. In response to the need for cost-effective sensors, [20] presents a low-priced chipless temperature threshold sensor, demonstrating the potential for affordable yet efficient biomedical applications. Radio frequency (RF) sensors, due to their nonintrusive nature and ease of installation, are preferred in various applications [25]. In [26], a partial discharge detecting system employing RF sensors is developed to detect dielectric insulation defects and prevent potential damage to electrical facilities. The applications of chipless RFID sensors are vast and diverse. Some critical applications include gesture recognition [27], enhancing security and authentication [28], defect detection in coatings and corrosion prediction [29], angle-of-arrival detection [30], and monitoring food/drug packaging integrity [31]. These examples emphasize the versatility and immense potential of chipless RFID sensors across different domains, paving the way for innovative solutions in various industries. Further exploration and research in this field will undoubtedly uncover even more applications and push the boundaries of what chipless RFID can achieve in sensing and identification.

5.3 Geometry and Designs

In this section, different geometries and designs will be described based on chipless RFID frequency domain tags. Mostly, FR4 substrate is used here to create copper resonant designs without the use of a ground plane to make a low-cost analysis with a fabricated prototype. The mathematical analysis of the structural geometry and the method of analysis is beyond the scope of this chapter.

5.3.1 Modified Square Split Ring Resonator Based Chipless Sensor: Design 1

With a physical dimension of 39 mm × 39 mm, modified square split ring resonators (MSSRRs) form the foundation of small sensor design. Five bits are produced by three resonating structures that are put into place on the substrate, as shown in Figure 5.4. To improve RCS and the number of bits, the slit widths, resonator width, and resonator spacing are all selected appropriately. The dimensions of the provided tag are shown in Table 5.2.

5.3.2 Rectangular Resonator-Based Chipless Sensor: Design 2

The six resonators on the FR4 substrate of this chipless miniature sensor, which measures 39 mm × 20 mm physically, add to its 6-bit encoding capabilities. As depicted in Figure 5.5, the occurrence of resonance peaks at particular frequencies is greatly influenced by variations in resonator length, slit width, the distance between neighboring resonators, and resonator width. The dimensions of the suggested sensor are provided in Table 5.3.

5.4 Experimental Assessment and Discussion

The RCS of a chipless RFID tag is a critical parameter that characterizes the tag's ability to scatter incident electromagnetic waves. In chipless RFID systems, RCS

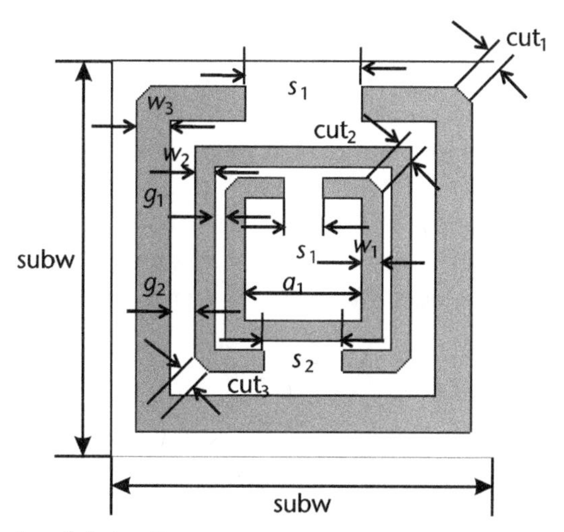

Figure 5.4 The geometry of design 1.

Table 5.2 Dimensional Parameters of Design 1

Parameters	*Dimensions (mm)*
a_1	12.01
s_1	4
s_2	8
s_3	12
w_1	2
w_2	2
w_3	3.5
g_1	1
g_2	2.5
subw	39.01
cut_1	1.55
cut_2	1.44
cut_3	1.33

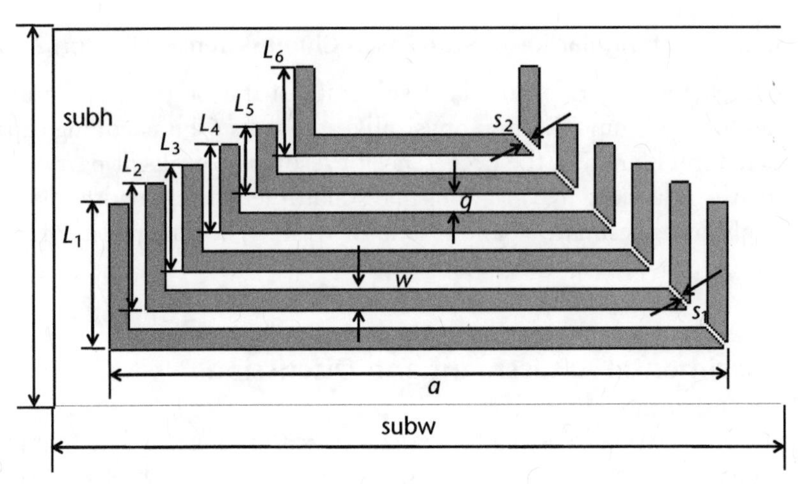

Figure 5.5 The geometry of design 2.

is vital for tag detection as it directly influences the tag's detectability and read range. A higher RCS implies a stronger reflected signal, enhancing the tag's visibility to the reader. Designing chipless RFID tags with optimized RCS patterns allows for efficient backscattering of interrogating signals, enabling robust detection and accurate identification in various applications, from inventory management to healthcare monitoring. Understanding and controlling RCS in chipless RFID tags is essential for maximizing system performance and ensuring reliable data retrieval, making it a key consideration in tag design and implementation. The goal is struck by a plane wave that is linearly polarized, producing RCS or the target's reflected wave. Figures 5.6 and 5.7 display the simulated RCS response of designs 1 and 2, respectively.

The curves show that design 1 can encode 5 bits because there are five resonant signatures between 2 and 8 GHz, while design 2 can encode 6 bits since there are six resonant peaks between 2 and 7 GHz. RCS peaks for suggested designs 1 and 2

Table 5.3 Dimensions of Design 2

Variables	*Values (mm)*
subw	39
subh	20
L_1	15
L_2	13
L_3	11
L_4	9
L_5	7
L_6	9
s_1	0.3
s_2	0.5
g	1
w	1
a	33

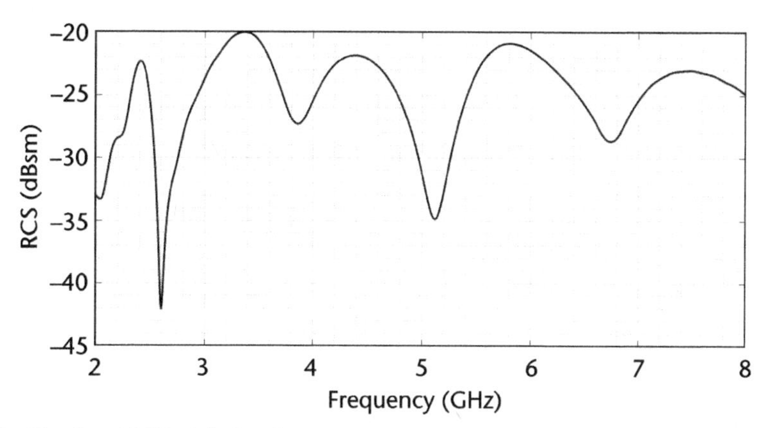

Figure 5.6 Simulated RCS of design 1.

are shown in Tables 5.4 and 5.5, along with their corresponding frequency points. In the first design, the maximum RCS is –20.02 dBsm, while in the second, it is –22.63 dBsm.

5.5 Current Distribution Patterns

The current distribution patterns in a chipless RFID tag play a pivotal role in its detection and functionality. These patterns describe how the electric currents are distributed throughout the tag's structure in response to an incident electromagnetic signal. The arrangement and distribution of these currents determine the tag's unique electromagnetic signature, enabling its identification and communication with the RFID reader. In the context of tag detection, understanding these patterns is crucial as they directly influence the tag's resonant behavior and backscattering characteristics. The design of the tag is meticulously engineered to achieve desired current distributions, optimizing resonant frequencies and enhancing the efficiency

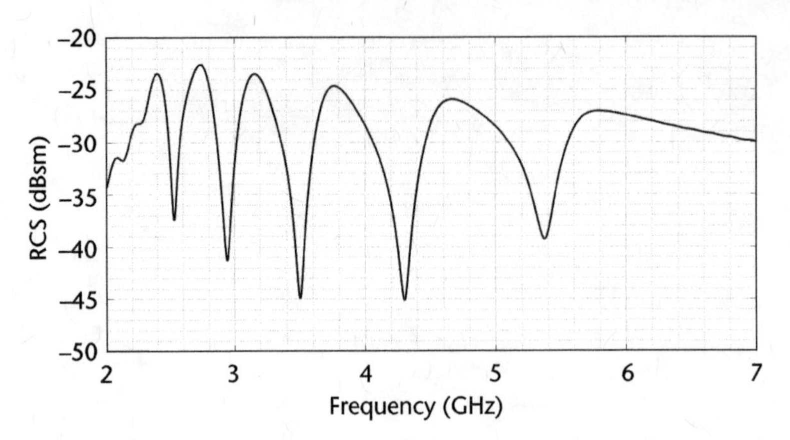

Figure 5.7 Simulated RCS of design 2.

Table 5.4 Simulated RCS Values of Design 1

Frequency (GHz)	*RCS (dBsm)*
2.40	–22.32
3.37	–20.02
4.38	–21.85
5.80	–20.88
7.48	–23.00

Table 5.5 Simulated RCS Values of Design 2

Frequency (GHz)	*RCS (dBsm)*
2.40	–23.44
2.73	–22.63
3.15	–23.5
3.77	–24.64
4.68	–25.92
5.80	–27.05

of backscattered signals. By controlling the current distribution, engineers can tailor the tag's response to specific frequencies, improving its detection accuracy and reliability in a chipless RFID system. Thus, a comprehensive understanding of current distribution patterns is fundamental in advancing chipless RFID technology and its diverse applications. A chipless RFID tag's operation is explained by the surface current density. Figure 5.8 exhibits the surface current density distribution for design 1. Figure 5.9 shows the surface current density of design 2. It is shown that the inner resonators reach the peak current density as the frequency rises gradually. This is owing to the fact that, in both cases, the maximum frequency is resonant with the least resonator.

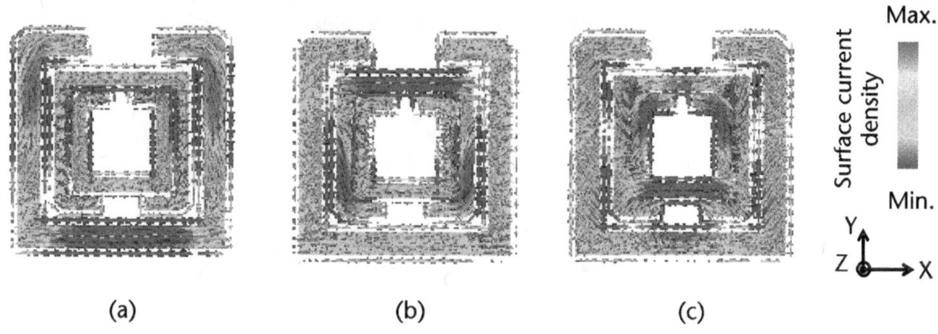

Figure 5.8 Surface current distribution of design 1 at (a) f = 2.40 GHz, (b) f = 5.80 GHz, and (c) f = 7.48 GHz.

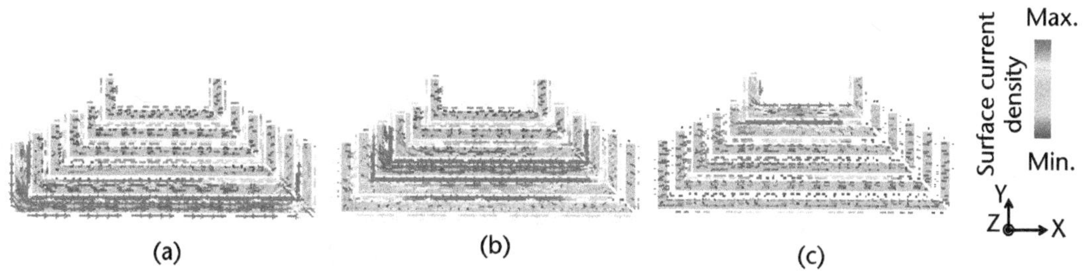

Figure 5.9 Surface current distribution of design 2 at (a) f = 2.40 GHz, (b) f = 3.15 GHz, and (c) f = 5.80 GHz.

5.6 Analysis of the Fabricated Prototypes

On the FR4 substrate, Figures 5.10 and 5.11, respectively, show the physical execution of the two passive chipless sensors that are being discussed. Figure 5.12 depicts the measurement arrangement. The anechoic chamber maintains a 25-cm gap between the tag and the reader. The comparisons of the simulated and measured results of designs 1 and 2, shown in Figures 5.13 and 5.14, respectively, are in good agreement. Tables 5.6 and 5.7 show the RCS peak magnitude in dBsm for the relevant frequencies in GHz for the measured results of designs 1 and 2, respectively.

5.7 Comparison Between Design 1 and Design 2

In Table 5.8, resonator-based chipless tags that have been previously published are compared to the suggested designs based on a number of factors, with the physical size of the tag, the number of bits encoded, the type of resonator, the frequency band, and the applications.

5.8 Chipless Tag Sensors for Noninvasive Applications

This section highlights two chipless RFID sensors resonating in the UHF band and is suitable for on-body applications. Designs 3 and 4 are implemented utilizing modified spiral resonators and MSSRRs, respectively.

Figure 5.10 Fabricated prototype of design 1.

Figure 5.11 Prototype of design 2.

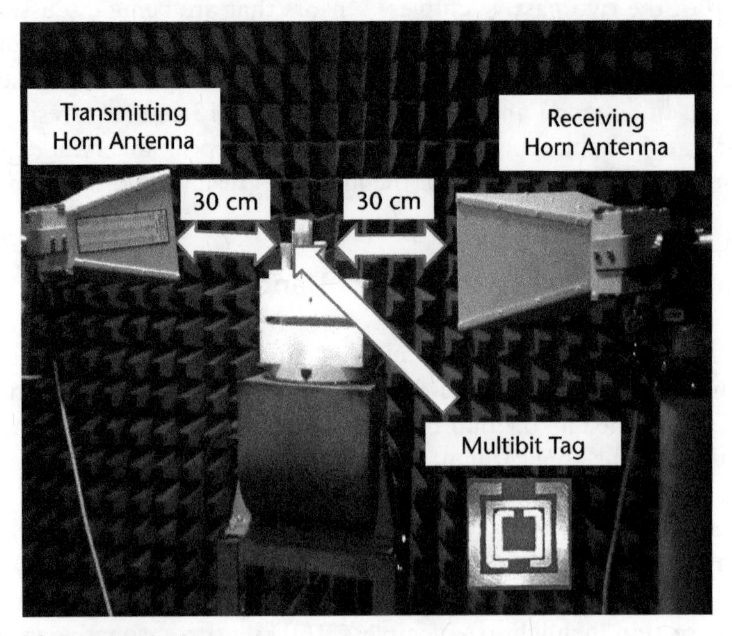

Figure 5.12 Measurement setup.

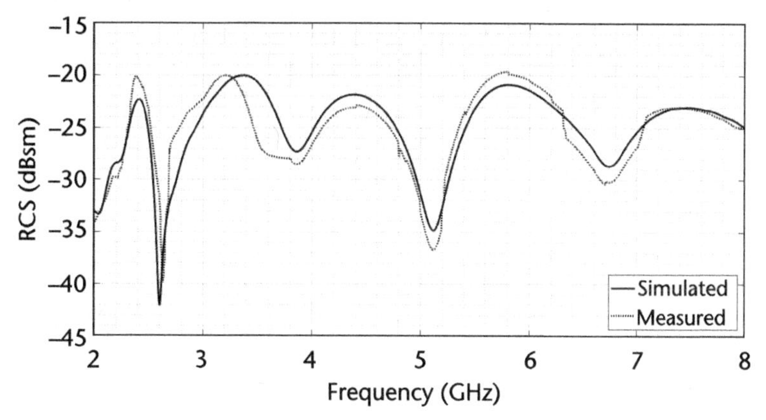

Figure 5.13 Simulated versus measured RCS of design 1.

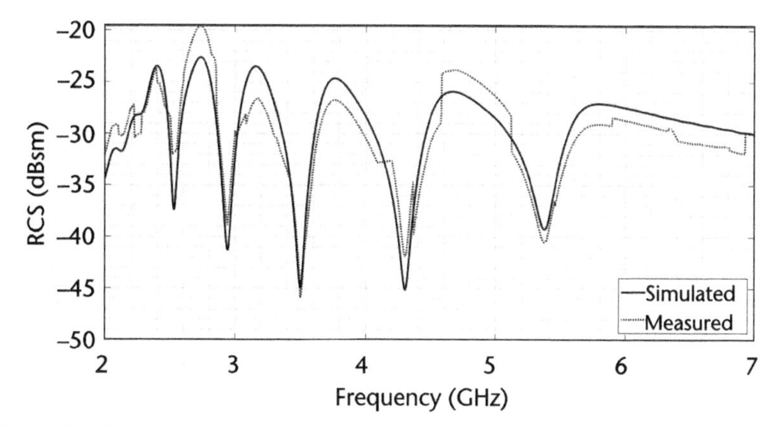

Figure 5.14 RCS of design 2.

Table 5.6 Measured RCS Peaks and Consistent Frequencies of Design 1

Frequency (GHz)	*RCS (dBsm)*
2.39	–20.16
3.22	–20.02
4.40	–22.86
5.80	–19.64
7.40	–23.01

5.8.1 Geometry and Design

Chemical etching using ferric chloride is a common and effective technique employed in the fabrication of chipless RFID tags. The process involves several key steps to create intricate and precise patterns on the tag's substrate. First, a suitable substrate, often a dielectric material like FR4 or specialized polymer, is selected. A mask containing the desired tag pattern is carefully aligned and applied to the

Table 5.7 Measured RCS Peaks and Corresponding Frequencies of Design 2

Frequency (GHz)	*RCS (dBsm)*
2.39	–23.66
2.74	–19.57
3.18	–26.60
3.77	–26.74
4.69	–23.80
5.9	–28.42

Table 5.8 Comparison of Resonator-Based Tags and Designs 1 and 2

Reference	*Tag Dimensional Parameters (mm²)*	*Bits Encoded*	*Resonator Type*	*Frequency Band (GHz)*	*Applications*
[32]	15 × 35	2	SIR	2–7	Product identification
[33]	24 × 24	3	Square SRR	1.5–6.7	Object tracking
	16.5 × 16.5		Circular SRR	1.2–6.5	
[34]	16.5 × 16.5	3	Circular SRR	0–1	Retail and healthcare
[35]	24 × 24	3	Square SRR	0.86–0.96, 2.4	Sleep disorder monitoring and human movement detection
[36]	—	2	Coupled square SRR	2–3	RF barcodes
Design 1	39 × 39	5	Modified square SRR	2–8	Biomedical
Design 2	39 × 20	6	Rectangular resonator	2–7	Biomedical

substrate. The prepared substrate is then immersed in a ferric chloride etchant solution, where the ferric chloride reacts with the exposed areas of the substrate not protected by the mask, effectively etching away the material. The etching duration is closely monitored to achieve the desired pattern depth and precision. The mask is then removed, and the substrate is thoroughly cleaned to halt the etching process. This procedure results in a chipless RFID tag with a specific pattern that determines its resonant frequencies and electromagnetic response. The flexibility and efficiency of chemical etching using ferric chloride make it a valuable technique for fabricating chipless RFID tags, allowing for precise control over tag designs and characteristics. The copper resonators of 0.035 mm thickness are designed (by chemical etching procedure) on a groundless FR4 substrate of a height of 1.6 mm.

5.8.2 Modified Spiral Resonator-Based Chipless Sensor: Design 3

Design 3 is constructed on altered spiral resonators with a maximum overall dimension of 71 mm × 71 mm. The 3-bit sensor is composed of three spirals and is shown in Figure 5.15. The parameters, such as the resonator width and the gap

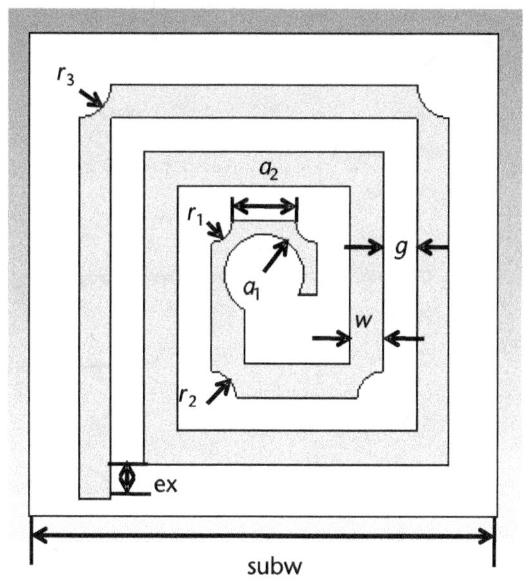

Figure 5.15 The structure of design 3.

between the resonators, have an excessive influence on the RCS. The dimensional parameters are displayed in Table 5.9.

5.8.3 Modified CSRR-Based Chipless Sensor: Design 4

The 5-bit tag sensor has a compact physical dimension of 30 × 30 mm^2. Design 4 is comprised of two resonant structures developed on the substrate, as shown in Figure 5.16. In addition, the slit widths, resonator width, and the separation between adjacent resonating elements determine the proper RCS at specific resonant frequencies. The presented tag is a hybrid of the circular split ring and stepped impedance resonators, which contribute to its high bit density with less resonating elements. Table 5.10 describes the dimensional parameters of design 4.

Table 5.9 Dimensional Parameters of Design 3

Parameters	*Dimensions (mm)*
a_1	6
a_2	16
r_1	8
r_2	18
r_3	28
w	5
g	5
ex	5
subw	71

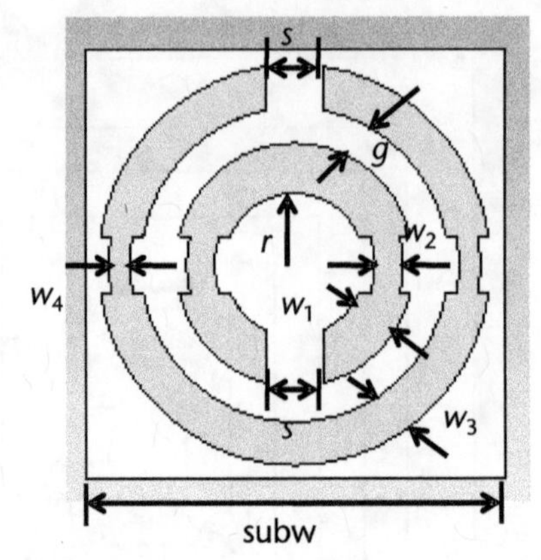

Figure 5.16 The geometry of design 4.

Table 5.10 Dimensional Parameters of Design 4

Parameters	*Dimensions (mm)*
r	5
w_1	3.48
w_2	1.5
w_3	2
w_4	1
g	2
s	2.5
subw	30

5.8.4 Experimental Results and Discussion

Using a CST Microwave Studio Suite 2020, the sensor designs are simulated. The correct RCS signal is provided by the transponder being imposed with a linearly polarized plane wave as a result of the backscattering from the tagged item when a probe is excited on the geometry's center. Figures 5.17 and 5.18, respectively, show the simulated RCS curves for designs 3 and 4.

Design 3 has an operational frequency from 0.3–3 MHz, covering the entire UHF band that encodes 3 bits as three resonant peaks are detected. Design 4 works over a frequency range of 0.4–6 GHz, covering the UHF and UWB bands. It has a data capacity of 5 bits as it has five resonant peaks. The maximum RCS is attained as –13.05 dBsm in design 3 and at –23.55 dBsm in design 4. The simulated RCS peaks and corresponding frequency of designs 3 and 4 are displayed in Tables 5.11 and 5.12, respectively.

The surface current distribution makes it easy to determine how an RFID tag without a chip works. The surface current density distribution for design 3 is shown in Figure 5.19 for the resonant frequencies of 0.86, 1.33, and 2.40 GHz. Figure 5.20 shows the surface current distribution for geometry design 4 at 0.86,

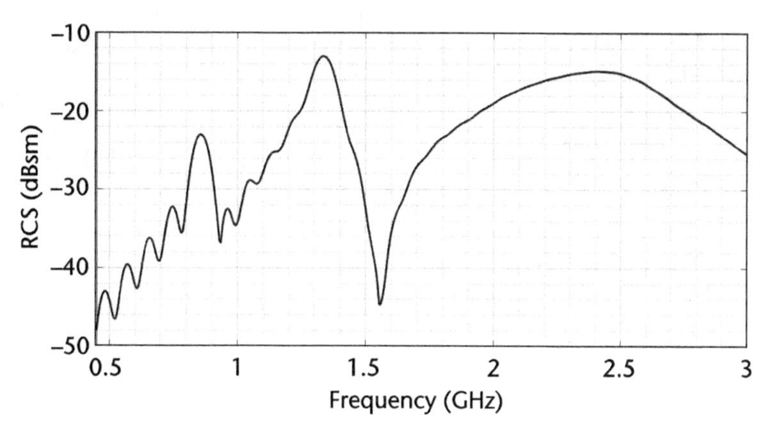

Figure 5.17 Simulated RCS of design 3.

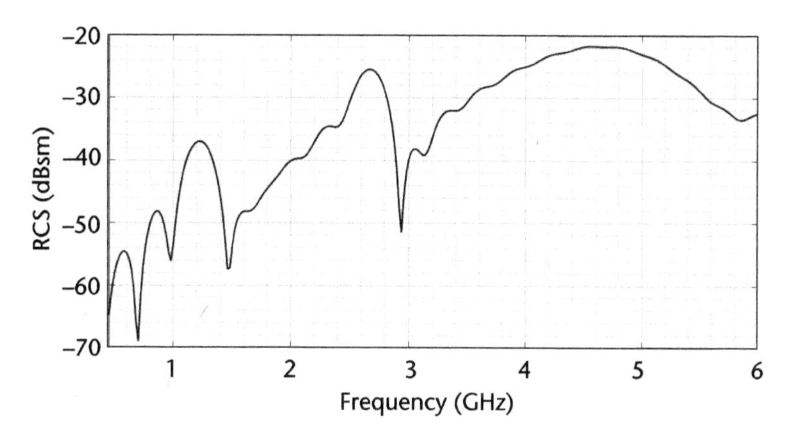

Figure 5.18 Simulated RCS of design 4.

Table 5.11 Simulated RCS Values of Design 3

Frequency (GHz)	*RCS (dBsm)*
0.86	–22.73
1.33	–13.05
2.40	–14.52

Table 5.12 Simulated RCS Values of Design 4

Frequency (GHz)	*RCS (dBsm)*
0.58	–54.63
0.860	–48.19
1.22	–37.21
2.66	–25.56
4.55	–21.72

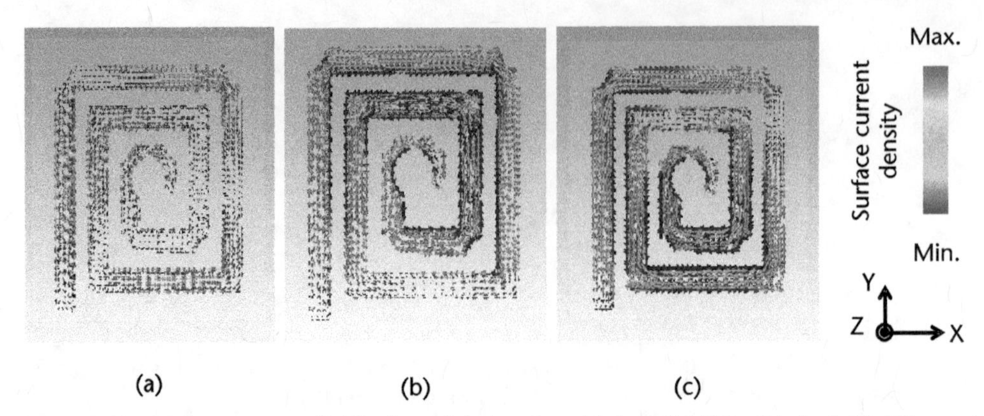

Figure 5.19 Surface current distribution of design 3 at (a) $f = 0.86$ GHz, (b) $f = 1.33$ GHz, and (c) $f = 2.40$ GHz.

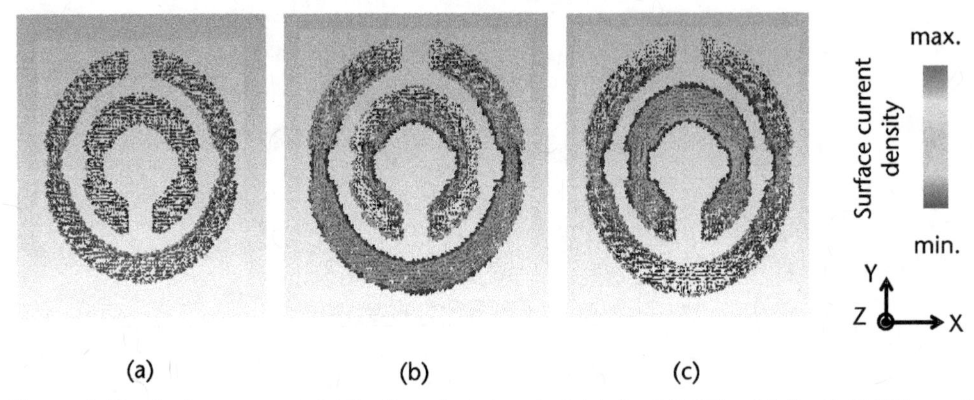

Figure 5.20 Surface current distribution of design 4 at (a) $f = 0.8$ GHz, (b) $f = 2.40$ GHz, and (c) $f = 5.36$ GHz.

2.66, and 4 GHz. It is shown that as the frequency gradually increases, the lesser resonant geometries reach the maximum current densities.

5.9 Analysis of the Fabricated Prototype

Figures 5.21 and 5.22 show the constructed prototypes of tag sensor designs 3 and 4 on the FR4 substrate, respectively. The measurement is done in an acoustically quiet environment with a 25-cm tag reader spacing. Measurements are done in bi-static mode using the PNA-N5222B. A comparison of the measured and simulated plots for designs 3 and 4 is shown in Figures 5.23 and 5.24, respectively, and it demonstrates good agreement. Tables 5.13 and 5.14 show the measured results of designs 3 and 4, respectively.

5.9.1 Comparison

An evaluation of the above-described resonator-based chipless tags and the designed structures based on various parameters such as type of resonator, tag physical

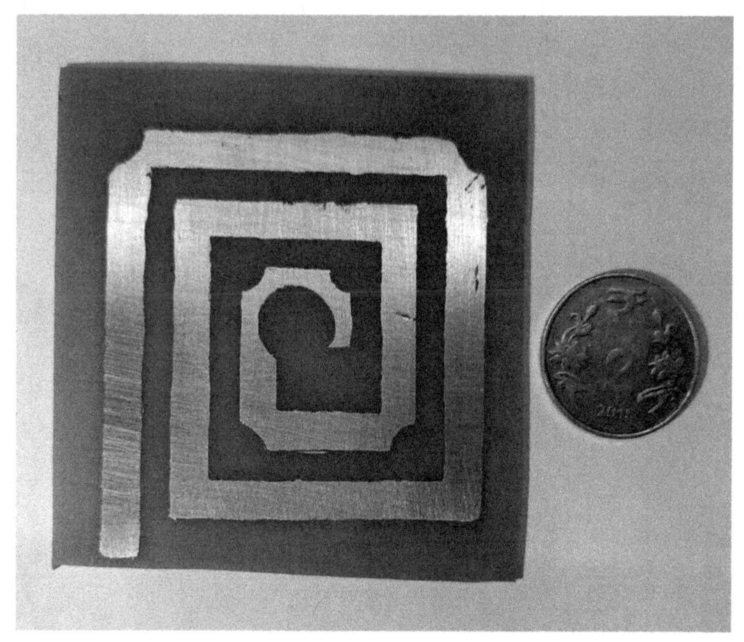

Figure 5.21 Fabricated prototype of design 3.

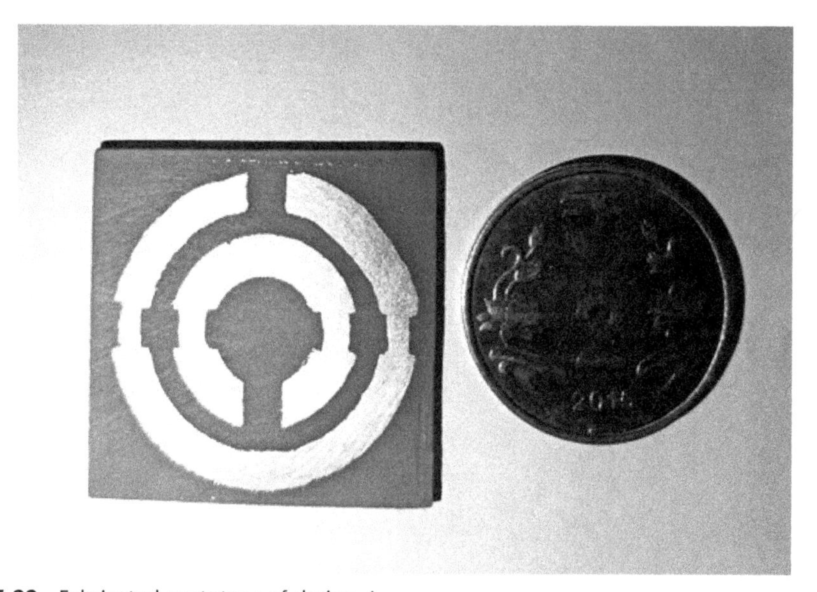

Figure 5.22 Fabricated prototype of design 4.

dimension, encoding capability, material, and application scenario is presented in Table 5.15.

In Table 5.15, we compare various RFID tag designs based on distinct resonator types, their physical dimensions, encoding capabilities, materials, and target application scenarios. The octagon resonator design, referenced in [38], features a unique octagonal shape with physical dimensions of 52 × 82 mm^2 and an encoding capability of 5 bits. Constructed using PET, this tag is ideal for orientation-independent

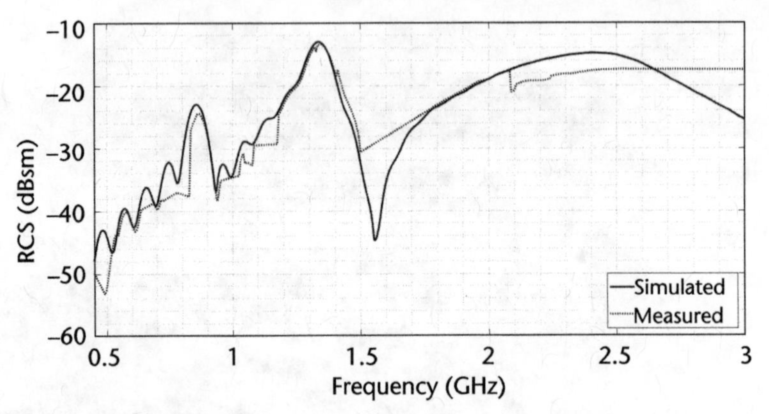

Figure 5.23 RCS of design 3.

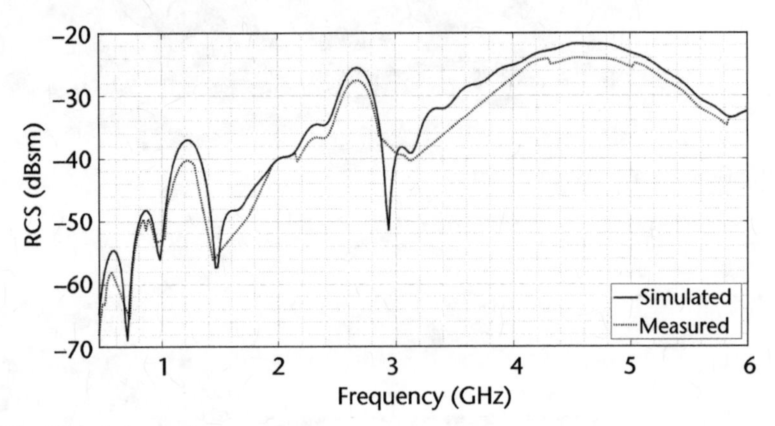

Figure 5.24 RCS of design 4.

Table 5.13 Measured RCS Peaks and Frequencies of Design 3

Frequency (GHz)	*RCS (dBsm)*
0.88	–24.44
1.36	–13.56
2.1	–17.34

object identification. The L-shaped slot resonator, denoted in [39], has a compact size of 20 × 20 mm², encoding up to 8 bits of information. Made from Rogers RT5880 material, it is well-suited for applications in smart retail. The square split ring resonator, presented in [40], offers a balanced trade-off between size and encoding capability (4 bits) with dimensions of 44 × 44 mm², making it suitable for IoT applications. The hairpin resonator design in [41] (40 × 40 mm²) has an encoding capability of 3 bits and utilizes pile as its material, serving purposes such as anticounterfeiting. Additionally, [42] introduces a C-shaped resonator with

Table 5.14 Measured RCS Peaks and Corresponding Frequencies of Design 4

Frequency (GHz)	*RCS (dBsm)*
0.56	–57.97
0.861	–51.45
1.24	–40.32
2.68	–27.58
4.7	–24.08

Table 5.15 Comparison of Distinct Resonator-Based Tags and the Designed Structure

Reference	*Type of Resonator*	*Physical Dimension of Tag (mm²)*	*Encoding Capability*	*Material*	*Application Scenario*
[38]	Octagon	52 × 82	5	PET	Orientation independent object identification
[39]	L-shaped slot	20 × 20	8	Rogers RT5880	Smart retail applications
[40]	Square split ring	44 × 44	4	FR4	IoT applications
[41]	Hairpin	40 × 40	3	Pile	Anticounterfeiting systems, laundry labels
[42]	C-shaped	35 × 15	7	Rogers RT/duroid 6010.2LM	Structural health monitoring
Design 3	Modified spiral	71 × 71	3	FR4	On-body applications
Design 4	Modified circular split ring	30 × 30	5	FR4	On-body applications

dimensions 35 × 15 mm^2, encoding 7 bits and employing Rogers RT/duroid material, making it apt for structural health monitoring. The designs, labeled 3 and 5, showcase modifications like a modified spiral and a modified circular split ring, respectively. These designs cater to on-body applications with physical dimensions of 71 × 71 mm^2 and 30 × 30 mm^2, encoding 3 and 5 bits, respectively. All designs use FR4 as the material, enhancing their applicability for on-body RFID applications and showcasing the versatility of chipless RFID tags in a variety of scenarios. More details of measurements can be observed in the literature [43–53].

5.10 Summary

This chapter introduced several small chipless sensor tags with different kinds of resonator-based miniaturized chipless sensors for biomedical uses. We presented several designs. Design 1 has a physical footprint that is 39 mm × 39 mm and operates in the 2- to 8-GHz frequency range, whereas design 2 has a smaller footprint that is 39 mm × 20 mm when compared to the first design and operates in the 2- to 7-GHz frequency range. In this context, critical design factors such as physical dimensions, radar cross section, current distribution, and encoding capacity are examined across different chipless RFID applications. Heartbeat, sleep, blood glucose level, or body temperature can all be monitored using the 2.4- and 5.8-GHz ISM bands. Real-time low-cost applications are a good fit for the tag because of their flexible design, lack of maintenance, and lower cost. Moreover, the lack of a ground plane may result in a cheaper manufacturing cost. The UHF band is safer for on-body applications such as heartbeat monitoring. Therefore, chipless tag sensors operating in the UHF band were also presented in this chapter; two chipless sensor tags operating in the UHF band were designed in which the modified spiral resonator-based tag covers the entire UHF band from 0.3–3 GHz and encodes 3 bits. Design 3 has an overall physical dimension of 71 sq. mm. The 5-bit design 4 has a compact physical size of 30 mm^2 and covers UHF and UWB from 0.4–6 GHz. Designs 3 and 4 also have a resonant peak at 860 MHz. All two tag designs are well-suited for on-body applications as the sensors operating in the UHF band are less harmful to the human body. The simulated and the fabricated prototypes produce similar results, which are clearly observed from the RCS comparison curves. Moreover, the inexpensive substrate without a ground plane further reduces the price of the tag. Using printing technology, which is more efficient than traditional etching or micromachining methods, it is possible to implement the resonators in the structure without a ground plane. An essential aspect of these tags is the analysis of the current distribution and RCS. Current distribution patterns within the tag are crucial as they determine its electromagnetic response. Understanding and optimizing these patterns allows for the design of tags with specific resonant frequencies, improving their detectability and readability. Additionally, analyzing the RCS, which quantifies the tag's ability to scatter incident electromagnetic waves, is fundamental in assessing tag visibility to the reader. In the context of biomedical applications, chipless RFID tags offer immense potential. By engineering these tags to have specific current distribution patterns and optimizing RCS, they can be tailored for biomedical sensing applications. For instance, designing chipless RFID tags to resonate at frequencies associated with certain biological indicators like glucose levels or body temperature enables nonintrusive and real-time monitoring. The ability to seamlessly integrate these tags into clothing or wearable accessories further enhances their utility for continuous health monitoring in a comfortable and unobtrusive manner. In conclusion, chipless RFID tags, with their customizable electromagnetic properties, hold great promise for revolutionizing biomedical applications, particularly in the domain of wearable health monitoring and diagnostics.

References

[1] Rezaeieh, S. A., K. S. Bialkowski, and A. M. Abbosh, "Folding Method for Bandwidth and Directivity Enhancement of Meandered Loop Ultra-High Frequency Antenna for Heart Failure Detection System," *IET Microwaves, Antennas & Propagation,* Vol. 8, No. 14, pp. 1218–1227.

[2] Behera, S. K., "Chipless RFID Sensors for Wearable Applications: A Review," *IEEE Sensors Journal*, Vol. 22, No. 2, January 15, 2022, pp. 1105–1120.

[3] Subrahmannian, A., and S. K. Behera, "Chipless RFID: A Unique Technology for Mankind," *IEEE Journal of Radio Frequency Identification,* Vol. 6, 2022, pp. 151–163.

[4] Dey, S., J. K. Saha, and N. C. Karmakar, "Smart Sensing: Chipless RFID Solutions for the Internet of Everything," *IEEE Microwave Magazine,* Vol. 16, No. 10, November 2015, pp. 26–39.

[5] Costa, F., S. Genovesi, M. Borgese, A. Michel, F. A. Dicandia, and G. Manara, "A Review of RFID Sensors, the New Frontier of Internet of Things," *Sensors,* Vol. 21, 2021, p. 3138.

[6] Dey, S., R. Bhattacharyya, S. E. Sarma, and N. C. Karmakar, "A Novel 'Smart Skin' Sensor for Chipless RFID-Based Structural Health Monitoring Applications," *IEEE Internet of Things Journal,* Vol. 8, No. 5, March 1, 2021, pp. 3955–3971.

[7] Corchia, L., G. Monti, and L. Tarricone, "A Frequency Signature RFID Chipless Tag for Wearable Applications," *Sensors,* Vol. 19, No. 3, 2019, p. 494.

[8] Karmaker, N. C., "Tag, You're It Radar Cross Section of Chipless RFID Tags," *IEEE Microwave Magazine,* Vol. 17, No. 7, July 2016, pp. 64–74.

[9] Behera, S. K., and N. C. Karmakar, "Wearable Chipless Radio-Frequency Identification Tags for Biomedical Applications: A Review [Antenna Applications Corner]," *IEEE Antennas and Propagation Magazine,* Vol. 62, No. 3, June 2020, pp. 94–104.

[10] Athauda, T., and N. Karmakar, "Chipped Versus Chipless RF Identification: A Comprehensive Review," *IEEE Microwave Magazine,* Vol. 20, No. 9, September 2019, pp. 47–57.

[11] Khaliel, M., A. El-Awamry, A. Fawky, and T. Kaiser, "Long Reading Range Chipless RFID System Based on Reflectarray Antennas," in *2017 11th European Conference on Antennas and Propagation (EUCAP),* 2017, pp. 3384–3388.

[12] Borgese, M., S. Genovesi, G. Manara, and F. Costa, "Radar Cross Section of Chipless RFID Tags and BER Performance," *IEEE Transactions on Antennas and Propagation,* Vol. 69, No. 5, May 2021, pp. 2877–2886.

[13] Behera, S. K., and N. C. Karmakar, "Chipless RFID Printing Technologies: A State of the Art," *IEEE Microwave Magazine,* Vol. 22, No. 6, June 2021, pp. 64–81.

[14] Karmakar, N. C., E. M. Amin, and J. K. Saha, *Chipless RFID Sensors*, Hoboken, NJ: John Wiley & Sons, 2015.

[15] Bäuerle, D. W., *Laser Processing and Chemistry,* Berlin: Springer-Verlag, 2011.

[16] Su, W., Qi Liu, B. Cook, and M. Tentzeris, "All-Inkjet-Printed Microfluidics-Based Encodable Flexible Chipless RFID Sensors," in *2016 IEEE MTT-S International Microwave Symposium (IMS),* 2016, pp. 1–4.

[17] Purushothama, J. M., S. Lopez-Soriano, A. Vena, B. Sorli, I. Susanti, and E. Perret, "Electronically Rewritable Chipless RFID tags Fabricated Through Thermal Transfer Printing on Flexible PET Substrates," *IEEE Transactions on Antennas and Propagation,* Vol. 69, No. 4, April 2021, pp. 1908–1921.

[18] Lim, N., J. Kim, S. Lee, N. Kim, and G. Cho, "Screen Printed Resonant Tags for Electronic Article Surveillance Tags," *IEEE Transactions on Advanced Packaging,* Vol. 32, No. 1, February 2009, pp. 72–76.

[19] Björninen, T., J. Virkki, L. Sydänheimo, and L. Ukkonen, "Possibilities of 3D Direct Write Dispensing for Textile UHF RFID Tag Manufacturing," *Proceedings of the IEEE APS Symposium,* Vancouver, Canada, July 2015, pp. 1316–1317.

[20] Amin, E. M., J. K. Saha, and N. C. Karmakar, "Smart Sensing Materials for Low-Cost Chipless RFID Sensor," *IEEE Sensors Journal,* Vol. 14, No. 7, July 2014.

[21] Nauroze , S. A., et al., "Additively Manufactured RF Components and Modules: Toward Empowering the Birth of Cost-Efficient Dense and Ubiquitous IoT Implementations," in *Proceedings of the IEEE,* Vol. 105, No. 4, April 2017, pp. 702–722.

[22] Attaran, A., and R. Rashidzadeh, "Chipless Radio Frequency Identification Tag for IoT Applications," *IEEE Internet of Things Journal,* Vol. 3, No. 6, December 2016, pp. 1310–1318.

[23] Borgese, M., F. A. Dicandia, F. Costa, S. Genovesi, and G. Manara, "An Inkjet Printed Chipless RFID Sensor for Wireless Humidity Monitoring," *IEEE Sensors Journal,* Vol. 17, No. 15, August 1, 2017, pp. 4699–4707.

[24] Andriamiharivolamena, T., A. Vena, E. Perret, P. Lemaitre-Auger, and S. Tedjini, "Chipless Identification Applied to Human Body," in *2014 IEEE RFID Technology and Applications Conference (RFID-TA),* 2014, pp. 241–245.

[25] Patre, S. R., "Passive Chipless RFID Sensors: Concept to Applications—A Review," *IEEE Journal of Radio Frequency Identification*, Vol. 6, 2022, pp. 64–76.

[26] Yang, Z., K. Y. See, M. F. Karim, and A. Weerasinghe, "Chipless RFID-Based Sensing System for Partial Discharge Detection and Identification," *IEEE Sensors Journal*, Vol. 21, No. 2 January 15, 2021, pp. 2277–2285.

[27] Monti, G., G. Porcino, and L. Tarricone, "Textile Chipless Tag for Gesture Recognition," *IEEE Sensors Journal*, Vol. 21, No. 16, August 15, 2021, pp. 18279–18286.

[28] Jiejun, Z., W. Junhong, C. Meie, and Z. Zhan, "RCS Reduction of Patch Array Antenna by Electromagnetic Band-Gap Structure," *IEEE Antennas and Wireless Propagation Letters,* Vol. 11, August 2012, pp. 1048–1051.

[29] Deif, S., and M. Daneshmand, "Multiresonant Chipless RFID Array System for Coating Defect Detection and Corrosion Prediction," *IEEE Transactions on Industrial Electronics,* Vol. 67, No. 10, October 2020, pp. 8868–8877.

[30] Alhaj Abbas, A., M. El-Absi, A. Abualhijaa, K. Solbach and T. Kaiser, "Dielectric Resonator-Based Passive Chipless Tag with Angle-of-Arrival Sensing," *IEEE Transactions on Microwave Theory and Techniques,* Vol. 67, No. 5, May 2019, pp. 2010–2017, doi: 10.1109/TMTT.2019.2901447.

[31] Fathi, P., N. C. Karmakar, M. Bhattacharya, and S. Bhattacharya, "Potential Chipless RFID Sensors for Food Packaging Applications: A Review," *IEEE Sensors Journal*, Vol. 20, No. 17, September 1, 2020, pp. 9618–9636.

[32] Nijas, C. M., S. Suseela, U. Deepak, P. Wahid, and P. Mohanan, "Low Cost Chipless Tag with Multi-Bit Encoding Technique," in *IEEE MTT-S International Microwave and RF Conference,* 2013, pp. 1–4.

[33] Mishra, D. P., and S. K. Behera, "A Novel Technique for Dimensional Space Reduction in Passive RFID Transponders," in *2021 2nd International Conference on Range Technology (ICORT),* 2021, pp. 1–4.

[34] Mishra, D. P., T. K. Das, and S. K. Behera, "Design of a 3-Bit Chipless RFID Tag Using Circular Split-Ring Resonators for Retail and Healthcare Applications," in *2020 National Conference on Communications (NCC),* 2020, pp. 1–4.

[35] Mishra, D. P., T. Kumar Das, P. Sethy, and S. K. Behera, "Design of a Multi-Bit Chipless RFID Tag Using Square Split-ring Resonators," in *2019 IEEE Indian Conference on Antennas and Propagation (InCAP),* 2019 pp. 1–4.

[36] Herraiz-Martinez, F. J., F. Paredes, G. Zamora Gonzalez, F. Martin, and J. Bonache, "Printed Magnetoinductive-Wave (MIW) Delay Lines for Chipless RFID Applications," *IEEE Transactions on Antennas and Propagation,* Vol. 60, No. 11, November 2012, pp. 5075–5082.

[37] Betancourt, D., K. Haase, A. Hübler, and F. Ellinger, "Bending and Folding Effect Study of Flexible Fully Printed and late-Stage Codified Octagonal Chipless RFID Tags," *IEEE Transactios on Antennas and Propagation,* Vol. 64, No. 7, July 2016, pp. 2815–2823.

[38] Sharma, V., S. Malhotra, and M. Hashmi, "Slot Resonator Based Novel Orientation Independent Chipless RFID Tag Configurations," *IEEE Sensors Journal,* Vol. 19, No. 13, July 1, 2019, pp. 5153–5160.

[39] Sai, D. M., D. P. Mishra, and S. K. Behera, "Frequency Shift Coded Chipless RFID Tag Design using Square Split Ring Resonators," in *2021 2nd International Conference on Range Technology (ICORT),* 2021, pp. 1–4.

[40] Corchia, L., et al., "Radio-Frequency Identification Based on Textile, Wearable, Chipless Tags for IoT Applications," in *2019 II Workshop on Metrology for Industry 4.0 and IoT (MetroInd4.0&IoT),* 2019, pp. 1–5.

[41] Javed, N., M. A. Azam, and Y. Amin, "Chipless RFID Multisensor for Temperature Sensing and Crack Monitoring in an IoT Environment," *IEEE Sensors Letters,* Vol. 5, No. 6, June 2021, pp. 1–4, Art No. 6001404.

[42] Ingle, V. K, and J. G. Proakis, *Digital Signal Processing Using MATLAB,* Third Edition, Cengage Learning, 2011.

[43] Mishra, D. P., and S. K. Behera, "Multibit Coded Passive Hybrid Resonator Based RFID Transponder with Windowing Analysis," *IEEE Transactions on Instrumentation and Measurement*, Vol. 71, 2022, pp. 1–7, Art No. 8005907.

[44] Mishra, D. P., and S. K. Behera, " Resonator Based Chipless RFID Sensors: A Comprehensive Review," *IEEE Transactions on Instrumentation and Measurement*, Vol. 72, pp. 1–16, 2023, Art No. 5500716.

[45] Jonnalagadda, S. V. S. T., D. P. Mishra, and S. K. Behera, "Chipless RFID Sensors for Vital Signs Monitoring–A Comprehensive Review," *IEEE Microwave Magazine,* Vol. 24, No. 11, November 2023, pp. 53–70, doi: 10.1109/MMM.2023.3303668.

[46] Mishra, D. P., T. K. Das, S. K. Behera, and N. C. Karmakar. "Modified Rectangular Resonator Based 15-Bit Chipless Radio Frequency Identification Transponder for Healthcare and Retail Applications," *International Journal of RF and Microwave Computer-Aided Engineering,* Vol. 32, No. 6, 2022, pp. e23127.

[47] Mishra, D. P., and S. K. Behera, "Modified Rectangular Resonators Based Multi-Frequency Narrow-Band RFID Reader Antenna," *Microwave and Optical Technology Letters,* Vol. 64, No. 3, 2022, pp. 544–551.

[48] Mishra, D. P., A. Subrahmannian, and S. K. Behera, "Design of Multi-Bit Chipless RFID Tag Using Hybrid Resonator for Retail and Healthcare Applications," *Journal of Microwave Engineering & Technologies,* Vol. 8, No. 2, 2021, pp. 34–42.

[49] Mishra, D. P., S. V. S. T. Jonnalagadda, and S. K. Behera, "Passive RFID Transponder using Open-Stubs for Bit-Capacity Enhancement," in *2023 3rd International Conference on Range Technology (ICORT),* Chandipur, Balasore, India, 2023, pp. 1–5.

[50] Mishra, D. P., and S. K. Behera, "Analysis of Chipless RFID Tags Based on Circular-SRR and Koch Snowflake Fractal for Space-Reduction and Bit-Coding Improvement," in *2021 IEEE Indian Conference on Antennas and Propagation (InCAP),* 2021, pp. 621–624.

[51] Mishra, D. P., and S. K. Behera, "Passive RFID Transponders Based on SRR and Koch-Island Fractal for Bit-Coding Enhancement," *2021 IEEE Advanced Communication Technologies and Signal Processing (ACTS),* 2021, pp. 1–5.

[52] Mishra, D. P., I. Goyal, and S. K. Behera, "A Comparative Study of Two Different Octagonal Structure-Based Split Ring Resonators," in *2022 IEEE Microwaves, Antennas, and Propagation Conference (MAPCON)*, Bangalore, India, 2022, pp. 1007–1012.

[53] Mishra, D. P., I. Goyal, and S. K. Behera, "A Technique to Improve RCS in Passive Chipless RFID Tags by Incorporating Array Structure," in *2022 IEEE Microwaves, Antennas, and Propagation Conference (MAPCON),* Bangalore, India, 2022, pp. 1478–1483.

CHAPTER 6

Passive Printable Chipless RFID Tag for Body Area Network

6.1 Introduction

Due to its numerous applications and enormous potential, chipless RFID technology has entered a highly urbanized era with the help of the IoT. Chipless RFID is a subsection of the RFID field that does not use a power source or an IC. However, the data is obtained from the physical structure of the tag using frequency signature. Here, the frequency-coded, design-based tags convey a detailed frequency response (e.g., RCS and S-parameters versus frequency) over the transmitted bits. Based on the various purpose and tag dimensions and design specifications, these frequency-coded signals or binary bits can be allocated in several ways. Identification and sensing are two common chipless RFID applications. Our focus is on the frequency domain approach (FDA) and FDA-based coding techniques in the chipless RFID domain. Frequency-coded bits are allotted to the tag output result for localization-based tags, and it is preferable so that more IDs can be generated and assigned. Meanwhile, these tags should be space-efficient (i.e., less form-factor) for real-time application. To achieve tagging efficiency, the bit capacity or the number of bits per square inch is taken care of. This is termed the bit density, where the capacity of the tag is defined. The techniques behind the sensing-based tags are associated with the physical parameter, which is humidity, temperature, and pressure utilizing changing frequency. This is because the resonance frequency signifies the data bits once resonance is used. The measured sensitivity, which measures how much the measured parameter varies with the parameter being observed (for example, what voltage change takes place with temperature for a temperature sensor), is also employed occasionally for tag sensors. ID-based tags from sensing-based tags are essentially used for application-specific tasks: both types of tags are often created for sensing and data storage. In a few instances, a sensing tag with the ID bits added or an ID-based tag with the sensing bits added. However, in each of these instances, this is done to enhance tag functionality for real-time applications. The combination of application specialization and a dearth of resonator variation limits adaptability and practicability. With a wider range of resonant frequencies due to higher permittivity, it becomes possible to design more intricate and unique resonant structures.

This allows for finer discrimination between different frequency patterns, enabling a higher information density and more bits to be encoded.

However, there are practical limits to increasing dielectric permittivity. Excessively high permittivity can also introduce challenges such as increased signal losses and decreased read range due to absorption and scattering of the RF signal within the high permittivity material. A design can only have a particular number of one kind of resonator before the area is full, for instance, if the tag has a specific dimension (area) requirement. The tag is, therefore, limited to a particular bit density. This can be avoided by making the tag work at higher frequencies, which would enable the use of smaller resonators. This method, however, makes the resonators very small and sometimes difficult to construct, making it harder to quantify the tag characteristic response. The customizability of the response is restricted by this tag design technique to a single type of resonator and a particular tag region, thereby limiting the tag's possible applications. Combining different resonator types allows for the minimization of the chipless tag's dimensions. However, caution must be exercised to ensure the enhancement of sensing quality, as a reduction in area may impact the overall performance of the tag. As a result, the tag's potential uses are constrained. Alternatively, combining different resonator types, larger bit densities, and better application diversity (i.e., use in a variety of applications, like ID and sensing applications) can be accomplished. This chapter will provide examples of these benefits. Spiral resonators are frequently employed in the radio frequency regime to achieve reasonably high inductance values with high Q-factors and compact electrical dimensions. Spiral resonators can also provide circuit solutions with greater performance and contribute to a higher level of integration in radio frequency integrated circuit (RFIC) applications. The electromagnetic properties of the spiral resonator have been examined using a variety of methods, according to reports in the literature [1–8]. To create a thorough and straightforward analysis guide, the single-model of the spiral resonator is designed [1–3]. The moment method is also used to obtain precise capacitance values. The modified Wheeler approach is suitable for better inductance values in the modeling of spiral resonators. Additionally, by taking into consideration the alternating current resistance value based on a ladder network and the coupling capacitance and resistance of the lateral substrate, the equivalent circuit model of the spiral resonators offers an important feature for yielding the wideband model [6–8].

Recently, the chipless transponder based on a spiral structure and the energy bandgaps with structures containing spiral form have been used to decrease the sudden swapping noise and distinguish between mixed-signal and power plane. The EM characteristics of coplanar resonators with the spiral structure are predicted, and accuracy can be confined by using the empirical formula; regrettably, these sorts of prototypes offer inconvenient accuracy results for printed circuit board (PCB) applications. This is due to lumped models exhibiting a relatively small range of frequency, and the structure having a spiral pattern on PCB is higher in electrical nature [8]. One of the recent advancements in chipless RFID is the application of a wireless body area network (WBAN). It can be classified as in-body and on-body networks. This chapter focuses on the resonators that may be applied as on-body RFID applications [8–10]. In the FDA, the resonant frequencies must satisfy the application frequencies at WBAN so that it can be claimed that the tag designs may be applied in the domain of body area networking RFID.

Figure 6.1 suggests the real-time implementation of the chipless RFID tag. Here, the tags are connected in several domains of applications. In between the tag and reader, there is active communication that takes place. Each data is recorded and stored for desired analysis. The different domains of RFID-integrated applications are as follows:

1. Deforestation;
2. Wildlife biology;
3. Personal data monitoring;
4. Handheld reader;
5. Packaging of goods;
6. Monitoring industrial automation.

Environment data can be observed and collected by the implementation of tag sensors in multiple applications, such as measuring humidity, temperature, pressure, and various gases. The forest related data is also recorded by RFID, and the data is sent to the cloud for further analysis. Various wildlife monitoring survey of wildlife habitat and their movement is done by integrating RFID tag. Personal data is also recorded and monitored using chipless RFID. RFID technology is also used for packaging goods and industrial automation.

The literature is discussed in Section 6.2, and the mathematical formulas for spiral resonators are shown in Section 6.3. The design and analysis of resonators are shown in Sections 6.4 and 6.5, and Section 6.6 concludes the discussion. The end of this chapter includes the essential references.

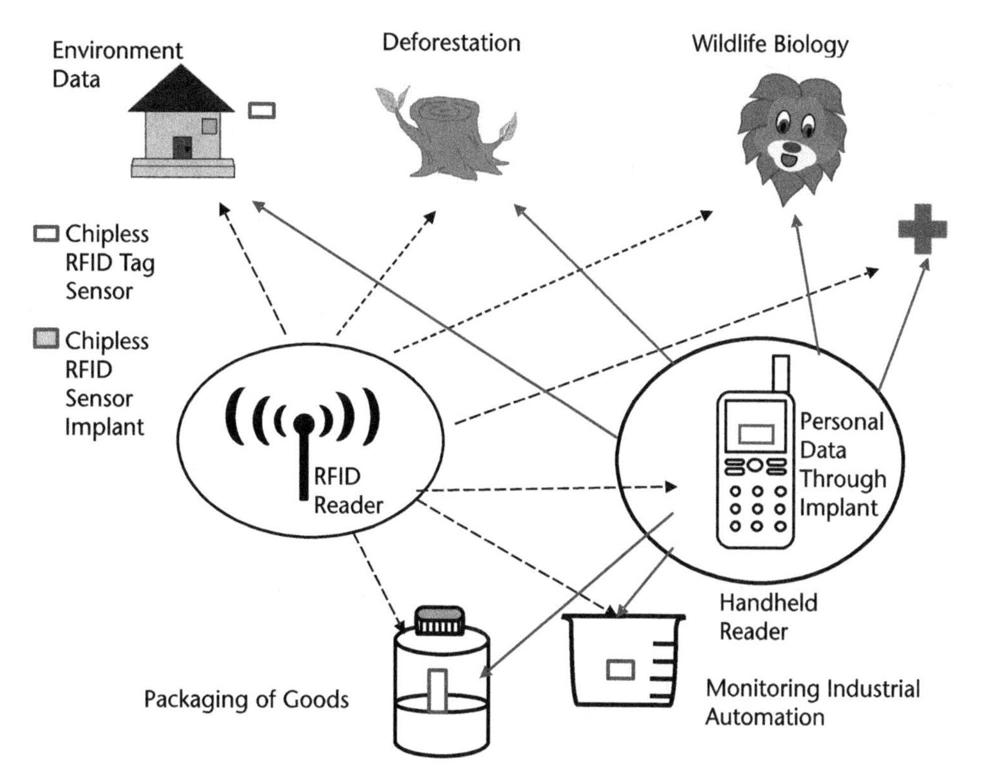

Figure 6.1 Utilization of chipless RFID in various applications.

6.2 Modified Spiral Resonator-Based Chipless Resonator

After establishing the purposes, aims, and basic requirements in the previous section, this chapter proposes the element size pattern of the chipless tag. The word "modified" signifies that the regular spiral resonator can be structurally modified for various frequency applications, which is advantageous in the chipless RFID domain. Here are some advantages of using spiral resonators in chipless RFID systems:

- *Compact size*: Spiral resonators can be designed to be compact in size, making them suitable for integration into small and constrained spaces such as product packaging, labels, or even wearable devices.
- *Frequency tunability*: The resonant frequency of a spiral resonator can be controlled and tuned by adjusting its dimensions, making it versatile and adaptable to different frequency bands. This tunability allows for compatibility with various RFID systems.
- *High data capacity*: Spiral resonators can encode a relatively large amount of data due to the multiple frequency resonances. This high data capacity is useful for applications requiring extensive information storage.
- *Low cost*: The fabrication of spiral resonators is often cost-effective compared to traditional RFID microchips. This can lead to cost savings in large-scale deployments, making chipless RFID more economically viable.
- *Compatibility with existing infrastructure*: Spiral resonators can be designed to operate within existing RFID frequency bands, ensuring compatibility with the infrastructure and RFID readers already in use.
- *Non–line-of-sight reading*: Like other RFID technologies, chipless RFID with spiral resonators can be read through obstacles and at a distance, providing convenience and flexibility in reading and tracking items.
- *Enhanced security and anticounterfeiting*: The unique patterns and encoding capabilities of spiral resonators can enhance security and help prevent counterfeiting. The intricate designs can be challenging to replicate without authorized knowledge.
- *Integration with printing technologies*: Spiral resonators can be integrated with printing technologies, enabling the incorporation of RFID functionality directly into labels or packaging during the printing process.
- *Environmental resilience*: Spiral resonators can be designed to withstand various environmental conditions, ensuring the RFID tags remain functional even in challenging or harsh environments.
- *Customizable designs*: The design of spiral resonators can be customized to meet specific requirements, including size, shape, and resonant frequency, allowing for tailor-made solutions for different applications.

Hence, spiral resonators in chipless RFID offer advantages such as compact size, tunability, high data capacity, cost-effectiveness, compatibility, non–line-of-sight reading, security, integration with printing technologies, environmental resilience, and customization, making them a promising choice for various RFID applications. This chapter also focuses on the spiral resonator magnetic resonance with mathematical variables. In general, copper spiral resonator structure can be investigated within a waveguide, specifically designed on a Rogers RT/duroid 5880 substrate. The analysis may be conducted through a numerical simulation utilizing the high frequency structure simulator (HFSS), which is based on finite element method (FEM). The results of this analysis make it feasible to determine the resonant frequency of metamaterials. This understanding is crucial for achieving mass metamaterial production with negative permeability through strategic resonator patterning. Maxwell's equations characterize the electromagnetic radiation's effect on any material and relate it to material parameters such as permeability and permittivity. The response of conventional materials to electromagnetic waves can be obtained through correlating limited charge and current densities with material resonant and evaluating Maxwell-Heaviside equations for EM fields. To improve a material's response, its effective electromagnetic parameters must be changed. The effective electrical permittivity and effective magnetic permeability of such customized material can be represented. So far, all naturally occurring materials have positive permittivity and permeability values. V. G. Vassalage, however, proved this incorrect in 1968 with his work [8–12], whereby he described his theoretical study of a monochromic uniform propagation of plane-wave in a medium along both negative permeability and negative permittivity. Owing to the left-handed triad made by the vectors E, H, and k, he named such medium materials left-handed materials (LHMs). LHMs are created artificially by placing miniperiodic metal structures on a substrate host. These matters obtain effectual properties by their structures rather than by their components. These materials' phase and group velocity are not in the same direction, which is opposite to natural materials. LHMs have a lattice parameter that is under the wavelength of the incident radiation [8, 13, 14]. Metamaterials are regarded as a broader level material than LHMs. It is a material level with either negative permittivity or negative permeability. LHM is made up of two kinds of cell elements: thin wires (TWs), split ring resonators (SRRs), or a composite of the two. MTMs' exotic features can be manipulated through the pattern of their building blocks. Pendry's metamaterial was a regular form of small metallic inclusions intended for use in electromagnetic fields. As a result, there has been a renewal of effort in developing structures with new material properties [13, 14]. The size and spacing of these inclusions should be reduced to a fraction of the incident signal's electromagnetic wavelength. Pendry, et al. proposed two cell elements, thin wires and split ring resonator structures, in 1998 and 1999, respectively, and concluded that thin wires can be utilized to obtain negative effective permittivity and split rings can be utilized to obtain negative effective permeability. The LHM can be realized using TW arrays and SRR over a similar frequency band [13, 14]. The standard SRR unit cell is made up of a pair of circular coplanar rings of a high-conductive metal. These rings have a split that is replaced by 180 degrees. Two rings are separated by a small space and positioned on a dielectric alienated from a similar center. SRRs of various shapes have been reported in the literature, including rectangular, square, circular, omega shape, s-shaped, and symmetrical ring, among others. These resonators are used in a wide range of applications, including antenna structures, cloaking, microwave absorbers, high-impedance surfaces, and sensors. SRRs have undergone several abridged refinements since their inception to optimize the resonance held up by the structure. As a result, these parameters can alter the effective permeability and permittivity of MTMs [8, 13, 14].

RFID reader system-level design is ensured by system integration and field trials. This section provides a detailed examination of the spiral resonator, which is a key element of the chipless RFID tag. The suggested chipless RFID tag presented in Chapter 6 encodes data using a multiresonator. The multiresonator is made up of cascaded microwave spiral resonators that are connected by a microstrip transmission line. For application in the printable chipless tag, microwave spiral resonators should be fully planar, have a narrow bandwidth (high Q-factor), and be compressed in size.

These sensors, which are based on the layouts of metamaterial unit cells, can operate without a connection between the reader and the sensor and can provide real-time monitoring of a variety of physical parameters. The spiral resonator (SR) geometry is another popular geometry in wireless sensors. Spiral resonators have found use in multiple areas, including wearable technology, biomedicine, strain sensing, eddy current sensing, and displacement sensing [8, 12, 15–18]. Numerous publications in the literature about the evaluation of the equivalent lumped circuit parameters have developed equivalent circuit models for microwave sensor geometries [8, 17, 18]. However, the unique implementation itself, which is built and constructed for a particular application with a rigorous set of limitations and constraints regarding the operating frequency or sensor reading range, is frequently the focus of these works. This is a negative since it makes it difficult to duplicate or modify the implementation to suit new operating parameters or application needs. According to the needs and limitations of the application at hand, the designer must make concessions and trade-offs when solving the design problem. Examples of such design targets in SR-based distance sensors include the maximum sensor range, operating frequency, or robustness to lateral displacements. The primary goal of this article is to give the designer a thorough grasp of how sensors work, as well as the implications of changing design parameters and how to best use them for a given application. Based on an SR tag imprinted on a dielectric substrate, the distance sensor suggested in this paper measures distance. A microstrip probing loop that serves as the reader's antenna queries the tag. By measuring the real part of the input impedance at the loop terminals, the sensor may determine the distance between the probe and the tag in real-time. This study uses numerical simulations to examine the impact of the proposed sensor's parameters, which are backed by approximations of analytical formulas.

The near-field inductive coupling between the SR and the microstrip probing loop is the active methodology of the planar-based sensor SR, which plays a role as an antenna of the reader. It has been demonstrated that the intensity of this coupling is inversely proportional to their distance from one another. This idea was used to extract a precise distance estimate by merely measuring the Z_0 at the probe terminals, where its value depends on the power of the coupling factor between the two elements.

6.3 Mathematical Formula of Spiral Resonator

The mathematical formula of the proposed tag can be formulated as follows. In this particular spiral shape design or normal spiral shape, there is no central part, and it is mostly called the planar spiral resonator (PSR). The resonance frequency of any spiral can be written by the expression [6–8]

$$f_1 = \frac{c}{2L} \tag{6.1}$$

where c is the velocity of the EM wave in free space, and L is the length of the transmission line.

Normally, these structures are otherwise called monofillar Archimedean spiral having no central part, and the radius (R) is taken as same here. So, it can be written as

$$L \approx 2\pi RN \tag{6.2}$$

where N is the number of turns.

For n number of resonances, the frequency can be written as

$$f_1 \approx \frac{c}{4\pi RN} \text{ and } \approx f_1 \frac{nc}{4\pi RN} \tag{6.3}$$

where $n = 1,2,3, \ldots.$

Similarly, the Archmedian spiral equation can be simplifying to

$$\rho(\varphi) = R_e(1 - \alpha\varphi) \tag{6.4}$$

where $\rho(\varphi)$ is the angle in polar coordinate and (φ) varies from 0 to $2\pi N$ and $\alpha = \frac{d}{2\pi r_e} = \frac{R_e - R_i}{2\pi R_e N} \ll 1$. Here, R_e *and* R_i are the external and internal radius, respectively.

6.4 Modified CSRR-Based Chipless Resonator-1

The spiral-based chipless RFID is shown in Figure 6.2. In this proposed simulated design, a 40 × 40 mm^2 dimension of substrate is taken. The patch is designed over an FR4 substrate, and the spiral width is taken as 2 mm (Table 6.1). It is a multi-resonator with a spiral shape. The proposed tag is designed using CST microwave studio. This tag sensor's sensing capability depends upon the varying characteristics of the RF field. This RF sensor can be used for IoT and sensor applications. The selection of FR4 substrate over here is to make the tag a less expensive and more practical application.

6.4.1 Simulated and Measured RCS

The simulated response is shown in Figure 6.3, where the varying RF frequency can be observed and resonates at 1.9 GHz, 2.4 GHz, and 3.01 GHz. The RCS is also

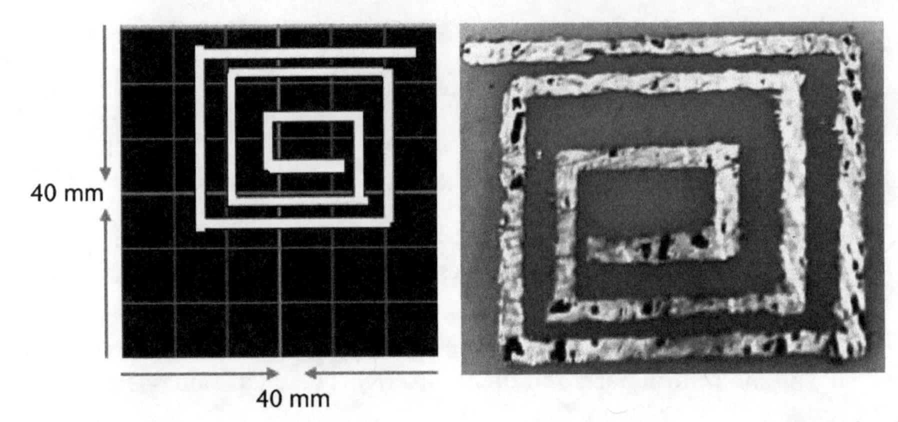

Figure 6.2 Simulated and fabricated prototype (not up to perfection by manual chemical etching method) of the spiral resonator-1.

Table 6.1 Parameters and Value

Parameter	*Width (mm)(W)*	*Length (mm)(L)*	*Height (mm)*
Substrate	40	40	1.6

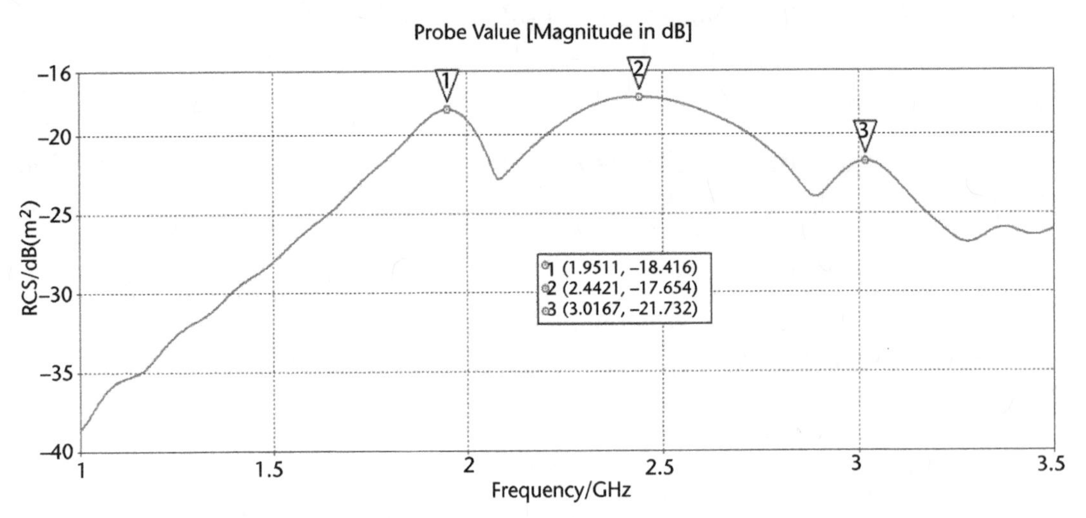

Figure 6.3 Simulated RCS result of the spiral resonator-1.

observed at these resonating frequencies as –18.416 dBsm, –17.65 dBsm, and –21 dBsm, respectively.

The RCS plot is investigated with consecutive bits. Then, the acquired response is again analyzed for the preferred domain (i.e., the FDA). In this chapter, the designed tag has three distinct bits with valid current distribution patterns. The obtained 2.4 GHz ISM band is very efficient for industrial applications. The simulated and the measured results are shown with a similarity gap, which is due to fabrication errors. Similarly, the current distribution is also taken for various resonant frequencies to validate the tag efficiency. The current distribution pattern is distinct as the frequency changes. The inner resonator starts conducting as the frequency increases. From frequencies 1.9 GHz, 2.4 GHz, and 3.0 GHz, the current distribution pattern changes. The measurement is carried out by a performance network analyzer (PNA) in bistatic mode of operation. The anechoic chamber is used for this spiral resonator measurement. There are two standardized horn antennas that were taken to carry out measurements. The measuring range of these two horn antennas is up to 6 GHz. The entire setup for the measurement of the proposed spiral resonator is shown in Figure 6.4. From 84 cm distance, the measurements are carried out. This distance is optimized after a series of measurements [19–24]. The measured response is shown in Figure 6.5.

6.4.2 Current Distribution

The RF characteristics of spiral resonators are also validated from the current distribution. The current distribution in a chipless RFID tag is a fundamental aspect that significantly impacts the tag's performance and functionality. Chipless RFID tags rely on unique patterns of currents within resonant structures to represent specific information or bits. The current distribution influences the resonant frequencies of the tag's structures. By manipulating the current distribution, the RFID tag can be tuned to specific frequencies, allowing for compatibility with various RFID systems and regulatory frequency bands. The current distribution is pivotal in determining the antenna performance of the RFID tag. Efficient current distribution ensures that the tag can effectively receive and reflect radio frequency signals, improving read range, sensitivity, and reliability of communication with the RFID reader. Designing an appropriate current distribution allows for spatial optimization of the tag, enabling its integration into various form factors such as labels, packaging, or other objects. This flexibility in spatial design is essential for the widespread adoption of chipless RFID technology across diverse applications. Complex and unique current distributions can be designed to enhance security and anticounterfeiting measures. Irregular or proprietary current patterns make it challenging for counterfeiters to replicate the RFID tag, adding a layer of security to the tagged products.

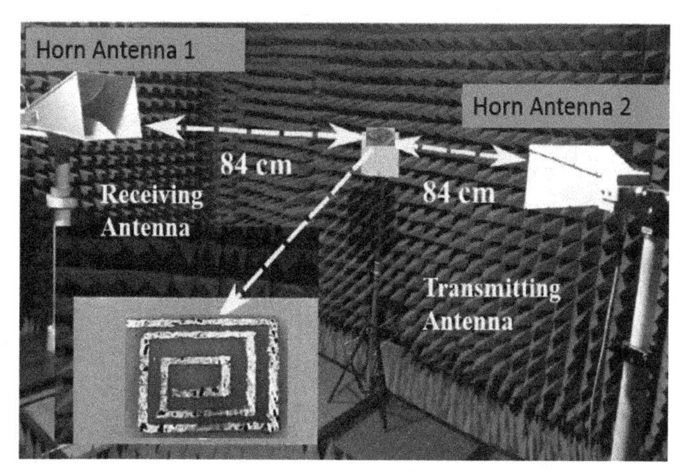

Figure 6.4 Measurement setup of the spiral resonator-1.

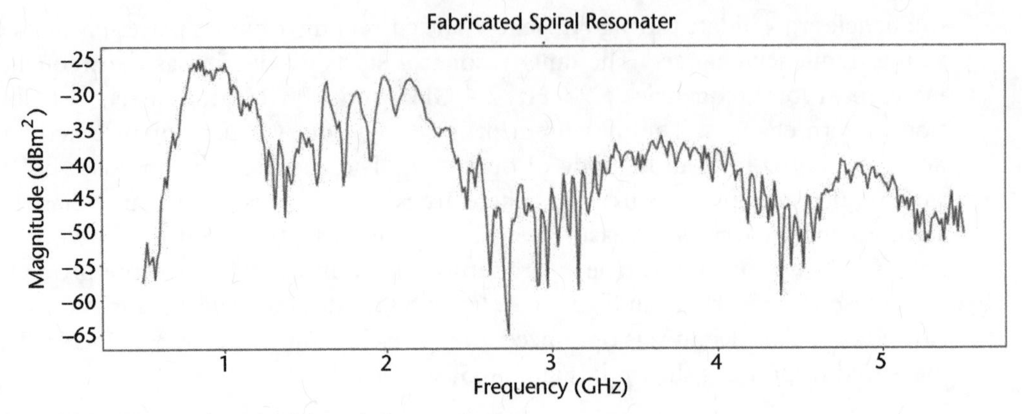

Figure 6.5 Measured result of the spiral resonator-1.

The current flowing through the patch signifies the RF characteristics of the spiral resonator sensor. The flowing current offers a direct relation with the frequency. When the frequency increases, the inner resonator starts to conduct. From the current distribution pattern (Figure 6.6) at 1.9 GHz, it can be observed that the current density is more at the outer resonator. In the case of 2.4 GHz, it is seen that the density is more at the inner resonator with respect to the standard scale.

6.5 Modified CSRR-Based Chipless Resonator-2

Similarly, another circular spiral resonator is taken as an example, in which a concentric circle with different arc lengths is proposed. Here, the arc length is taken using some standard formula.

$$\text{arc length} = 2\pi r \frac{\theta}{360} \tag{6.5}$$

A common center is found, and then using this formula, standard arc length is obtained.

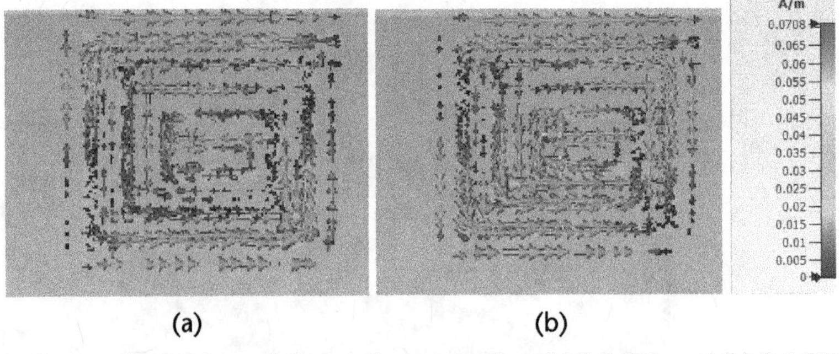

Figure 6.6 Current distributions of the spiral resonator-1 at (a) 1.9 GHz and (b) 2.4 GHz.

The 60 × 60 mm^2 substrate is taken to design the circular-shaped resonator with a height of 1.6 mm (Table 6.2). The design is chosen here, to make a comparison with the previously proposed spiral resonator (Figure 6.7). It is important to propose a convenient tag for the body area application. For this reason, the chipless RFID sensing capability should be more. Here, the sensing capability is enhanced by modifying the structure and its RCS response. The RCS of an RFID tag determines how much of the incident RF signal is reflected back towards the RFID reader. A higher RCS means more signal is reflected, which can potentially improve the read range. The RCS contributes to the effective interaction between the RFID tag and the reader. A larger RCS can enhance the effective interaction, allowing the RFID reader to detect the tag from a greater distance, thereby improving the read range. The material properties of the RFID tag significantly influence its RCS. Conductive or reflective materials used in the tag can increase RCS, leading to better reflection of RF signals and potentially extending the read range. The design of the tag, which is a significant factor in RCS, can be optimized to maximize the reflection of RF signals towards the reader, thus positively impacting the read range. The system's configuration, including reader power, reader sensitivity, and the surrounding environment, also affects how RCS contributes to the read range.

The second design is more capable of covering a wide band, though it has the maximum number of peaks covering from 990 MHz to 3.2 GHz. This band contains the ultrawideband capable of IoT applications like body area and biomedical domain. The RFID tag is again simulated on CST microwave studio with the standard procedure. Here six resonant peak frequencies are obtained.

Table 6.2 Parameter and Values for Chipless Resonator-2

Parameter	*Width (mm) (W)*	*Length (mm) (L)*	*Height (mm)*
Substrate	60	60	1.6

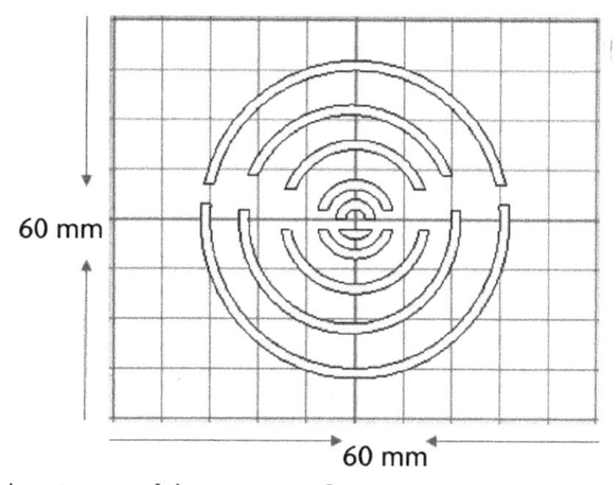

Figure 6.7 Simulated prototype of the resonator-2.

6.5.1 Radar Cross Section

The simulated RCS is shown in Figure 6.8. Here, the resonance frequency can be obtained at 998 MHz, 1.1 GHz, 1.5 GHz, 2.2 GHz, 2.6 GHz, and 3.2 GHz. The resonant frequency over here is more distinct and covers more bands. In this case, the standard formula is taken. RCS of the spiral resonator can be established with the help of the following equation:

$$\sigma = A \times \left|\Gamma^2\right| \times D \tag{6.6}$$

where A is the cross-sectional area of the circular-shaped spiral resonator, $|\Gamma^2|$ is reflectivity, D is the maximum directive gain, and σ is the RCS. The reflectivity can be found by the following equation:

$$|\Gamma|^2 \left|\frac{(Z_L - Z_S^*)}{(Z_L - Z_s)}\right|^2 \text{ where } 0 \le |\Gamma|^2 \le 1 \tag{6.7}$$

where $|\Gamma|$ = reflection coefficient, Z_L= load impedance, Z_S= source impedance, and (*) implies the complex conjugate.

$$\sigma = A_t \times |\Gamma|^2 \times D \tag{6.8}$$

σ = RCS, A_t = area of cross section of the target in m^2 $|\Gamma|^2$ = reflectivity, and D = directivity.

The RCS of the small plate is defined as the

$$\sigma_{\text{plate}} = \pi w^2 H^2 / \lambda^2 \tag{6.9}$$

where w = width, H = thickness of the plate, and λ = operating wavelength.

For chip-based tags, the RCS can be written as

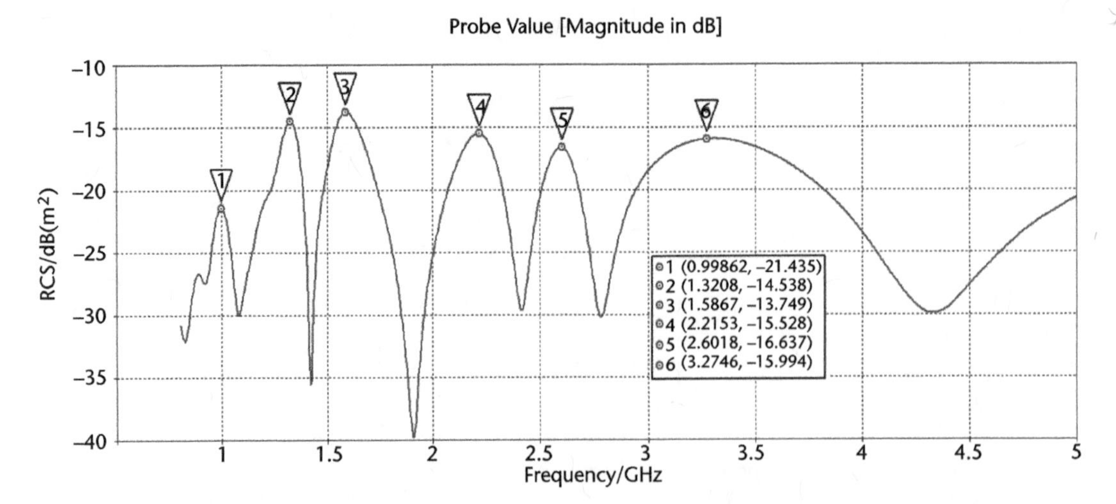

Figure 6.8 Simulated result of the resonator-2.

$$\sigma_{tag} = \lambda^2 R_a G_{tag}^2 / \pi |Z_a + Z_c| \tag{6.10}$$

R_a = Real part of the antenna input impedance;
Z_c = RFID tag chip impedance;
G_{tag} = Gain of the tag antenna;
$Z_c = 50\Omega$ for a chipless tag.

The read range equation of the tag can be written as

$$R = \frac{\lambda}{4\pi}\sqrt{\frac{P_t G_t G_r \left(1 - |\Gamma|^2\right)}{P_r}} \tag{6.11}$$

where R is the separation between the receiving and transmitting antenna, and P_t and G_t are the transmitted power and the gain of the transmitting antenna. P_r and G_r hold the corresponding parameter of the tag antenna used at receiver, and $|\Gamma|$ is the reflection coefficient.

6.5.2 Current Distribution

The current distribution factor of this resonator (Figure 6.9) is more prominent and distinguished in this case. It can be noted that at 998 MHz, there is a certain current distribution at the outer circle. Similarly, at 1.3 GHz next to the outer circle, the inner resonator starts conducting. In this way, at 998 MHz, 1.3 GHz, 1.5 GHz, 2.2 GHz, and 2.6 GHz, the distinct current patterns are observed. The current

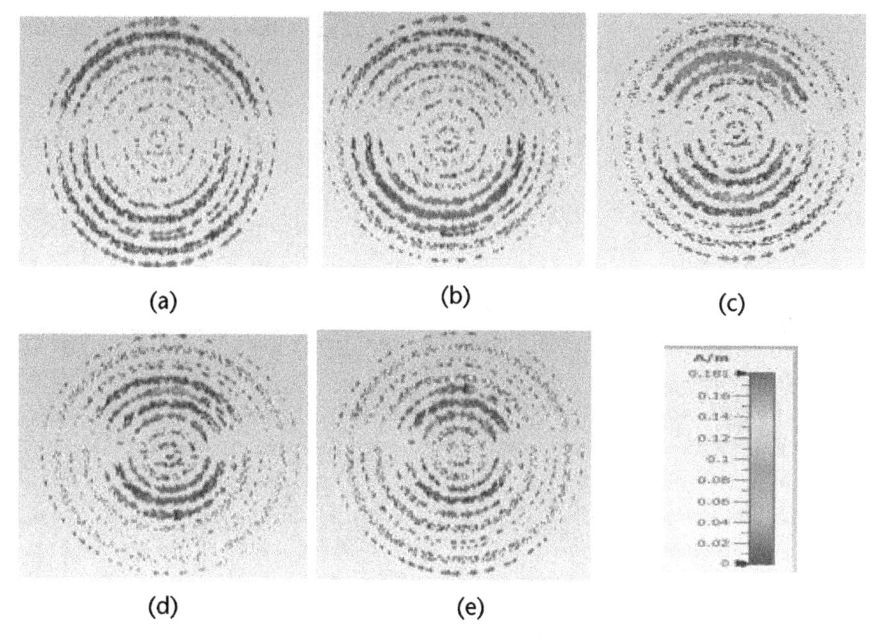

Figure 6.9 Current distributions of the resonator-2 at (a) 998 MHz, (b) 1.3 GHz, (c) 1.5 GHz, (d) 2.2 GHz, and (e) 2.6 GHz.

distribution pattern signifies the tag validation whether it is actually satisfying resonant frequency characteristics or not.

6.6 A High Bit Density Chipless RFID Tag Using Modified Square Split Ring Resonators

Chipless RFID technology has the potential to overrule the existing technologies of identification and will be a game-changer shortly. Nevertheless, one of the critical constraints of this advancement is the low bit density. This section presents a compact, high-bit capacity chipless tag with an encoding capability of 10 bits operating over a frequency range of 0.5 MHz to 6 GHz. The tag based on modified square split ring resonators (MSSRR) having a physical footprint of 38.4 mm × 38.4 mm covers both the UHF and the UWB. The proposed tag is a suitable candidate for product identification and has the capability of identifying 2^{10} (i.e., 1,024) objects without structural encoding if windowing techniques are utilized. The resonating elements developed on an ungrounded, readily available FR4 substrate material ($\varepsilon_r = 4.4$, $\tan\delta = 0.02$) contribute to a fall in the overall cost when compared to a chipped tag.

6.6.1 Proposed Tag Design Geometry

The proposed tag design consists of six square split ring resonators chamfered at all the inner and outer corners. The top view of this copper-based MSSRR developed on a low-cost FR4 substrate is displayed in Figure 6.10. The copper resonators of 0.035-mm thickness are etched out on the 1.6-mm thick substrate with a dielectric constant and loss tangent of 4.4 and 0.02, respectively.

The overall physical size of this compact tag is 38.4 mm × 38.4 mm, where the six resonating elements contribute to 10 distinct resonant peaks. The split ring resonators have the property of encoding more bits with a smaller number of resonators when compared to the simple resonant structures and therefore utilized in this high-bit density design. The size of resonators and slit gaps, as observed from the tag geometry in Figure 6.10, are optimized in order to attain the resonant peaks at specific frequencies with considerable RCS values. The physical dimensions of the presented tag design are tabulated in Table 6.3.

6.6.2 Simulated Insights of the MSSRR-Based Tag

To investigate the tag response in a specified frequency range, the simulations are carried out in a CST Microwave Studio Suite 2020. The illumination of the target occurs with the aid of an LP plane wave at the center of the tag structure, and the backscattered RCS is obtained. By the frequency domain approach, every resonant peak in the RCS plot corresponds to one bit. Therefore, the proposed design can be utilized as a 10-bit tag. The presented tag works in a frequency band covering a part of UHF (0.3 GHz to 3 GHz) and UWB (3.1 GHz to 10.6 GHz) bands. More precisely, it spans over the L-band (1 GHz to 2 GHz), the S-band (2 GHz to 4 GHz), and a section of the C-band (4 GHz to 8 GHz). The RCS values at a particular frequency are arranged in Table 6.4, where it is clear that the maximum RCS is

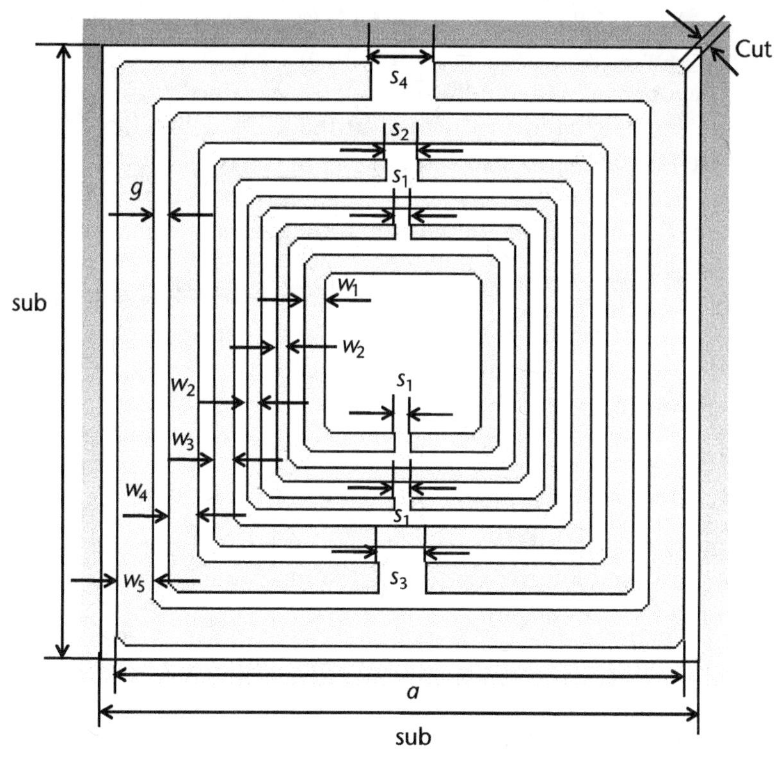

Figure 6.10 Chipless tag based on modified square split ring resonator.

Table 6.3 Dimensional Parameters of the MSSRR-Based Tag

Parameters	*Values (mm)*	*Parameters*	*Values (mm)*
sub	38.4	s_2	2
w_1	1	s_3	3
w_2	0.8	s_4	4
w_3	1.3	a	36
w_4	1.8	g	1
w_5	2.3	cut	0.71
s_1	1	—	—

Table 6.4 Simulated RCS Values at Resonant Frequencies of the Proposed Design

Frequency (GHz)	*RCS (dBsm)*	*Frequency (GHz)*	*RCS (dBsm)*
0.80	-39.15	2.66	-19.20
1.08	-30.47	2.94	-20.70
1.37	-25.41	3.50	-19.58
1.73	-27.66	4.06	-21.48
2.46	-17.83	5.35	-19.96

obtained as –17.83 dBsm at 2.46 GHz. The resonator with minimum physical dimension gives rise to the highest resonant frequency and vice versa.

The working of a chipless tag is well-explained from its surface current distribution, as depicted in Figure 6.11 at 0.8 GHz, 2.46 GHz, and 5.35 GHz for the proposed design. It is observed that when the frequency slightly soars, the highest current density is experienced by the smallest resonating element. This is due to the fact that the resonator with the largest dimension resonates at the lowest frequency and vice versa.

A comparison of various resonator-based chipless tags and the presented tag design are compared according to different parameters like resonator type, frequency band, tag dimension, encoding capability, substrate material, and applications is concluded in Table 6.5.

6.6.3 Measured Insights of the Fabricated Prototype

The proposed tag is fabricated on an ungrounded FR4 substrate by wet chemical etching utilizing a ferric chloride solution, as illustrated in Figure 6.12. The measurement setup, as displayed in Figure 6.13, comprises the Keysight PNA-N5222B, the transmitting and receiving horn antennas (bistatic mode of measurement), and the device under test (DUT), that is, the fabricated prototype. The measurement is taken place in an anechoic chamber where the interrogation is observed to be 30 cm. Figure 6.14 provides a comparison of simulated and measured RCS of the proposed MSSRR-based tag, and they show good agreement. Table 6.6 discloses the measured RCS peak values with respective frequencies.

6.7 Koch-Island Fractal-Based Tag

Chipless RFID has emerged as a prominent means of object identification and detection. What sets this tagging technology apart from optical barcodes is its ability to operate without the constraint of requiring a direct line of sight for communication. Within this section, we delve into the intricate aspects encompassing the design, simulation, and fabrication of a chipless RFID transponder ingeniously structured around the third iteration of the Koch-island fractal, showcasing a sophisticated

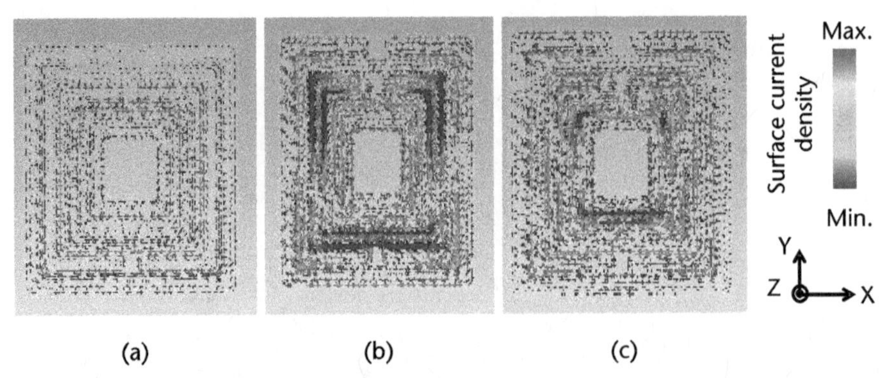

Figure 6.11 Surface current distribution of the MSSRR-based tag at (a) $f = 0.8$ GHz, (b) $f = 2.46$ GHz, and (c) $f = 5.35$ GHz.

Table 6.5 A Comparison of the Proposed Tag with Several Resonator-Based Tags

Resonator Type	*Frequency Band (GHz)*	*Tag Dimension (mm^2)*	*Encoding Capability*	*Substrate Material*	*Application*
Concentric square ring + plus shaped	1 to 12	60 × 60	5	FR4	Biomedical [25]
Double circular split ring	1 to 6	28 × 28	1	FR4	Wound monitoring (pH sensing) [26]
U-shaped	23 to 27	60 × 20	4	RT/Duroid 5880	Healthcare (remote monitoring) [27]
Diagonal coupled line	4 to 7	—	7	Taconic TLX-8	Item tagging [28]
Stepped impedance	2 to 2.6	53.92 × 26.05	6	Taconic TLX-8	IoT [29]
Spiral	4.15 to 8	13 × 20	6	Taconic TLX-8	Long range industrial [30]
Modified square split ring	0.5 to 6	38.4 × 38.4	10	FR4	Product identification

Figure 6.12 Fabricated prototype of the MSSRR-based tag.

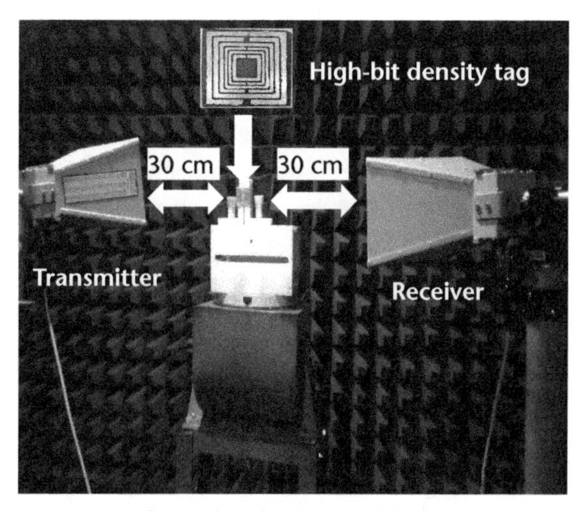

Figure 6.13 Measurement setup for testing the proposed 10-bit tag.

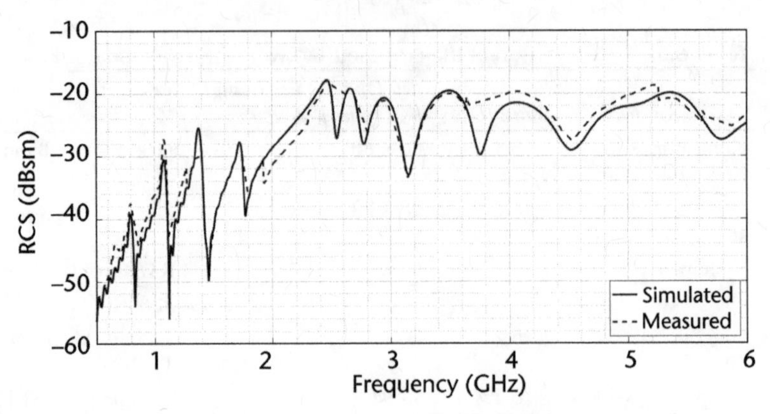

Figure 6.14 Simulated versus measured RCS of the MSSRR-based tag.

Table 6.6 Measured RCS Peaks and Corresponding Frequencies of the Proposed Design

Frequency (GHz)	*RCS (dBsm)*	*Frequency (GHz)*	*RCS (dBsm)*
0.8	−37.52	2.66	−19.20
1.08	−27.22	2.94	−21.16
1.40	−29.92	3.50	−20.04
1.73	−27.75	4.06	−19.74
2.45	−17.95	5.25	−18.67

and comprehensive analysis. The proposed RFID tag has been subjected to simulation, employing the economical FR4 substrate with a dielectric constant (ε_r) of 4.4 and a nominal loss tangent of 0.02. These simulations were meticulously executed within the CST Microwave Studio Suite 2020. In the first iteration, each segment of the curve is (1/3) of its length. Four of these portions are to be chosen from the initiator. This intricate division implies that for each subsequent curve iteration, its unfolded or extended length becomes $(4/3)^n$, where n stands for the particular iteration level. This unfolding characteristic is of paramount significance when crafting antennas with such geometric intricacy. To probe the simulation further, the bistatic mode was employed. Figure 6.15 elegantly showcases the first iteration of the Koch-island fractal, while the subsequent iterations are illustrated in Figure 6.16.

The tag's maximum dimension encompasses an area of 24 square mm, as depicted in Figure 6.17. Within this context, it is noteworthy that for the first iteration, the tag reaches a maximum length of 15 mm, whereas the second iteration maximum dimension is 5 mm, and the third iteration's maximum length is 1.65 mm. Notably, the smallest segment's length in the third iteration measures 0.556 mm. To gain insights into the RCS response, the tag was simulated across all iterations and the results are illustrated in Figure 6.18. The highest RCS observed was −25.6 dBsm for the third iteration. Consequently, it becomes evident that the third

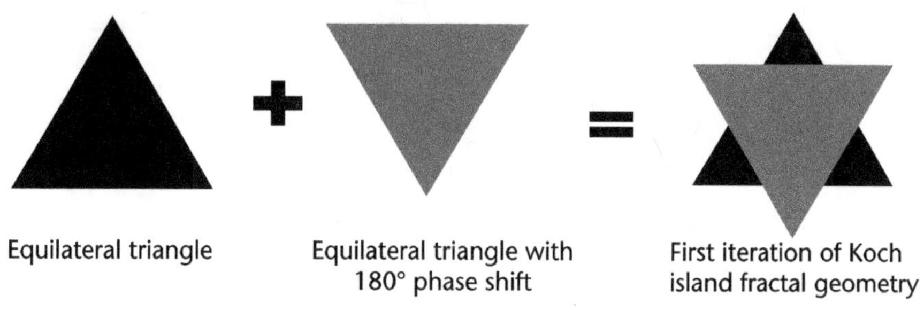

Figure 6.15 Koch-island fractal geometry (1st iteration).

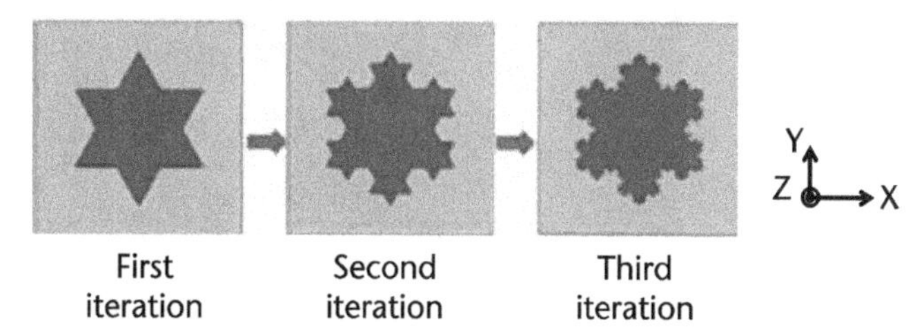

Figure 6.16 Koch-island fractal design steps.

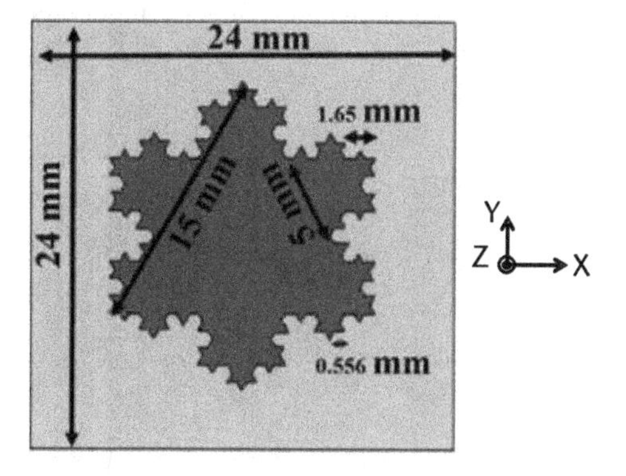

Figure 6.17 Koch-island fractal layout with dimensional details.

iteration of the Koch-island fractal surpasses the first and second iterations in terms of suitability based on the specified design dimensions. This transponder adheres to the backscattering principle [31–45]. Even when we explore higher iterations, the number of bits remains constant, as indicated in Figure 6.18, resulting in a solitary resonance, a key finding as per frequency domain analysis. This, in turn, leads to a single bit tag within the frequency domain.

Figure 6.19 displays the current distribution at the resonant peak frequency of the tag, which is 5.8 GHz. The distinctive patterns observed in this distribution strongly suggest that the transponder can be effectively detected by the reader [30].

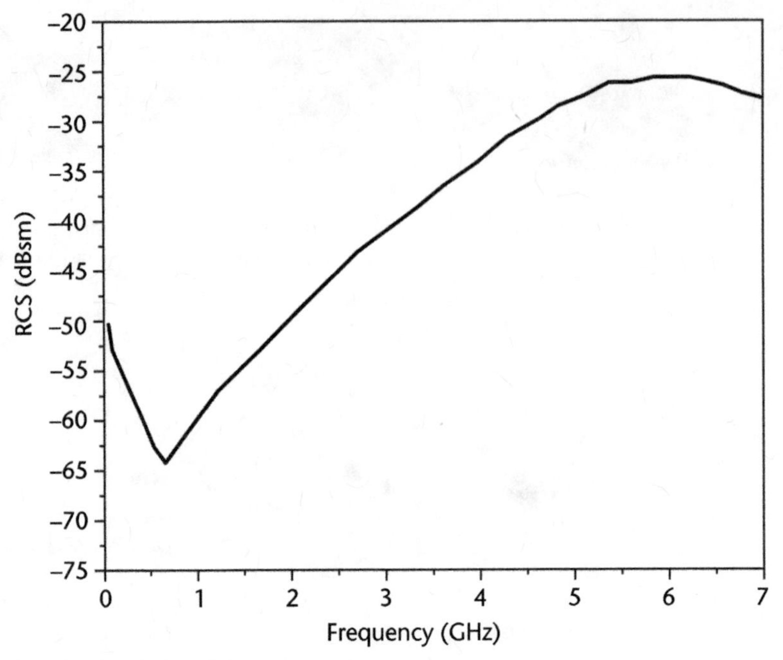

Figure 6.18 Simulated RCS of the proposed tag.

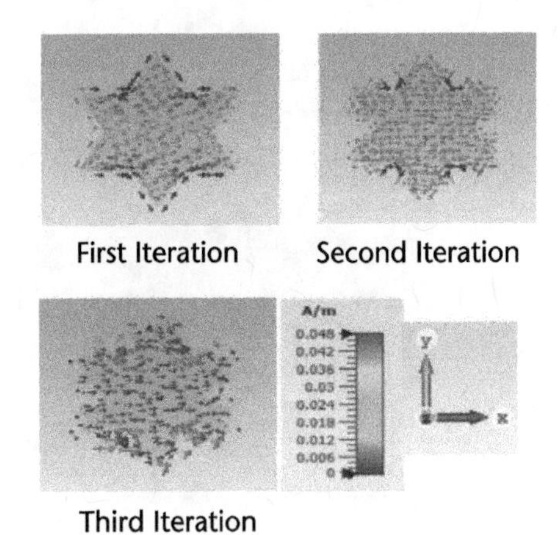

Figure 6.19 Surface current distributions of the tag at 5.8 GHz at $\Phi = 0°$.

6.8 Summary

In this chapter, the experimental analysis of the spiral resonator was proposed and its sensing behavior was investigated. In wireless body area networking, the placement of the chipless RFID tag should be efficient enough to capture the maximum data of the patients in order to establish uninterrupted communication. The proposed resonators are the kind of possible analysis of RFID tags in the domain of on-body chipless RFID networking. The frequency response of the simulated and fabricated transponders can be concluded for a good candidacy in real-time body area networking. The resonators of different shapes can vary the frequency signature and storage capacity, which were explained in this chapter for a good understanding of the readers to visualize the operation of the tag and its real-time applications. It is required to analyze more and more compact resonator-based tags that resonate to the matched frequencies that can be used in wireless body area networks as well as other varying RFID applications like retail, logistics, or traffic toll-fee collections. For any upcoming scopes, the various structures, dimensions, and resonance frequencies can be examined. The UWB and ISM band frequency found in this instance clarifies the viability of the tag.

References

[1] Tu, M.-T., P. Cheong, and W.-W. Choi, "Defected Ground Structure with Half-Wavelength Spiral Resonator of Ultra Wide Band Chipless RFID Tag," *IEEE Journal of Radio Frequency Identification*, Vol. 3, No. 3, September 2019, pp. 121–126, doi: 10.1109/JRFID.2019.2924146.

[2] Elgeziry, M., F. Costa, and S. Genovesi, "Wireless Monitoring of Displacement Using Spiral Resonators," *IEEE Sensors Journal*, Vol. 21, No. 16, August 15, 2021, pp. 17838–17845, doi: 10.1109/JSEN.2021.3085191.

[3] Paredes, F., G. Zamora, F. J. Herraiz-Martinez, F. Martin, and J. Bonache, "Dual-Band UHF-RFID Tags Based on Meander-Line Antennas Loaded with Spiral Resonators," *IEEE Antennas and Wireless Propagation Letters*, Vol. 10, 2011, pp. 768–771.

[4] Brinker, K. R., M. Vaccaro, and R. Zoughi, "Application-Adaptable Chipless RFID Tag: Design Methodology, Metrics, and Measurements," *IEEE Transactions on Instrumentation and Measurement*, Vol. 69, No. 6, June 2020, pp. 3882–3895.

[5] Deif, S., and M. Daneshmand, "Multiresonant Chipless RFID Array System for Coating Defect Detection and Corrosion Prediction," *IEEE Transactions on Industrial Electronics*, Vol. 67, No. 10, October 2020, pp. 8868–8877.

[6] Fathi, P., S. Bhattacharya, and N. C. Karmakar, "Dual-Polarized Keratin-Based UWB Chipless RFID Relative Humidity Sensor," *IEEE Sensors Journal*, Vol. 22, No. 3, February 1, 2022, pp. 1924–1932.

[7] Abouyoussef, M. S., A. M. El-Tager, and H. El-Ghitani, "Quad Spiral Microstrip Resonator with High-Quality Factor," *2018 35th National Radio Science Conference (NRSC)*, 2018, pp. 6–13.

[8] Karmakar, N. C., *Handbook of Smart Antennas for RFID Systems*, John Wiley & Sons, 2011.

[9] Vena, E. P., and S. Tedjni, "A Depolarizing Chipless RFID Tag for Robust Detection and its FCC Compliant UWB Reading System," *IEEE Transactions on Microwave Theory and Techniques*, Vol. 61, No. 8, 2013, pp. 2982–2994.

[10] Kiourti, A., "RFID Antennas for Body-Area Applications: From Wearables to Implants," *IEEE Antennas and Propagation Magazine*, Vol. 60, No. 5, 2018, pp. 14–25.

[11] Jayakrishnan, M. P., A. Vena, B. Sorli, and E. Perret, "Solid-State Conductive-Bridging Reconfigurable RF-Encoding Particle for Chipless RFID Applications," *IEEE Microwave and Wireless Components Letters*, Vol. 28, No. 6, 2018, pp. 506–508.

[12] Finkenzeller, K., *RFID Handbook: Fundamentals and Applications in Contactless Smart Cards, Radio Frequency Identification and Near-Field Communication*, John Wiley & Sons, 2010.

[13] Kundtz, N. B., D. R. Smith, and J. B. Pendry, "Electromagnetic Design with Transformation Optics," *Proceedings of the IEEE*, Vol. 99, No. 10, October 2011, pp. 1622–1633, doi: 10.1109/JPROC.2010.2089664.

[14] Iyer, A. K., and G. V. Eleftheriades, "Free-Space Imaging Beyond the Diffraction Limit Using a Veselago-Pendry Transmission-Line Metamaterial Superlens," *IEEE Transactions on Antennas and Propagation*, Vol. 57, No. 6, June 2009, pp. 1720–1727, doi: 10.1109/TAP.2009.2019890.

[15] Vena, E. P., and S. Tedjini, "Design of Compact and Auto Compensated Single-Layer Chipless RFID Tag," *IEEE Transactions on Microwave Theory and Techniques*, Vol. 60, No. 9, 2012, pp. 2913–2924.

[16] Zheng, F., Y. Chen, T. Kaiser, and A. H. Vinck, "On the Coding of Chipless Tags," *IEEE Journal of Radio Frequency Identification*, Vol. 2, No. 4, 2018, pp. 170–184.

[17] Behera, S. K., and N. C. Karmakar, "Wearable Chipless Radio-Frequency Identification Tags for Biomedical Applications: A Review [Antenna Applications Corner]," *IEEE Antennas and Propagation Magazine*, Vol. 62, No. 3, June 2020, pp. 94–104.

[18] Ramos, Z. A., A. Vena, M. Garbati, and E. Perret, "Single-Layer, Flexible, and Depolarizing Chipless RFID Tags," *IEEE Access*, Vol. 8, 2020, pp. 72929–72941.

[19] Mishra, D. P., and S. K. Behera, "Multibit Coded Passive Hybrid Resonator Based RFID Transponder with Windowing Analysis," *IEEE Transactions on Instrumentation and Measurement*, Vol. 71, Art no. 8005907, 2022, pp. 1–7.

[20] Mishra, D. P., and S. K. Behera, "Resonator Based Chipless RFID Sensors: A Comprehensive Review," *IEEE Transactions on Instrumentation and Measurement*, Vol. 72, Art no. 5500716, 2023, pp. 1–16.

[21] Jonnalagadda, S. V. S. T., D. P. Mishra, and S. K. Behera, "Chipless RFID Sensors for Vital Signs Monitoring – A Comprehensive Review," *IEEE Microwave Magazine*, pp. 2–19, Online ISSN: 1557-9581.

[22] Mishra, D. P., T. K. Das, S. K. Behera, and N. C. Karmakar. "Modified Rectangular Resonator Based 15-Bit Chipless Radio Frequency Identification Transponder for Healthcare and Retail Applications," *International Journal of RF and Microwave Computer-Aided Engineering*, Vol. 32, No. 6, 2022, e23127.

[23] Mishra, D. P., and S. K. Behera, "Modified Rectangular Resonators Based Multi-Frequency Narrow-Band RFID Reader Antenna," *Microwave and Optical Technology Letters*, Vol. 64, No. 3, 2022, pp. 544–551.

[24] Mishra, D. P., A. Subrahmannian, and S. K. Behera, "Design of Multi-Bit Chipless RFID Tag Using Hybrid Resonator for Retail and Healthcare Applications," *Journal of Microwave Engineering & Technologies*, Vol. 8, No. 2, 2021, pp. 34–42.

[25] Arpitha, M. M., P. Sethy, and S. K. Behera, "A 5-Bit Rectangular Resonator Array for Biomedical Applications," *2022 3rd International Conference for Emerging Technology (INCET)*, 2022, pp. 1–4.

[26] Hasan, M. M., and N. Pala, "Cross-Polarized RCS Based Chipless RFID Tag for Wound Monitoring Through pH Sensing," *2021 IEEE Texas Symposium on Wireless and Microwave Circuits and Systems (WMCS)*, 2021, pp. 1–6.

[27] DiCarlofelice, A., E. DiGiampaolo, and P. Tognolatti, "mm-Wave Chipless RFID Tag for Healthcare Applications," *2020 XXXIIIrd General Assembly and Scientific Symposium of the International Union of Radio Science*, 2020, pp. 1–3.

[28] Babaeian, F., and N. Karmakar, "A Cross-Polar Orientation Insensitive Chipless RFID Tag," *2019 IEEE International Conference on RFID Technology and Applications (RFID-TA)*, 2019, pp. 116–119.

[29] Abdulkawi, W. M., and A.-F. A. Sheta, "Design of Chipless RFID Tag Based on Stepped Impedance Resonator for IoT Applications," *2018 International Conference on Innovation and Intelligence for Informatics, Computing, and Technologies (3ICT)*, 2018, pp. 1–4.

[30] Babaeian, F., and N. C. Karmakar, "A Semi-Omnidirectional Resonator for Chipless RFID Backscattered Tag Design," *2020 27th International Conference on Telecommunications (ICT)*, 2020, pp. 1–5.

[31] Karmakar N. C., *Handbook of Smart Antennas for RFID Systems*, John Wiley & Sons, 2011.

[32] Jalil, M. E. B., et al., "High Capacity and Miniaturized Flexible Chipless RFID Tag Using Modified Complementary Split Ring Resonator," *IEEE Access*, Vol. 9, 2021, pp. 33929–33943.

[33] Mishra, D. P., S. V. S. T. Jonnalagadda, and S. K. Behera, "Passive RFID Transponder using Open-Stubs for Bit-Capacity Enhancement," *2023 3rd International Conference on Range Technology (ICORT)*, Chandipur, Balasore, India, 2023, pp. 1–5.

[34] Mishra, D. P., T. Kumar Das, P. Sethy, and S. K. Behera, "Design of a Multi-Bit Chipless RFID Tag using Square Split-Ring Resonators," *IEEE Indian Conference on Antennas and Propagation (InCAP),* 2019, pp. 1–4.

[35] Mishra, D. P., T. K. Das, and S. K. Behera, "Design of a 3-bit Chipless RFID Tag using Circular Split-Ring Resonators for Retail and Healthcare Applications," *2020 National Conference on Communications (NCC),* 2020, pp. 1–4.

[36] Mishra, D. P., and S. K. Behera, "A Novel Technique for Dimensional Space Reduction in Passive RFID Transponders," *2021 2nd International Conference on Range Technology (ICORT),* 2021, pp. 1–4.

[37] Mishra, D. P., and S. K. Behera, "Analysis of Chipless RFID Tags Based on Circular-SRR and Koch Snowflake Fractal for Space-Reduction and Bit-Coding Improvement," *2021 IEEE Indian Conference on Antennas and Propagation (InCAP),* 2021, pp. 621–624.

[38] Mishra, D. P., and S. K. Behera, "Passive RFID Transponders Based on SRR and Koch-island Fractal for Bit-Coding Enhancement," *2021 IEEE Advanced Communication Technologies and Signal Processing (ACTS), 2021*, pp. 1–5.

[39] Mishra, D. P., I. Goyal, and S. K. Behera, "A Comparative Study of Two Different Octagonal Structure-Based Split Ring Resonators," *2022 IEEE Microwaves, Antennas, and Propagation Conference (MAPCON)*, Bangalore, India, 2022, pp. 1007–1012.

[40] Mishra, D. P., I. Goyal, and S. K. Behera, "A Technique to Improve RCS in Passive Chipless RFID Tags by Incorporating Array Structure," *2022 IEEE Microwaves, Antennas, and Propagation Conference (MAPCON)*, Bangalore, India, 2022, pp. 1478–1483.

[41] Jonnalagadda, S. V. S. T., D. P. Mishra, and S. K. Behera, "Frequency Coded Orientation Insensitive Chipless RFID Tags for Vital Signs Monitoring," *2023 3rd International Conference on Range Technology (ICORT)*, Chandipur, Balasore, India, 2023, pp. 1–5.

[42] Jonnalagadd, S. V. S. T., D. P. Mishra, and S. K. Behera, "Frequency Coded High Density Complementary U – shaped Chipless RFID Tag Design," *2023 3rd International Conference on Range Technology (ICORT)*, Chandipur, Balasore, India, 2023, pp. 1–5.

[43] Jonnalagadda, S. V. S, D. P. Mishra, M. M. Arpitha, P. Sethy, and S. K. Behera, "Frequency Coded Chipless RFID Tag Design using L and U - Shaped Resonators," *2022 IEEE Microwaves, Antennas, and Propagation Conference (MAPCON)*, Bangalore, India, 2022, pp. 301–305.

[44] Subrahmannian, A., D. P. Mishra, and S. K. Behera, "Multi-Bit Passive RFID Tag Design using Concentric Rectangular Strip Resonators for 2.4 GHz and 5.8 GHz ISM Band," *2022 IEEE Microwaves, Antennas, and Propagation Conference (MAPCON)*, Bangalore, India, 2022, pp. 1434–1438.

[45] Sethy, P., D. P. Mishra, T. Kumar Das, S. K. Behera, P. P. Sahu, and M. Acharya, "Design of Chipless RFID tag using a Modified Swastik-Shaped Resonator for UWB Band Applications," 2020 *IEEE International Students' Conference on Electrical, Electronics and Computer Science (SCEECS)*, Bhopal, India, 2020, pp. 1–3.

CHAPTER 7

Resonators in Chipless RFID Technologies and Signal Processing Techniques

7.1 Introduction

Similar to optical detection, RFID is a wireless method of tracking and detecting people and other objects. Even though RFID has many benefits over optical systems, the additional features are readability without the need for LOS, higher information storage capacities, and the capacity to simultaneously read numerous tags. The price of conventional RFID transponders (i.e., tags with chips or silicon integrated circuits) is high for numerous applications, including inexpensive goods, compared to barcodes and chipless tags. Passive tags are powered by inquiry signals rather than batteries and work in the backscattering principle. RFID uses electromagnetic waves to send and receive data from a reader to a tag or transponder. The ability to embed RFID tags into objects, faster data transport, and a wider scanning range of tags are some of the benefits of this technology over conventional methods of identification [1]. The chain of supplies for stores, the military service network, pharmakon management and tracking, access management, applications for metering and sensing, tracking of parcels and documents, recurring billing solutions, tracking of assets, real-time navigational aids, automated identification of vehicles, and tracking of pets or livestock are just a few of the current uses of RFID. Because the demand for automatic detection is growing to track and keep an eye on the various industries stated above, there is an increased need for flexible RFID tags. Unlike lower-frequency tags, which currently have constrained reading distances of 1 ft to 2 ft, UHF band RFID transponders feature a greater data transfer rate and a wider scanning range of over 10 ft [2–4]. The price of the RFID tags must be exceedingly low for RFID technology to realize a fully universal system and be commercially viable in large quantities. Affordable substrate for RFID and other RF applications due to a number of factors holds importance in deciding the cost factor. A substrate made of flexible, printable materials that are readily available may solve the issue up to a certain extent. Normally, copper printed on FR4 sheet is the most affordable material as a chipless tag for its great demand and widespread manufacture [1, 2]. Thus, it becomes more practical to produce RFID inlays on copper in bulk. Copper sheets have a thin surface profile as well. Instead of using conventional metal etching methods, it can be used for quick printing procedures like direct-write methodology with the right coating.

Chipless RFID tags can be of time and frequency domain based on the type of operation. The main focus here is on the frequency domain, which was explained in the previous chapter. The time domain tags also utilize unique patterns or structures to encode information in the time domain. Chipless RFID tags use resonant elements as the core components, which resonate at specific frequencies when exposed to an electromagnetic field. Information is encoded into the tag by utilizing the resonant frequencies and bandwidths of these resonant elements. Different patterns or configurations of resonant elements represent specific data bits or characters. Time domain encoding involves measuring the time taken for a signal to travel through the resonant elements or structures. Information is encoded based on the delay or phase shift in the received signal. One of the common techniques for time domain encoding involves sending a pulse or short-duration signal to the tag. The tag responds by reflecting the pulse after a certain delay, and this delay represents the encoded information. The presence or absence of a reflected pulse in a particular time span represents a binary value (0 or 1). Operations of time-domain chipless RFID tags can be shortly expressed as follows. When exposed to an electromagnetic signal, the resonant elements within the RFID tag absorb energy and resonate at their characteristic frequencies. An interrogator or reader device sends out an interrogation signal, typically a short-duration pulse, to the RFID tag. The resonant elements in the tag absorb the energy from the interrogation signal and resonate at their respective frequencies. The tag reflects the signal after a certain delay, which depends on the resonant frequency of the elements. The reader detects the reflected signal and measures the time delay between the transmitted and reflected pulses. The measured time delay is divided into time bins and, based on the presence or absence of a reflected pulse in each bin, the reader decodes the encoded binary information. The decoded binary values are combined to retrieve the encoded data, which may represent a unique identifier, product information, or any other relevant data. Time domain chipless RFID tags enable data encoding and retrieval without a microchip, providing a low-cost and efficient solution for identification and tracking in various applications such as inventory management, supply chain, and asset tracking.

Therefore, the frequency domain and time domain are two distinct approaches to implementing chipless RFID tags, each with its own set of advantages. The choice between these approaches depends on specific application requirements and priorities. Frequency domain chipless RFID tags (FDCL-RFID tags) typically have simpler structures and designs compared to time domain chipless tags. This simplicity often translates to lower production costs, making FDCL-RFID tags more cost-effective for certain applications. They may require less complex manufacturing processes, facilitating mass production and integration into a variety of items or materials. The simpler structure of these tags allows for easier scaling and mass production. This makes them more suitable for high-volume applications where many tags need to be produced and deployed. The FDCL-RFID tags can achieve high data capacity and encoding flexibility by utilizing a wide range of resonant frequencies and spectral responses. This enables the encoding of a rich set of data, providing a versatile identification and tracking solution. These are often designed to be compatible with existing RFID systems and infrastructure, allowing for seamless integration and interoperability. This compatibility is crucial for upgrading or enhancing existing RFID systems. The FDCL-RFID tags can demonstrate better robustness to environmental changes such as temperature, humidity, and nearby materials. This robustness contributes to reliable and consistent performance in diverse operating conditions. These tags can achieve longer detection ranges and higher read rates compared to the time domain. This is advantageous in applications where rapid and accurate identification of multiple tags is essential. The FDCL-RFID tags can be designed to work in conjunction with other communication systems, enabling multimodal communication and information exchange. This versatility can be beneficial for complex applications.

These tags offer flexibility in antenna design, allowing for various antenna configurations to optimize performance and meet specific application requirements. Frequency domain chipless tags can achieve extended memory and multilayer encoding capabilities, enabling storage of a larger amount of data and enhancing the complexity of the encoding scheme. Understanding these advantages helps in choosing the appropriate RFID tag technology based on the specific needs of the application, budget considerations, and the desired level of complexity and performance.

Printing electronics on or into paper substrates can be done quickly and effectively using an inkjet printer with copper ink; as a result, it is also possible to quickly incorporate or place components like antennas, electronics, memories, batteries, or sensors on or in paper modules. In addition, various fabrics might be applied to the paper substrate to make it hydrophobic and fire resistant. This can readily fix any problems with fiber-based goods [4]. One of the ecofriendly products is paper, which now acts as a substrate for chipless RFID. Densities, coatings, thicknesses, textures, dielectric constant, and dielectric loss tangent are among the distinctions in chipless RFID applications. Paper substrates must first undergo dielectric RF evaluation in order to be used in any RF "on-paper" designs. It should be emphasized that fabrication and even assembly methods that are compatible with printed circuit boards can be used to make paper boards that resemble printed wiring boards at a reasonable cost. In a multilayer, three-dimensional platform, these are capable of supporting passives, cables, RFID, sensors, and other components. Small liquid droplets, as small as a few, are ejected by modern inkjet printers in order to print. By using conductive paste and the novel inkjet printing technology, prototype circuits can be quickly created without the need for numerous rounds of photolithographic mask design or conventional etching methods that have been widely employed in industry. Printing may be fully managed from the designer's computer without the need for a clean room. The resolution of the printer is dependent on the size of a droplet. The cartridge is a heater and reservoir-equipped piezodriven jetting device [3, 4]. The inkjet printer used for this study was a Dimatix Materials Printer DMP-2800. Using a DMC-11610 print head or cartridge, inkjet printing fires a single ink droplet from the nozzle in the direction of the desired region, as opposed to subtractive etching which removes undesired metal from a substrate's surface. As a result, there is no waste produced, leading to a cost-effective production solution. To obtain good metal conductivity, silver nanoparticle inks are frequently used in inkjet printing. Once the droplet of silver microparticle has been pushed through the nozzle, the removal of additional fluid and substance contaminants from the depositions is found to be needed in a sintering procedure. The other benefit of the sintering procedure is that it strengthens the link between the paper substrate and the deposition. In relation to the temperature and time to dry, the conductive ink's conductivity ranges from 0.4 to 2.5 $\times 10^7$ S/m. A weak connection results from the bigger gaps between the particles at the lower temperature. Chipless RFID sensors in printing technology are studied in [5, 6]. Inkjet, paper-based substrate, wood substrate-based, brush painting, doctor blading, thermal, screen, gravure, and more are explained in [5, 6]. The details can also be referred to in Chapter 2 of this book.

Nowadays, the utilization of RFID technology is rapidly surging in the healthcare, biomedical, and IoT domains. The chipless RFID overcomes the shortcomings of a chipped RFID system and plays a crucial role in the fields of IoT, retail, and healthcare domain. Recently, there is a growth in the efficacy of additive manufacturing techniques (AMTs), which offer the benefit of chipless RFID to becoming popular in the Internet of Things and millimeter-wave fifth generation networks, storage of power, passive RF elements, and sensing systems. This chapter describes different resonators and their usage in designing the chipless RFID tags and certain signal processing approaches that are used for efficient tag detection at the reader end.

7.2 Comparison of Chipped and Chipless Sensors

The most attractive RFID technology is chipless due to its lower cost, simpler design, numerous real-time commercial applications, and improved read range. The major distinctions between chipped and chipless RFID, as well as their practical applications, are outlined in Table 7.1.

RFID tags have transformed product monitoring and identification over the last few decades. Since they don't require batteries, passive RFID tags are less expensive than active ones. Also, they live longer lives. But occasionally, the price of a tag may be nearly equal to the cost per unit of the item that has to be identified. Recent studies have focused on finding ways to lower the cost of tags to make them more affordable than optical barcodes, with a goal price of less than $0.01. A chipless tag has a read range of about 1m or greater. A chipless tag doesn't need a silicon chip or an ASIC. Hence, printing tags makes the realization process

Table 7.1 Chipped and Chipless RFID Features

Properties	*Chipped RFID*	*Chipless RFID*
ASIC chips	ASIC is embedded	No IC
Cost	Costly	Affordable
Range of frequencies	120 kHz to 5.8 GHz [1, 4]	More than 15 GHz [1, 4]
Range of power transmission	Greater than 3W to 4W	< 10 mW
View range distance	5m [5]	2m [5]
Effects of noise	No such effect as signal processing can be done due to the ASIC chip facility	Due to reduced power, noise has a greater negative influence
Physical structure	Brittle	Compatible and flexible

programmable. The cost of a chipless tag is typically comparable to an optical barcode because it can be printed directly on the packing material. An interrogator or reader antenna's electromagnetic wave provides electricity to a chipless tag. When the incident wave hits the item, the tag acts as an identification. Other times, a signature analysis is done to identify and distinguish a person [5–7]. Chipless RFID tags are substantially easier to manufacture than conventional RFID tags. The lack of an ASIC chip eliminates the need for microelectronic production, simplifies the connection between the tag and chip, and results in a strong and affordable tag antenna. In reality, creating a basic chipless tag only requires applying conducting ink to a dielectric substance.

The structural classification of RFID resonators is shown in Figure 7.1. Homogeneous/single resonance resonators have a single resonant frequency of operation, whereas hybrid/multiresonance resonators support more than one resonant frequency. The geometry of the resonator plays a very important role in multiple resonances [1, 2, 6].

7.2.1 Half and Quarter Wavelength Resonators

A half-wavelength resonator has a length of half a wavelength with open circuits at both ends. The fundamental resonant frequency (f_0) is given by (7.1).

$$f_0 = \frac{c}{2l\sqrt{\epsilon_{eff}}} \tag{7.1}$$

where c is the speed of light, l is the length of the microstrip line, and ε_{eff} is the effective dielectric constant of the microstrip. For $n = 2, 3$, and so forth, this resonator likewise resonates at $f = nf_0$. A stub of a quarter-wavelength having a grounded end and an open-circuited end is called a quarter wavelength resonator, having resonant frequency f_0:

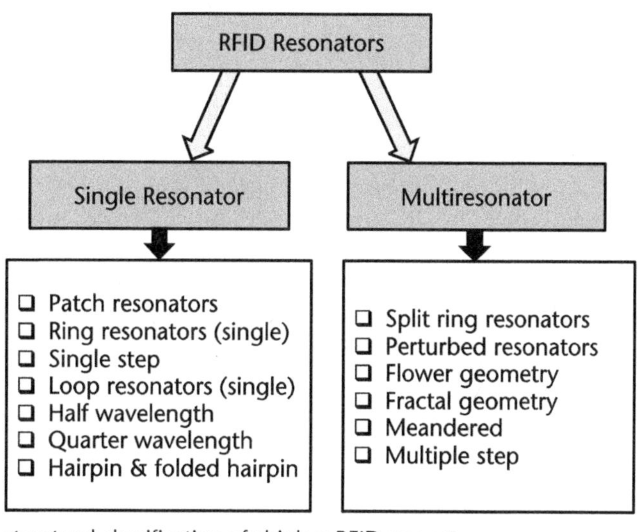

Figure 7.1 The structural classification of chipless RFID resonators.

$$f_0 = \frac{c}{4l\sqrt{\epsilon_{eff}}} \tag{7.2}$$

For $n = 2, 3$, and so forth, this resonator resonates at $f = (2n - 1)f_0$ for $n = 2, 3, \ldots, n$.

7.2.2 Patch Resonators

A patch resonator is used to increase the power-handling capability of the filter due to the larger size of the resonator. This resonator has lower conductivity loss compared to the narrow microstrip line resonator. However, it suffers from greater radiation loss. The common shapes of the patch resonator are square, rectangular, triangular, and hexagonal (Figure 7.2). It has gained the highest popularity because of its extensive use in chipless RFID applications as a key factor in obtaining the required RCS [1].

Comparing various microwave cavity resonators and microcantilever-based sensors, planar microwave sensors are favored for their small size, simplicity of manufacture, cheap cost, and lightweight characteristics [5–9]. Due to their high-quality factor and compact size, open-ended half-wavelength resonators are modified to become SRRs, which are extensively employed as sensing components.

7.2.3 Hairpin Resonators

To improve the compactness of the half wavelength resonator, it can be folded into other shapes, such as hairpin, folded hairpin, and meandered shapes. The term *hairpin resonator* refers to a U-shaped, open-circuited, half-wavelength microstrip transmission line (Figure 7.3). At frequencies where its length ($2l$) is an integer multiple of half the guided wavelength (λ_g), this resonator resonates. The fundamental frequency is given by (1). When developing filters, the resonators are also used for proper coupling. A square open-loop resonator is very adaptive for the Butterworth and Chebyshev filter designs. To lower the size of the transponder, the half-wavelength resonator may be folded into several forms, such as a hairpin, a folded hairpin, an open shape, and a meandering shape. A hairpin resonator is a half-wavelength microstrip open-transmission line twisted into a U form (Figure 2.5). At frequencies where its length ($2l$) is an integer multiple of half the guided wavelength, λ_g, this resonator can resonate. The fundamental frequency is provided by (1). Hairpin resonators are also intended to make it simple to couple the resonators together for the filter design. For the Butterworth and Chebyshev filter designs, a hairpin open-loop resonator is particularly adaptable [6].

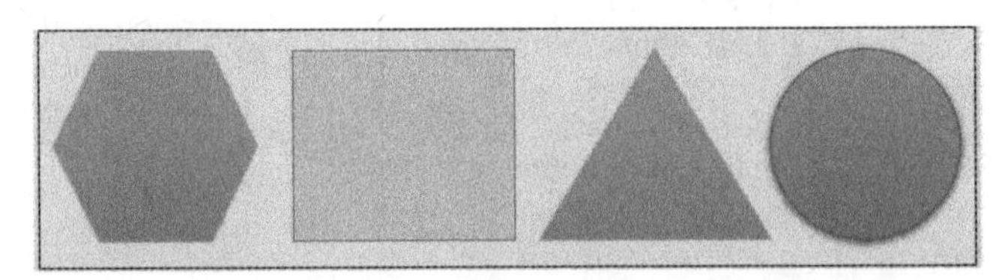

Figure 7.2 Hexagonal, square, triangular, and circular-shaped patch resonators.

Figure 7.3 Hairpin and folded hairpin resonator.

The following are the effects and benefits of using hairpin resonators in chipless RFID tags.

Resonance Frequency Encoding

Hairpin resonators are designed to resonate at specific frequencies determined by their dimensions. By employing hairpin resonators with varying sizes and shapes, chipless RFID tags can encode information into the resonant frequencies, allowing for data representation. Different configurations of hairpin resonators create unique frequency signatures or spectral responses. These signatures encode binary or multibit data patterns, enabling the representation of digital information without the need for an integrated microchip. Hairpin resonators, when properly designed and deployed in a tag, contribute to frequency domain-based encoding. Each resonator or a combination of resonators represents a unique frequency, forming the basis of the tag's identification and data encoding.

Compact Design and Versatility in Data Encoding

Hairpin resonators offer a compact and planar design, making them suitable for integration into chipless RFID tags. Their small footprint allows for efficient use of space, which is critical for designing compact and unobtrusive tags. The design and dimensions of hairpin resonators can be varied to achieve a wide range of resonant frequencies, allowing for versatile data encoding schemes. This versatility enhances the capacity and flexibility in representing information. Hairpin resonators can be configured in a manner that allows for encoding more bits of data per unit length. This enhances the data capacity of chipless RFID tags, enabling the storage of a larger amount of information.

Sensitivity to Perturbations

The resonant frequencies of hairpin resonators can be influenced by perturbations in the tag or the surrounding environment. Monitoring changes in resonant frequencies due to perturbations can provide additional information, enabling applications such as environmental sensing.

Frequency Shift Keying and Frequency Multiplexing

Hairpin resonators can be utilized to implement frequency shift keying (FSK) modulation, where different resonant frequencies represent distinct binary states. This modulation technique is commonly used for data encoding in chipless RFID tags. Hairpin resonators can offer stable and consistent resonant responses, ensuring robust data encoding. Their robust performance is important for reliable identification and data retrieval in chipless RFID systems. Hairpin resonators allow for frequency multiplexing, where multiple resonant frequencies can be used to encode data simultaneously. This further enhances the tag's data capacity and improves the efficiency of data retrieval. Their compact design, versatility in data encoding, and stability make them a valuable tool for achieving effective and reliable chipless RFID communication.

7.2.4 Ring Resonator

A ring resonator (Figure 7.4) is one guided-wavelength transmission line in a closed loop. Its higher resonant modes also occur at $f \approx nf_0$ (for $n = 2, 3...$). The ring structure can be bent into a square, triangle, or hexagonal shape for ease of coupling in the filter design. Apart from this, a cut from the ring results in a split structure (rectangular split-ring, square split-ring, circular split-ring, etc.), which results in multiple resonances with compact dimensions.

7.2.5 Stepped Impedance Resonator

Two or more microstrip line segments with varying impedance values make up a stepped impedance resonator. This resonator is less than a half-wavelength resonator in terms of physical length and also resonates at the same frequency and may be defined as:

$$\tan\theta_1 \tan\theta_2 = \frac{Z_2}{Z_1} \tag{7.3}$$

The electrical length of the two microstrip lines are denoted by θ_1 and θ_2, with the corresponding characteristic impedances Z_1 and Z_2 (Figure 7.5). This resonator also resonates at $f = nf_0$ (for $n = 2, 3, ...$). The resonator is folded into an open-loop shape or other shapes to form a compact structure and to ease the coupling with another resonator in the filter design. With the correct combination of different stepped impedance resonators, the spurious response can be controlled.

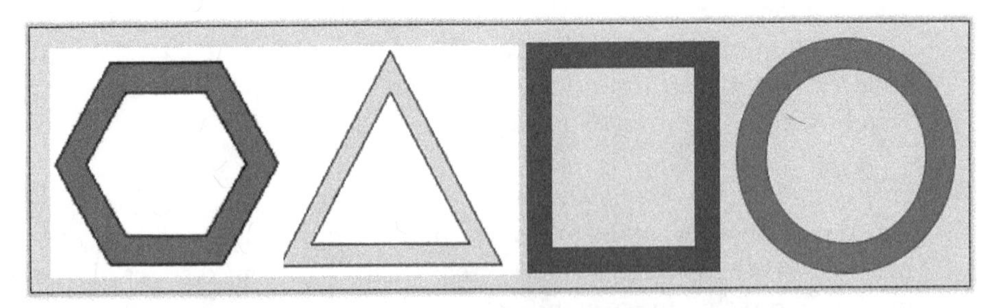

Figure 7.4 Common types of ring resonators: hexagonal, triangular, square, and circle.

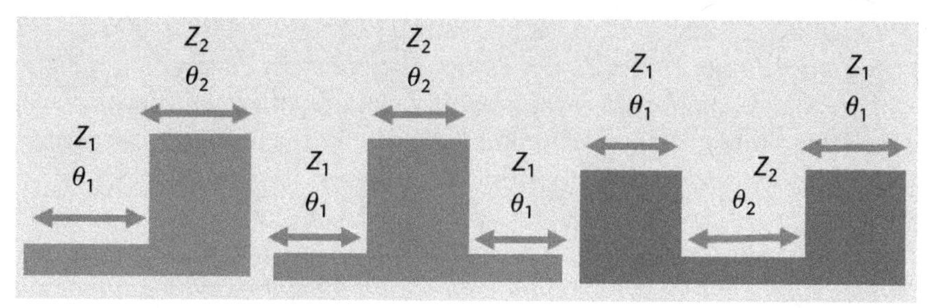

Figure 7.5 Various SIR resonators.

7.2.6 Resonators with Perturbation

The perturbation can give multiple resonances. It can be achieved by the addition of different patches with a regular geometry or by making slots or cuts (Figure 7.6). They can be of cut, stub, slot, and hybrid perturbation. The basic dual-mode resonator structures are: (1) patch resonator with stub perturbation, (2) patch resonator with cut, (3) ring resonator with stub perturbations, and (4) ring resonator with a notch. To reduce the resonator size, a meandered line resonator with stub perturbation with a fractal-shaped resonator or a ring of arrows resonator can be used. By layering two ring resonators with stub perturbation on top of one another in a multilayer arrangement, dual-band bandpass filters are created. The fundamental mode frequencies of the resonator are also changed by a dual-mode resonator that employs a composite right/left-hand transmission line with two pairs of carved slots. The dual-mode resonator can be loaded with capacitive stubs to exhibit multiple resonant frequencies.

Perturbations on a patch antenna can significantly impact performance by altering its resonance frequency, polarization, radiation pattern, impedance matching, and overall efficiency. A patch antenna is a popular type of antenna often used in wireless communication systems. Perturbations, or changes to its structure or environment, can cause variations in its EM characteristics. Perturbations can cause a shift in the resonance frequency of the patch antenna. Alterations in the antenna's dimensions impact the antenna's operating frequency. Perturbations can affect the polarization of the EM wave radiated by the patch antenna. Changes in the geometry or orientation of the antenna can lead to a shift in polarization from linear to circular or vice versa, impacting system compatibility and performance. Perturbations can modify the radiation pattern of the patch antenna. Changes in the antenna's structure or nearby objects can alter the radiation characteristics, including beamwidth, directivity, and radiation efficiency, affecting signal coverage and reception. It can introduce impedance mismatches in the patch antenna

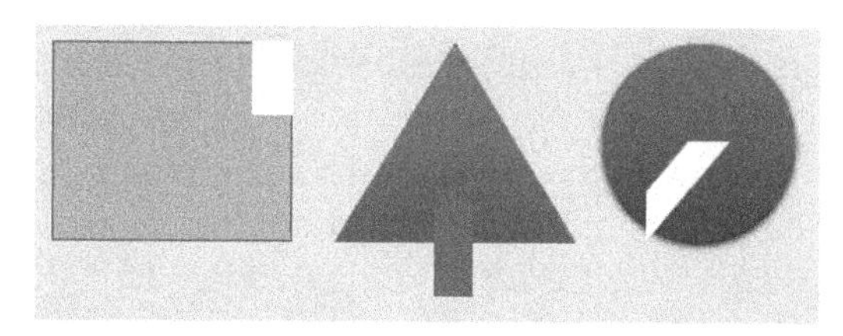

Figure 7.6 Cut, stub, and slot perturbations.

system. Modifications to the antenna's geometry or nearby objects can affect the antenna's input impedance, causing reflections and degrading the antenna's performance in terms of power transfer and signal quality. Perturbations can impact the bandwidth of the patch antenna. Changes in the antenna's structure can influence the bandwidth, leading to either an increase or decrease in the frequency range over which the antenna operates efficiently. Perturbations can result in near-field effects around the patch antenna. Nearby objects or structures can cause distortion, scattering, or absorption of the EM fields, altering the antenna's behavior and affecting performance. Perturbations can induce cross-polarization in the patch antenna system. Changes in the antenna's structure or surrounding environment can cause unintended polarization components, affecting signal purity and reception quality. Modifications to the antenna structure or placement can alter the distribution of electromagnetic energy and potentially influence SAR levels in the surrounding area. Understanding the effects of perturbations on patch antennas is crucial for antenna design, optimization, and deployment. Engineers and designers must consider these effects to ensure the antenna meets the desired specifications and performs effectively in its intended application.

7.2.7 Polarization Errors

Polarization errors in half and quarter-wavelength resonators in chipless RFID tags refer to the degradation of tag performance caused by the mismatch between the polarization of the incident electromagnetic wave and the orientation of the resonant structures within the tag. These resonators, which are crucial for encoding information, operate optimally when the incident electromagnetic field's polarization matches their orientation. However, in real-world scenarios, the incident field's polarization may not align perfectly with the resonators, leading to reduced tag efficiency and read accuracy. Half and quarter-wavelength resonators are designed to resonate at specific frequencies based on their physical dimensions and orientation. When the incident electromagnetic field has a polarization that aligns well with these resonators, efficient energy absorption and reflection occur. Polarization errors arise when the incident electromagnetic wave has a different polarization than what the resonators are optimized for. In chipless RFID tags, this mismatch can significantly reduce the efficiency of the resonators in absorbing and reflecting the energy, impacting the tag's read range and reliability. Resonators, such as dipole or loop structures, are sensitive to the orientation of the incident electromagnetic wave. When the polarization of the incident field is perpendicular or at an angle to the resonator's orientation, the efficiency of the resonance process decreases, affecting the tag's ability to reflect signals effectively. Polarization errors lead to a decrease in the effective read range of the chipless RFID tag. The tag may not reflect signals back to the reader as efficiently as expected, resulting in a reduced read range, and potentially causing misreads or failed detections. The encoding scheme of chipless RFID tags is based on the resonant behavior of the structures within the tag. When polarization errors occur, the accurate decoding of the encoded information may be compromised, affecting data integrity and reliability. The sensitivity of half and quarter-wavelength resonators to incident field polarization means that the tag's performance can vary based on the orientation of the tag concerning the incident electromagnetic wave. This can pose challenges in achieving consistent and reliable performance across different orientations.

7.2.7.1 Mitigation Strategies

To mitigate polarization errors, strategies such as polarization-diverse resonator designs, circularly polarized antennas, machine learning algorithms for adaptive adjustments, or educating end-users on optimal tag orientation are employed. These strategies aim to enhance the tag's robustness and performance in varying polarization conditions. Implementing the multiresonator design that comprises resonators with different orientations to accommodate various polarizations helps increase the probability of successful tag detection irrespective of the incident field's polarization. Random distribution of resonators within the tag mitigates the impact of polarization errors. This strategy helps ensure that some resonators align with the incident field's polarization, improving the tag's overall performance. Developing chipless RFID tags with the ability to dynamically adjust their resonance properties based on the incident field's polarization can be achieved through tunable materials or structures within the tag. Understanding and addressing polarization errors in chipless RFID tags are essential to optimize tag performance, improve read accuracy, and enhance the reliability of RFID systems in diverse real-world deployment scenarios.

The multiple resonances give multiple bits by increasing the storage capacity of chipless RFID tags, which is a useful feature in sensing applications. The use of resonators as sensors is described in the next section [6].

7.3 Categories of Chipless Sensors

In passive wireless sensors, there is no physical connection made; instead, the connection is made via a wireless radio transceiver. These sensors work with the RF signal that the reader has interrogated [7]. In general, there are two types of wireless sensors, chipped and chipless, based on how they operate and how well they can handle power. A chipped RFID passive sensor operates by altering impedance parameters to detect variations in a physical environment, which is based on the modulation of reflected RCS theory. Chipped sensors can be classified as electromagnetic or electrical. Sensors that use batteries are active sensors, and those that do not use chips, as well as the battery, are passive sensors. Similar distinctions can be made between passive, harmonic, and self-tuned electromagnetic sensors. Chipped sensors have the highest level of independence.

The performance of the transducer, which is influenced by the surroundings and the maximum interface range, can be used to determine the range of a chipless RFID. The mechanism monitors the motion of an acoustic wave over a piezoelectric material and falls under the category of SAW-based sensors. The propagation delay can be changed to alter the acoustic wave's velocity. A commercial sensor based on chipless RFID technology is the SAW-based sensor [3]. The main drawbacks of this sensor are that it is nonplanar, the manufacturing is difficult, and the interface distance is only about 10 meters [8]. This results from conversion loss when switching from an acoustic to an electromagnetic wave. An EM transduction-based chipless

RFID, on the other hand, can interrogate across a greater distance (30m) since it suffers from less EM conversion loss. For a high frequency application, this particular sensor is practical [8, 9].

7.4 Applications of Chipless Sensors

Chipless sensors have a number of benefits, including low cost, great sensitivity, enhanced security, and resistance to interference. The small weight of this type of sensor is a significant benefit as well. Additionally, it can provide a flexible, long-lasting sensor model with the highest degree of accuracy. Figure 7.7 shows a classification of chipless RFID passive sensors. Sensors for temperature, pressure, humidity, gas strain cracking, and biosensing are all included in the EM transduction sensor group [7]. Their utility in identifying structural deformation is increased by their scalability. For example, reliable knowledge of the spatial and temporal variations in soil and crop characteristics within a field is essential for precision agriculture [7]. In addition to the benefits of chipless sensors mentioned previously, EM transduction enables a chipless sensor to have a high read range of up to 30m. These components are used in many IoT devices and enable smart sensing.

7.5 Convolutional Codes

In this section, encoding and decoding architecture for convolutional codes are explored. The encoder can be expressed in a variety of equally effective ways. We'll also talk about the main convolutional code decoding technique based on the Viterbi algorithm. An in-depth understanding of convolutional coding can be found in [10].

Convolutional coding is a form of error-correcting code used in communication systems, including chipless RFID technology. It's employed to enhance data

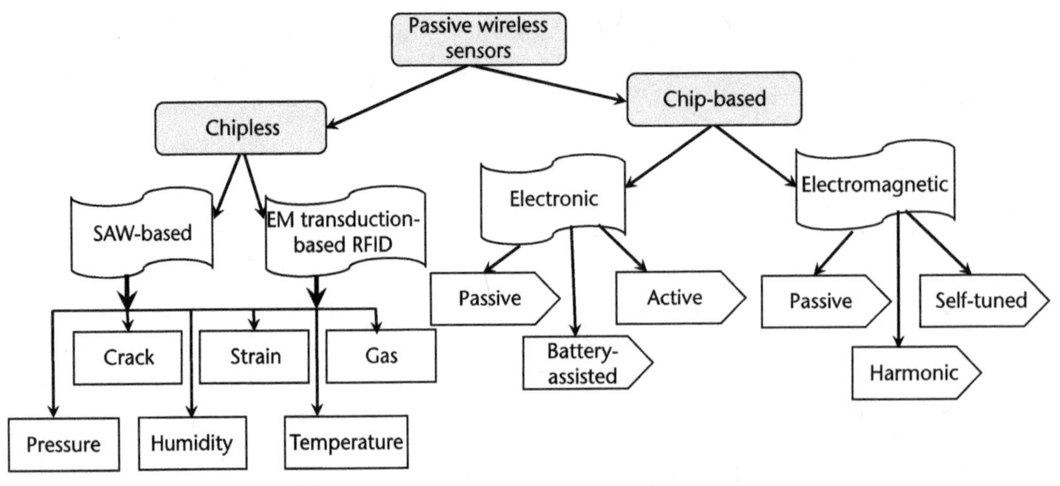

Figure 7.7 Classification of sensors.

integrity by adding redundancy to transmitted data, allowing for error detection and correction at the receiver. In chipless RFID, convolutional codes help in robust data encoding and decoding, especially in scenarios where tag responses are affected by noise, interference, or other impairments.

In chipless RFID, convolutional codes are used to encode information into the frequency or time domain responses of the tag. The encoded data is then decoded by the reader to retrieve the original information. Convolutional coding helps combat noise, interference, and other signal impairments, ensuring reliable and accurate data communication between the tag and the reader. Convolutional coding is a powerful tool in chipless RFID systems, enhancing the robustness of data encoding and decoding and ultimately improving the performance and reliability of the RFID technology in various applications.

7.5.1 Encoder and Decoder

A convolutional code consists of a shift register and a set of binary feedback connections. The shift register is a set of memory elements (flip-flops) that store the recent input bits. The connections determine how the input bits are combined and fed into the shift register to produce the output (encoded) bits. The connections and feedback are defined by generator polynomials, typically represented in octal or hexadecimal notation. These polynomials determine how the input bits affect the output bits. Constraint length (K) refers to the number of memory elements (stages in the shift register). K implies that the encoder considers K previous bits when generating the output. The encoding process involves initializing the shift register with all zero values. Then, the current input bit and the bits stored in the shift register according to the generator polynomials are checked and correlated. Thereafter, the result is written into the shift register, shifting out the oldest bit. The bits that are shifted out from the shift register are the encoded output bits.

Shift registers accept information bits as input, adding the input information bits (modulo-2 addition) to produce the output encoded bits and storing the shift register's contents. The linkages to the modulo-2 adders were heuristically constructed without any prior knowledge of algebra or combinatory, as shown in Figure 7.8.

The definition of the rate of transmission of the code in the context of convolution code is $r = k/m$ where k is the total number of parallel input information bits, and m is the total number of parallel output encoded bits at a given time [10–12]. For a convolutional code, the constraint length k is defined as $k = n + 1$, where n is the highest possible stage count (or memory size) for every shift register. The state data for the convolutional encoder is stored in shift registers, and the output bit size is constrained by the constraint length. The code rate is set to $r = 2/3$, the maximum memory size is set to n/3, and the constraint length is set to l/4 for the convolutional encoder depicted in Figure 7.9.

State Diagram

The state diagram serves as a representation of a convolutional encoder's state information. The shift registers include data that describes the convolutional encoder's current state. In Figure 7.10, the encoder's state diagram is shown [11, 12].

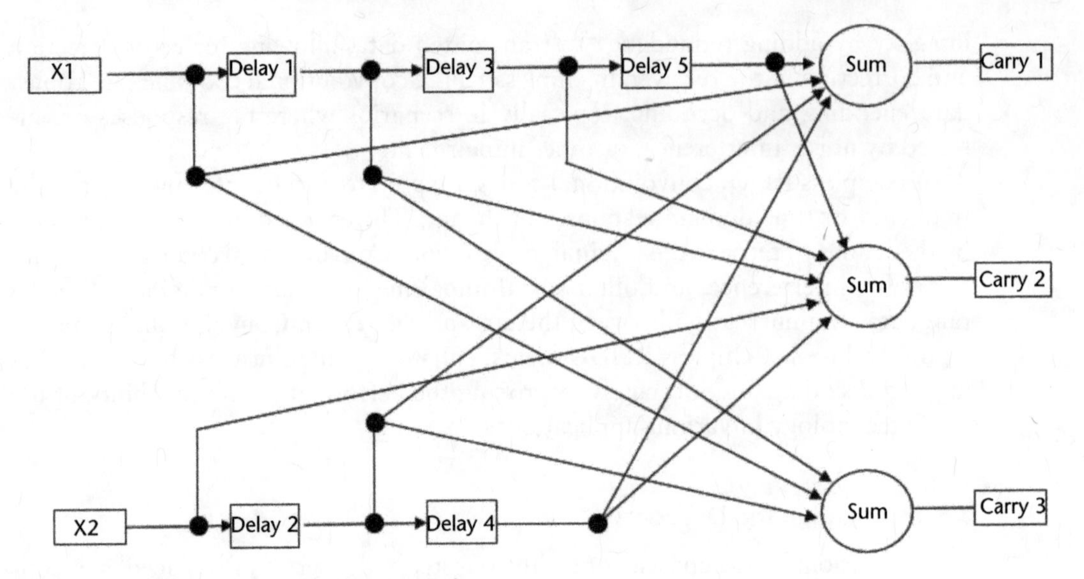

Figure 7.8 Structure of a delay-based encoder.

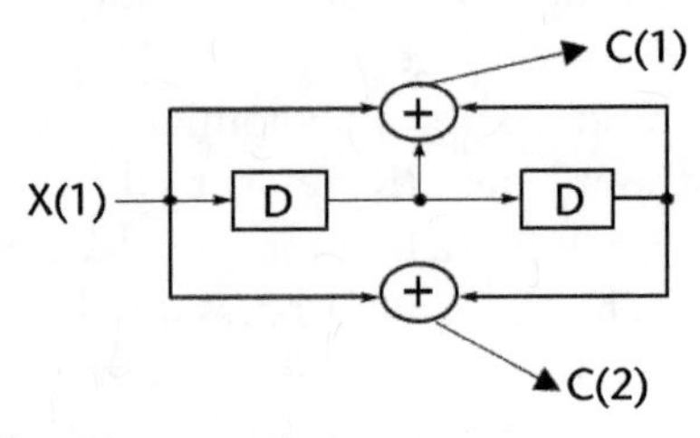

Figure 7.9 Signal interpretation of an encoder.

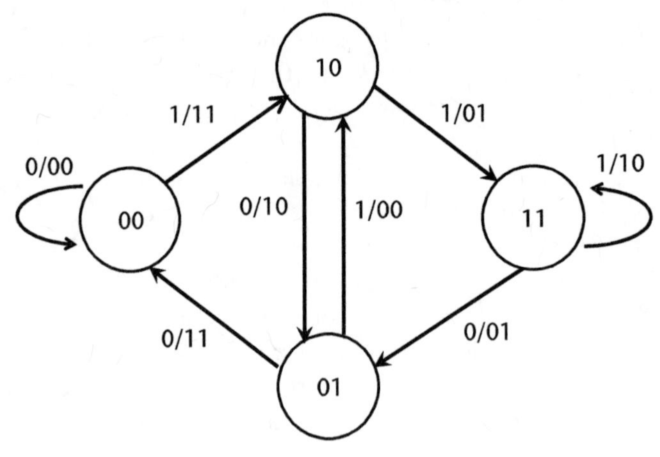

Figure 7.10 State diagram of the encoder.

Convolutional decoding uses algorithms to estimate the most likely transmitted sequence based on the received (possibly noisy) sequence. Two popular decoding algorithms are the Viterbi algorithm and maximum likelihood sequence estimation (MLSE). The following is a simplified decoding process using the Viterbi algorithm:

- *Initialization*: Initialize the state metrics and traceback paths;
- *Branch metric calculation*: Calculate the branch metrics, representing the difference between the received bit and the expected bits for each possible transition;
- *State metric update*: Update the state metrics for each state based on the branch metrics and the previous state metrics;
- *Survivor path selection*: Select the survivor path for each state, which is the most likely path based on the state metrics;
- *Trace back*: Trace back through the survivor paths to obtain the most likely transmitted bit sequence.

In a convolutional code, the encoded order sequence c is convoluted from input system X. The received order r is obtained after transmitting the sequence, covering a noisy channel. The estimated code sequence y from the sequence that was sent as r is determined using the Viterbi method utilizing an ML computation in order to maximize the probability that r is received in the given anticipated code order [13, 14]. The block diagram of the Viterbi decoder is shown in Figure 7.11. Sequence y cannot be any arbitrary sequence and must be one of the permitted coding sequences. The partial trail metric assigns a number to respectively of the trellis diagram's condition. The fractional route metric is computed from state $s = 0$ at time $t = 0$ to a specific state $s = 1$. From the pathways that terminate at a particular state, the best partial path metric is selected. Depending on whether a and b are selected in a traditional or alternative manner, the larger or smaller metric may be the optimum partial path metric. The rest of the metrics indicate the no survivor's paths, whereas the chosen measure reflects the journey of the survivor. In the trellis diagram, the pathways with survivors are saved, while the paths without survivors are deleted. The Viterbi algorithm selects the ML path as the solitary survivor path that is still extant at the conclusion of the procedure. The sequence would then be provided by tracing back the ML path on the trellis diagram.

7.6 Noise Reduction Algorithms

Several noises are discovered while measuring the RFID tag in a real-time context. To remove the noise, traditional signal processing techniques are used. Signals are typically altered by noise reduction techniques, either slightly or significantly. To prevent signal alterations, utilize the local signal-and-noise orthogonalization algorithm [6, 15].

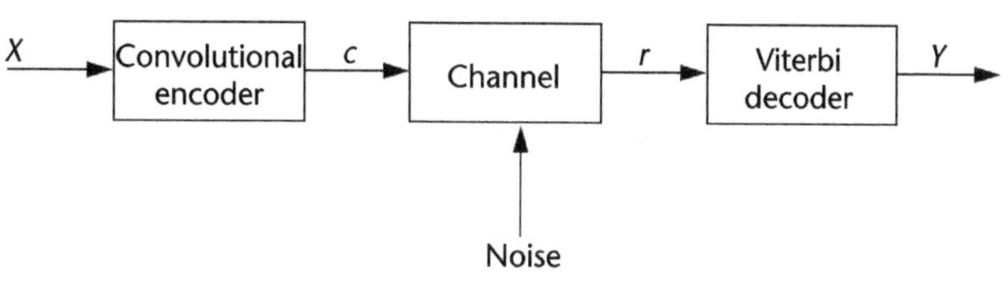

Figure 7.11 Block diagram of the Viterbi decoder.

7.6.1 Convolutional Codes' Decoding Complexity

The quantity of time intervals, or t, and the total number of concurrent bits of information, or k, make up the order of the input information k^*t with regard to a broad convolutional code. As a result, the trellis diagram exhibits $L + m$ phases. A thorough search for the ML sequence would demand computer power of the order of $O(2k^*t)$ due to the fact that there are precisely $2k^*t$ potential paths in the trellis diagram. By doing the ML search in the trellis one level at a time, the Viterbi algorithm lessens this complexity. Each trellis node (state) does $2k$ calculations. Each step of the trellis has a total of two million nodes. If either k or m rises, complexity will rise exponentially [14–16].

7.6.2 Weiner Filter

Signal processing techniques, such as the Wiener filter, combine additive noise, well-known stationary signal and noise spectra, and LTI filtering to infer a desired objective, random process, from an observed noisy process. By using this, the mean square error between the planned system and the predicted random progression is minimized [17–19]. The frequency domain is where wiener filters are typically applied. A damaged image $I(n, m)$ is transformed using the discrete Fourier transform (DFT) to get $I\ (x, y)$. One can calculate the original image spectrum by multiplying the product of $I\ (x, y)$ by the Wiener filter $G\ (u, v)$ as $\hat{S}(x, y) = G\ (x, y)\ I\ (x, y)$. The image's estimate is then derived from its spectrum using the inverse DFT. These spectra are used to define the filter as $H\ (x,\ y)$ = Fourier spectrum of the point spread function (PSF). By applying the Fourier transform to the signal autocorrelation, the power spectrum of the signal process is obtained. By applying the Fourier transform to the noise autocorrelation, the power spectrum of the noise process is obtained.

In the context of chipless RFID, the Wiener filter can be utilized for signal enhancement, noise reduction, and optimal recovery of chipless tag responses. This is a linear filter designed to minimize the mean square error between the desired signal and the filtered signal. It is employed to estimate an optimal reconstruction of a desired signal from a noisy or degraded version of that signal. The Wiener filter operates in the frequency domain, aiming to restore the original signal by utilizing the power spectral density of both the desired and observed signals. In chipless RFID, tag responses can be affected by various factors such as noise, interference, or weak signal strength during communication. The Wiener filter can enhance the received tag response by reducing noise and interference, improving the accuracy and reliability of the detected information. RFID systems, including chipless RFID, often encounter noise during signal transmission and reception. The Wiener filter can be applied to reduce the noise in the received signal, allowing for better identification and decoding of chipless tag responses. When a tag response is degraded or distorted due to environmental conditions or interference, the Wiener filter can be used to optimally recover the original tag response. This ensures that the RFID reader receives a more accurate representation of the tag's data, leading to improved data retrieval. By utilizing the Wiener filter to enhance tag responses, the decoding process becomes more accurate and efficient. The filtered response provides a cleaner representation of the encoded data, leading to fewer decoding errors and better overall performance. In dynamic RFID environments, signal conditions can change rapidly. The Wiener filter can be adapted in real-time to varying signal characteristics, ensuring optimal filtering and signal enhancement based on the current conditions. The Wiener filter's applications in chipless RFID are critical for overcoming challenges related to noise, interference, and signal degradation, ultimately improving the accuracy and efficiency of data communication between RFID readers and chipless tags.

7.6.3 Kalman Filters

The Kalman filter uses the desired signal as input, together with noise, and outputs the unknowables that are reasonably close to their real values. It takes advantage of velocity and positioning and selects the interesting signal from faulty observations. The state space representation of the Kalman filter is as follows:

$$M(n+1) = I(n)M(n) + J(n)U(n) + V(n) \tag{7.4}$$

where the state vector $M(n)$, the input vector $U(n)$, and the unknown noise with covariance $V(n)$ are each given, the state estimates are validated by the Kalman filter.

$X(n|n)$: estimations of $X(n)$, where $Y(n)$, $Y(n-1)$, and so on are the provided measurements.

$X(n+1|n)$: estimations of $X(n+1)$, where $Y(n)$, $Y(n-1)$, and so on are the provided measurements.

The covariance of $X(n)$, denoted as $P(n)$, represents the error covariance matrix associated with the state estimates at time n. $I(n)$ and $J(n)$ are the state transition matrix and input matrix at time n, respectively. The Kalman filter produces state estimates $X(n|n)$ based on given measurements $Y(n)$, $Y(n-1)$, and so forth. The error covariance matrix $P(n|n)$ verifies the accuracy of these estimates, utilizing the measurements $Y(n)$ and $Y(n-1)$. The quantities known to us are $X(n|n)$, $U(n)$, $P(n)$ and the new measurement $z(n+1)$.

$$V(n+1) = z(n+1) - \hat{z}(n+1|n) \tag{7.5}$$

$$X(n+1|n+1) = X(n+1|n) + H(n+1)V(n+1) \tag{7.6}$$

where $H(n+1)$ is Kalman gain. For real-time applications, the Kalman filter is utilized because the aforementioned equations can be simply implemented. The Kalman filter can be represented by mathematical equations that present a well-organized computational (recursive) way to estimate the state of a process, where the mean of the squared error is minimized. The filter supports estimations of past, present, and future states, and it can do so even when the precise characteristics of the modeled system are not known. These kinds of estimations are used in navigation and other processes to accurately track an object in a medium or free space [16]. The Kalman filter estimates a state variable x_k which is a discrete controlled process with the help of linear stochastic differential equation as depicted by (7.7).

$$x_k = \Phi x_{k-1} + Bu_{k-1} + w_{k-1} \tag{7.7}$$

with a measurement given by:

$$z_k = Hx_k + v_k \tag{7.8}$$

Where w_k and v_k process and measurement noises with a covariance of Q and R respectively. Φ is the state transition matrix and B is a control input that will affect the state variable when present in the equation. H is the measurement sensitivity.

$$\Phi = \begin{bmatrix} 1 & t & \cdots \\ 0 & 1 & t \\ \cdots & 0 & \cdots \end{bmatrix} \tag{7.9}$$

For a 2×2 matrix,

$$\Phi = \begin{bmatrix} 1 & t \\ 0 & 1 \end{bmatrix} \tag{7.10}$$

where $0 \le t \le \tau$ and $\tau > 0$. Let $t = 1$.

Therefore,

$$\Phi = \begin{bmatrix} 1 & 1 \\ 0 & 1 \end{bmatrix} \tag{7.11}$$

$$B = \begin{bmatrix} 0 & 0 \\ 0 & 0 \end{bmatrix} \tag{7.12}$$

$$H = [1 \quad 0] \tag{7.13}$$

$$R_k = E\langle v_k v_k^T \rangle = \sigma^2 I \tag{7.14}$$

where $I = \begin{bmatrix} 1 & 0 \\ 0 & 1 \end{bmatrix}$ and σ is the variance of AWGN random variable = 1.

Therefore, $R = 1$,

$$Q = \begin{bmatrix} 1 & 0 \\ 0 & 1 \end{bmatrix} \tag{7.15}$$

In Kalman estimation, we estimate a parameter posteriorly, $\hat{x}_{k(+)}$ with the help of a priori estimate, $\hat{x}_{k(-)}$.

The procedure for Kalman filtering is as follows:

Provide initial estimates for $\hat{x}_{k-1(+)}$ and $P_{k-1(+)}$.

Project the state ahead with $\hat{x}_{k(-)} = \Phi_{k-1}\hat{x}_{k-1(+)}$.

Project the error covariance ahead with $P_{k(-)} = \Phi_{k-1}P_{k-1(+)}\Phi_{k-1}^{T}$ where $P_{k(-)}$ is a priori covariance (the error covariance matrix before the update).

Computing the Kalman gain $\overline{K}_k = P_{k(-)}H_k^T\left[H_kP_{k(-)}H_k^T + R_k\right]^{-1}$.

Update the estimate with the measured value.

$\hat{x}_{k(+)} = \hat{x}_{k(-)} + \overline{K}_k\left(z_k - H_k\hat{x}_{k(-)}\right)$.

Update the error covariance $P_{k(+)} = \left(I - \overline{K}_kH_k\right)P_{k(-)}$ and these iterations will go on up to the end where $P_{k(+)}$ is a posteriori covariance.

Estimated measurement, $\hat{z}_{k(-)} = H_k\hat{x}_{k(-)}$.

7.7 Adaptive Control Algorithm

The adaptive filter that makes up the automatic noise control (ANC) system comprises two essential components. These two components consist of a control algorithm and an adaptable digital filter. The essential signal processing is handled by the digital filter, whose coefficients are modified by the control algorithm. Since they are frequently employed in ANC systems as an adaptive controller, adaptive filter algorithms play a significant role in these systems. Due to their importance, this chapter concentrates on the more complicated and sophisticated control algorithms that have been used in ANC systems as adaptive controllers. Well-defined specifications are necessary for the design of digital filters with static coefficients. Conversely, adaptive filters automatically adjust their filter coefficients in order to minimize a specified objective function [20, 21]. The essential design of an adaptive filter is shown in Figure 7.12.

When the adaptive filter's output, $y(m)$, is relative to a wanted signal $w(m)$, an error signal $e(m)$, which is transmitted back to the adaptive filter is generated. In this scenario, m is the repetition number. An adaptive method adjusts coefficients of the adaptive FIR filter on each repetition, in general, collecting facts from the error signal and reference signal. The following is how the error signal $e(m)$ is recognized:

$$e(m) = d(m) - y(m) \tag{7.16}$$

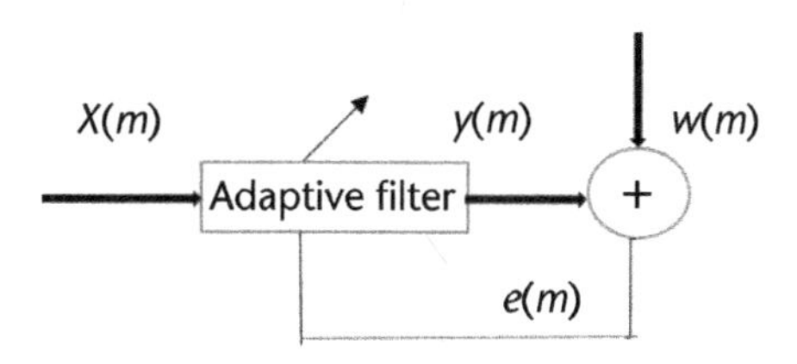

Figure 7.12 Basic block diagram of adaptive filter.

The adaptive method's coefficient adjustment algorithm basically updates the filter coefficients by combining the reference signal and error signal $e(m)$ with the goal of minimizing, for instance, the mean-square error [17]. The error signal shows how closely the adaptive filter's output signal resembles the target signal. When choosing adaptive filters, a number of critical factors come into play, including computational complexity, divergence issues, or the filter coefficients' average produces a biased result. Such problems will be affected by the adaptive filter algorithm chosen. Numerous adaptive filter applications exist, each with unique difficulties. Since this filter solution for a single application might not be appropriate for another, caution should be applied while choosing adaptive algorithms for a given application. For any adaptive application, variables, including computing cost, performance, and robustness, should be taken into account while choosing an adaptable algorithm [21].

The many different adaptive algorithms that are accessible include the LMS method, NLMS method, leaky LMS (LLMS) method, and RLS algorithms, to name just a few. Only a small number of their filtered-X versions have been taken into consideration in this chapter for the aim of duct noise reduction.

7.8 Least Mean Square Algorithm

This process shifts its filter coefficients against the rapid slope of the square error signal w.r.t the coefficient vector [18, 20] because it is a gradient-based approach. The LMS technique directly calculates the instantaneous square's gradient estimate from the mean square value of the error signal, which is then used to update the filter coefficients. The LMS algorithm update equation is:

$$y(m) = vT(m) \times (m) \tag{7.17}$$

$$e(m) = w(m) - y(m) \tag{7.18}$$

$$v(m+1) = vT(m) + \mu e(m) \times (m) \tag{7.19}$$

Where $w(m)$ is its coefficient vector, and $y(m)$ is the output signal of the adaptive FIR filter at time m, it can be defined as:

$$v(m) = [v1(m)v2(m)\ldots vL(m)]T \tag{7.20}$$

The definition of the reference signal vector $x(m)$ is as follows:

$$x(m) = [x(m), x(m-1)\ldots x(m-L+1)]T \tag{7.21}$$

where $e(m)$ is the error signal.

$$0 < \mu < \frac{1}{L^* E\left[x(m)^2\right]} \quad (7.22)$$

Where $E\ [x(m)^2]$ is the signal strength of the input signal $x(m)$, and L is the filter length.

7.9 Filtered-x Least Mean Square Algorithm

The LMS technique assumes that the erroneous signal results from a difference between the desired signal $w(m)$ and the filtered output signal $y(m)$. The resulting signal of the adaptive filter in ANC applications has not correspond to an estimation of the purported intended signal. In these scenarios, an estimate of the required signal is generated by a dynamic system whose input is the adaptive filter's output signal. The control path or forward path is a common name for this dynamic system. This path comprises an amplifier, error microphone, antinoise loudspeaker, low pass filter, A/D converter, and an acoustic path between the latter two [17–21]. For ANC applications, a typical adaptive filter algorithm that produces an approximation of the required signal as an output signal is not defined. In this scenario, the conventional adaptive filter algorithm's filtered-x variant must be employed to modify the coefficients in order to reduce errors like the least square error or the mean square error. One of the most well-liked adaptive filtering algorithms designed for ANC applications is the filtered-x LMS (F-x LMS) method. When there is a forward path, it is possible to use the filtered-x LMS algorithm for control applications. The reference sensor signal is typically filtered using an FIR-filter estimate of the forward path by this approach, and the gradient estimate is then computed using the reference signal, which is filtered. The filtered x-LMS algorithm produces a filtered reference signal xC′(m) by the forward path, which is taken into consideration by filtering the input signal $x(m)$ by an estimation of the ahead path. The weight adjustment algorithm LMS then receives this filtered reference signal as input. The required signal $d(m)$, travels along the main physical path P. The output of the adaptive filter is obtained by filtering $x(m)$ with the adaptive FIR-filter W. This result is represented as $y(m)$. By interference in the sound of yC(m), which is output $y(m)$ filtered by the forward path, the intended signal $d(m)$ is afterwards received in the error microphone. The output $y(m)$ is the input to the canceling loudspeaker and the $e(m)$ error signal. The filtered-x LMS method is depicted in block form in Figure 7.13. The error signal is displayed as,

$$e(m) = d(m) + yC(m) \quad (7.23)$$

The filtered-x LMS algorithm's vector notation is [18–20],

$$y(m) = vT(m) \times (m) \quad (7.24)$$

$$e(m) = w(m) + yC(m) \quad (7.25)$$

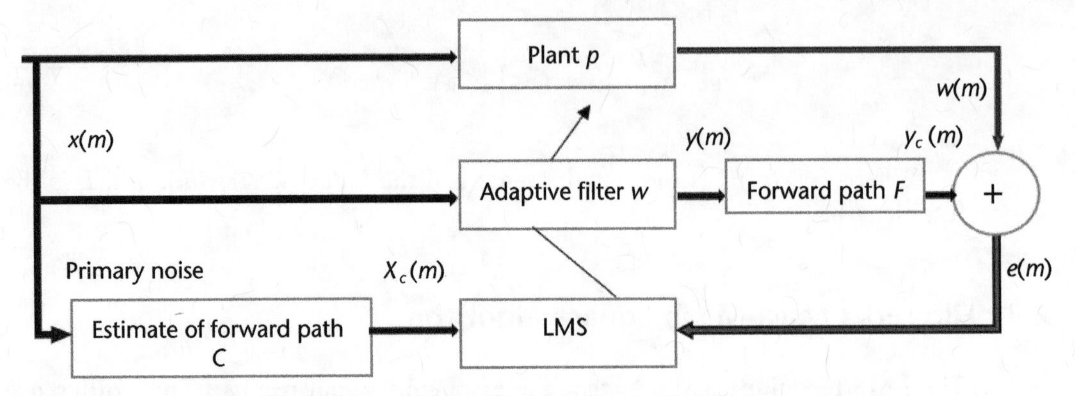

Figure 7.13 Structure of the filtered-x LMS algorithm used in the ANC system.

$$v(m+1) = v(m) - 2\mu e(m) \times C'(m) \tag{7.26}$$

The step size is chosen in accordance with [14] for this algorithm to reach mean square convergence.

$$0 < \mu < \frac{1}{E\left[x_{CF}^2(m)\right](L_W + \Delta)} \tag{7.27}$$

Here, L is the number of samples that make up the forward path's total delay and is the length of the adaptive filter.

7.10 Normalized Least Mean Square Algorithm

This algorithm is created by varying the step size of the LMS algorithm in response to the energy in the reference signal vector. It is one of the most crucial techniques for sustaining the steady state response and effective speed of convergence [17]. The LMS algorithm instability issue caused by variations in the reference signal's power is therefore resolved by the normalized LMS (NLMS) technique. The NLMS algorithm's coeffects correction is offered by:

$$w(m+1) = w(m) + \beta \frac{x(m)}{c + \mathrm{x(m)}^2} e(m) \tag{7.28}$$

Here, β is new step size, $x(m)$ is the norm, which affects the step size in a way that minimizes LMS sensitivity, and ε is a minimal real positive value that prevents division by zero in the event that $x(m)$ becomes zero. When β obeys the inequalities [17], the NLMS algorithm converges at:

$$0 < \beta < 2 \tag{7.29}$$

Figure 7.14 demonstrates a simple structure that exemplifies how the filtered-x NLMS algorithm for an ANC system operates.

The forward path of the NLMS algorithm within the context of the active noise control (ANC) system involves the filtration of the input signal $x(m)$, as illustrated in Figure 7.14, to produce a filtered reference signal $x_c'(m)$. The weight adjustment method NLMS uses this filtered reference signal as an input. The intended signal, $d(m)$, is transmitted along the main physical path P. The yield of the adaptive filter is obtained by filtering $x(m)$ with the adaptive FIR-filter W. $y(m)$ is used to represent this result. The aural interfering of $y_M(m)$, which is the loudspeaker output $y(m)$, was filtered by the forward path, and produces the intended signal $d(m)$ in the error microphone. The canceling loudspeaker and error signal $e(m)$ both take input from the output $y(m)$. The algorithm update will be represented by the following:

$$v(m+1) = v(m) + \beta \frac{x_{CF}(m)}{c + (x_{CF}(\mathrm{m}))^2} e(m) \tag{7.30}$$

7.11 Leaky Least Mean Square Algorithm

When the reference signal employed by the LMS method is not properly conditioned, bias accumulation in the adaptive filter's coefficients probably takes place [17]. As a result, the LMS algorithm is probably going to have a divergence problem [17, 18]. In order to solve the issue of a poorly conditioned reference signal and hence resolve LMS algorithm-related divergence issues, the LLMS algorithm is successful. A leakage technique is typically used to address the issue of bias accumulation in the adaptive filter coefficients. In essence, white noise is simulated using a modified adaptive algorithm that is applied to the reference signal. The coefficient adjustment for the F-x LMS algorithm is supplied by:

$$v(m+1) = v(m) + \mu e(m) \times (m) \tag{7.31}$$

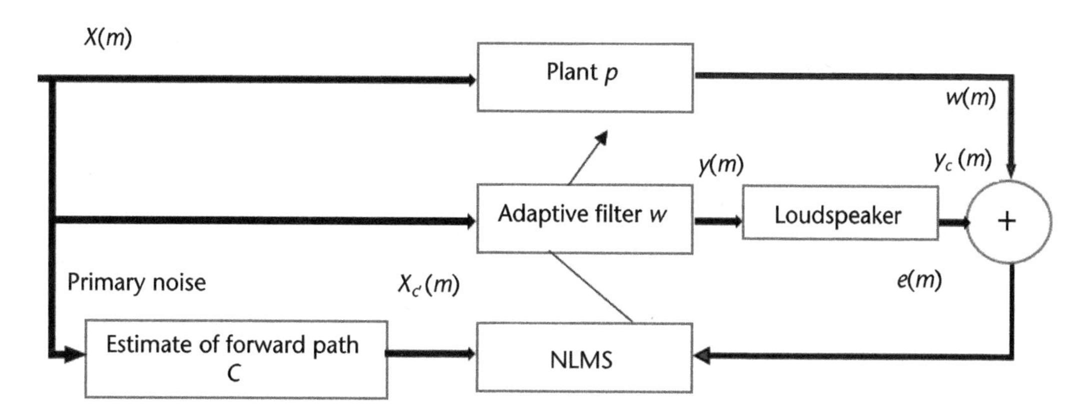

Figure 7.14 Structure of the filtered-x NLMS algorithm used in the ANC system.

where μ is the step size, $e(m)$ is the error signal, and $x(m)$ is the input signal. $v(m + 1)$ represents the value of the filter coefficient at time $(m + 1)$, which is the next iteration or time step. $v(m)$ represents the value of the filter coefficient at the current time step m. Adjusting the value of μ allows control over the rate at which the adaptive filter adapts to changes in the input signal, helping to prevent divergence issues and ensuring the algorithm's overall effectiveness. The step size μ determines the magnitude of the adjustment made to the filter coefficients during each iteration of the algorithm. It is a positive constant and plays a crucial role in balancing the convergence speed and stability of the adaptive filter. A smaller μ value results in slower convergence but increased stability, while a larger μ value may lead to faster convergence but with the risk of instability or divergence. The appropriate selection of μ depends on the specific characteristics of the system being modeled and the properties of the input signals [16]. In the generalized block diagram (Figure 7.15) of the filtered-x LLMS algorithm, the forward path initially filters the input signal $x(m)$, generating the filtered reference signal $x_c'(n)$. The LLMS weight adjustment algorithm processes this filtered reference signal. Simultaneously, the desired signal $d(m)$ travels through the primary physical path P. The adaptive filter, represented by W, filters $x(m)$ to produce the output $y(m)$. The interference from the loudspeaker, $y_c(m)$, undergoes forward path filtering and contributes to the intended signal $d(m)$ in the error microphone. Both the canceling loudspeaker and the error signal $e(m)$ receive input from the output $y(m)$.

7.12 The Exponentially Weighted Recursive Least Square Algorithm

Using a reference signal $x(m)$ and a target signal $d(m)$, the least-squares solution for the adaptive filter coefficients $w(m)$ is determined by the recursive least square (RLS) algorithm, a recursive adaptive filter at every repetition of m. The RLS algorithm offers an effective computational technique for computing the least squares results for the adaptive filter coefficients $w(m)$ in every repetition of m [18]. The RLS approach is more computationally complex than the LMS technique, but it has a reduced steady-state error and accelerates convergence [17]. The expressions in the RLS algorithm are as follows:

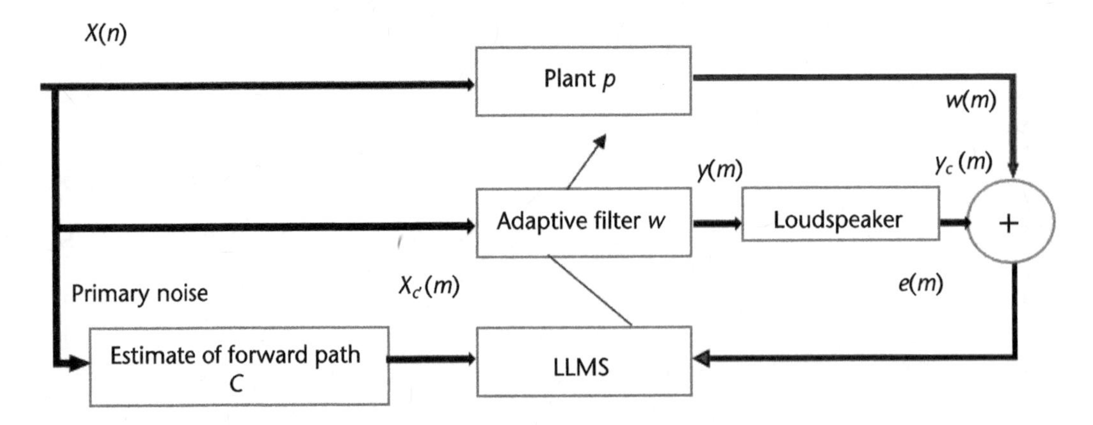

Figure 7.15 Structure of the filtered-x LLMS algorithm used in the ANC system.

$$k(m) = \frac{-1O(m-1)x'_C(m)}{x'T_C(m)^{-1}O(m-1)x'_C(m)+1} \quad (7.32)$$

where $k(m)$ denotes the gain factor, $O(m)$ is the iteratively calculated inverse of the input signal autocorrelation matrix, $x'_C(m)$ denotes the vector of the filtered input signal, filtered by C', $x'T_C(m)$ denotes the transposition of $x'_C(m)$, and $y_C(m)$ denotes the yield signal of forward path. $O(0) = I\delta{-}1$ where $w(0) = 0$ and delta is a tiny positive value.

7.13 Filtered–u Recursive Least Mean Square Algorithm

When active noise reduction is being implemented in a duct and acoustic feedback is present, infinite impulse response (IIR) filters offer a benefit. Among the many algorithms for ANC employing IIR filters is the Feintuch filtered-u recursive LMS (FuRLMS) algorithm [17]. The poles of the adaptive IIR filter eliminate the poles introduced by the auditory feedback [16, 17]. Since IIR filters' poles offer compatible qualities with a minor order structural order, less mathematical operations are required. Due to the presence of feedback, the IIR filter's poles function similarly to FIR filters but with a significantly lower order. With just a finite set of coefficients, this feedback results in an endless impulse response. IIR filters require less because of this. Computations are for FIR other than for each sample [18, 19]. In comparison to adaptive FIR filters, the convergence rate is slower, and the poles may cause instability issues. The error signal may not always be minimized at each iteration, and it might reach a local minimum [20, 21]. A basic structure of the adaptive IIR-based FuRLMS algorithm in use is shown in Figure 7.16. The yield of the FuRLMS controller is given by:

$$Y(m) = \sum_{i=0}^{N-1} a_i(m)x(m-i) + \sum_{j=1}^{M} b_i(m)y(m-j) \quad (7.33)$$

N and M are the orders of filters a and b, respectively, and $a_i(m)$ and $b_i(m)$ are the weight vectors of those filters, respectively. The residual error signal is given by:

$$e(m) = d(m) - c(m) * y(m) \quad (7.34)$$

where $c(m)$ is the filter C's impulse response. For both direct and feedback filters, the IIR filter adapts by using the same residual error. Both filters stop adjusting when the residual error is at its lowest value. At this time, filter A fully simulates the plant P, while filter B fully affects the feedback path. The feedback path and the forward path are both represented by C. After both A and B have joined, the measured fitting error $e(m)$ is calculated [21]. By using poles, the filtered-u recursive LMS algorithm does away with sound feedback, but it also has significant drawbacks. When compared to FIR filters, they have a slow convergence rate. In some situations, using a small step size results in sluggish convergence, which is undesirable [17–19].

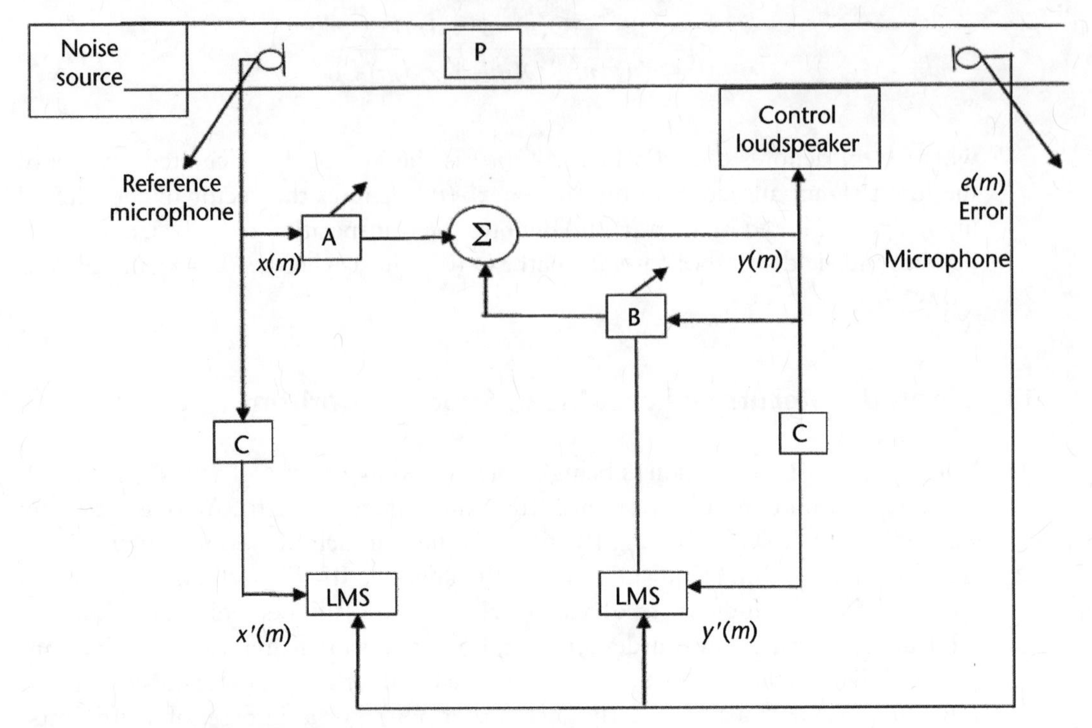

Figure 7.16 FuRLMS algorithm block diagram.

7.14 Windowing Technique for Chipless RFID

Signal processing algorithms can make use of a variety of window types, including the Hanning, Hamming, Bartlett, optimal window fractional Fourier transform, and others. We can examine their primary traits by analyzing their frequency responses. We'll examine how windows are used in spectrum analysis, where it's important to understand how well they detect closely spaced signal components. The ideal window for accurately obtaining the data from the chipless tags is also an important aspect. One of the most basic window functions is the rectangular window. Equation (7.35), where M is the window length, serves as its representation.

$$w(n) = \begin{cases} 1, 0 \leq n \leq M-1 \\ 0, \textit{otherwise} \end{cases} \tag{7.35}$$

Here, $w(n)$ is the rectangular window function of length M. The index point n is the variable that represents any position within the length of value 0 to $M - 1$ where the window function results in 1. If the value of n is outside the length 0 to $M - 1$, then the window $w(n)$ results in 0. This means that along the x-axis, the length is taken, including 0 to M, and along the y-axis, the magnitude is taken as 1. For example, if $M = 10$, then the rectangular window results in 1 from 0 to 9; for the rest of all other positions of length, the window results in zero [22–26].

Different windowing strategies such as the rectangular window, Bartlett window, Hamming window, and Hanning window can be applied at the reader end as a part of signal processing, but it is essential to study how the width of the window function influences signal detection [22–26]. To correctly retrieve the information encoded in the chipless tags, an adequate window must be determined. In MATLAB, we can analyze three basic evaluations of a window function of an information stream, as illustrated in Figure 7.17. The loss of information occurs when the window function is less than the fundamental period of a message signal, as shown in Figure 7.17(a). The use of a larger window [Figure 7.17(b)] confuses the information. Figure 7.17(c) shows the precise window function where the signal fits within the window. The major advantages associated with this window are:

1. It has a minimum stopband attenuation, which is suitable for RCS analysis as backscattered signal strength is very low.
2. The Gibbs phenomenon (due to sudden transition from 0 to 1 or 1 to 0) does not affect much in comparison to the windows as presented in Figure 4.6.
3. It is suitable for encoding multiple bits in the desired frequency bands, thereby successfully implementing the coding capability.

Algorithmic steps for rectangular windowing technique-based encoding are as follows:

1. Data extracted from the RCS response signal and stored in a table (RCS_TAB) along with the frequencies.
2. Fixing the attributes for signal processing. Here, these are "frequency" and "RCS."
3. Find the total number of instances and find their locations.
4. Find the desired frequencies (pick two particular frequencies) in between encoding to be done. Here, it is the lower frequency (F_L) and the higher frequency (F_H).
5. By padding zeros, remove the signal components except for the region between F_L and F_H [i.e., zeros (desired range)].
6. Generate an accurate rectangular window (rect) in the desired range and make the required amplitude. The window size should be five times smaller than the total signal frequency range to be processed for a better presentation of coding information.

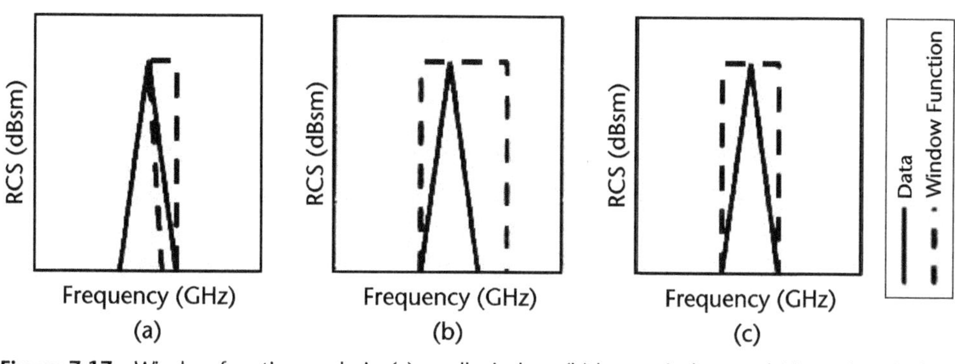

Figure 7.17 Window function analysis: (a) small window, (b) large window, and (c) precise window.

7. Pass the signal through the "rect" and observe whether it is well-fitted to the window.
8. Make a reference amplitude as per the desired measured signal level in the window to avoid signal transformations/clips.
9. Subtract (SUB) the window from the original signal (RCS). This will remove the signal component or the unwanted signal in the required rectangular window (SUB_RCS).
10. Repeat this for different frequency regions (create multiple windows) for further encoding using a for loop.
11. Adding the result window to the RCS. This will result in the desired encoded signal (CODED_RCS).

7.15 Summary

The printing technology of RFID is an important aspect of commercialization. Different resonators used in chipless RFID tags and various signal processing algorithms were the primary points of discussion in this chapter. Several coding techniques were followed in order to enhance the data handling capability of the tag. The tag noise can be reduced by multiple noise reduction techniques. This innovative approach utilizes various techniques and components to achieve robust data communication between RFID readers and chipless tags. We discussed key aspects related to chipless RFID, focusing on principles, components, encoding methods, resonators, and error correction techniques. Chipless RFID technology encodes data using diverse methods, including time domain and frequency domain approaches. Resonant elements and structures play a vital role in encoding information into the response of chipless RFID tags. Various resonators, such as hairpin resonators and other structures, are fundamental components in chipless RFID tags. These elements are carefully designed to resonate at specific frequencies, enabling the encoding of unique information based on their resonance characteristics. Error correction and mitigation strategies, such as convolutional coding and the Wiener filter, are employed to enhance data integrity, reduce noise, and optimize the decoding process. These techniques play a critical role in achieving reliable and accurate data communication. Chipless RFID technology finds applications in various domains, including inventory management, supply chain, asset tracking, and more. Ongoing research and advancements continue to enhance chipless RFID systems, making them more efficient, reliable, and suitable for a wider range of applications. The future of chipless RFID holds promising prospects, including improved encoding techniques, enhanced data capacity, integration with the IoT, and broader adoption across industries. Further research and development will drive innovation, making chipless RFID a fundamental tool in modern identification and tracking systems. Chipless RFID technology is continually evolving, offering a versatile and efficient solution for a diverse array of applications. By leveraging resonant elements, advanced encoding methods, error correction techniques, and real-world applications, chipless RFID is poised to shape the future of identification and data communication in an increasingly interconnected world. Several challenges arise during the real-time implementation of the chipless tag, which will be discussed in Chapter 9.

References

[1] Leng, T., X. Huang, K. Chang, J. Chen, M. A. Abdalla, and Z. Hu, "Graphene Nanoflakes Printed Flexible Meandered-Line Dipole Antenna on Paper Substrate for Low-Cost RFID and Sensing Applications," *IEEE Antennas and Wireless Propagation Letters*, Vol. 15, 2016, pp. 1565–1568.

[2] Rizwan, M., et al., "Possibilities of Fabricating Copper-Based RFID Tags with Photonic-Sintered Inkjet Printing and Thermal Transfer Printing," *IEEE Antennas and Wireless Propagation Letters,* Vol. 16, 2017, pp. 1828–1831.

[3] Horter, T., H. Ruehl, W. Yang, Y.-S. Chiang, K. Glaeser, and A. Zimmermann, "Image Analysis Based Evaluation of Print Quality for Inkjet Printed Structures," *Journal of Manufacturing and Material Processing*, https://doi.org/10.3390/jmmp7010020.

[4] Forouzandeh, M., and N. Karmakar, "Self-Interference Cancelation in Frequency-Domain Chipless RFID Readers," *IEEE Transactions on Microwave Theory and Techniques*, 2019, pp. 1–16.

[5] Behera, S. K., and N. C. Karmakar, "Chipless RFID Printing Technologies: A State of the Art," *IEEE Microwave Magazine*, Vol. 22, No. 6, June 2021, pp. 64–81.

[6] Mishra, D. P., and S. K. Behera, "Resonator Based Chipless RFID: A Frequency Domain Comprehensive Review," *IEEE Transactions on Instrumentation and Measurement*, Vol. 72, Art no. 5500716, 2023, pp. 1–16.

[7] Sethy, P., and S. K. Behera, "Chipless RFID Sensors for Bioimplants: A Comprehensive Review," *IEEE Microwave Magazine*, Vol. 24, No. 7, July 2023, pp. 41–60.

[8] Costa, F., S. Genovesi, and A. Monorchio, "Chipless RFIDs for Metallic Objects by Using Cross Polarization Encoding," *IEEE Transactions on Antennas and Propagation*, Vol. 62, No. 8, 2014, pp. 4402–4407.

[9] Jin-Kyu, B., C. Nak-Sun, and K. Dong-Hun, "Optimal Design of a RFID Tag Antenna Based on Plane-Wave Incidence," *IEEE Trans. Magn.*, Vol. 48, No. 2, February 2012, pp. 795–798.

[10] Kim, M.-G., "On Systematic Punctured Convolutional Codes," *IEEE Transactions on Communications,* Vol. 45, No. 2, February 1997, pp. 133–139.

[11] Forouzandeh, M., and N. Karmakar, "Compact Polarizer Chipless RFID Tag," *2020 4th Australian Microwave Symposium* (*AMS*), 2020, pp. 1–2.

[12] Dong, S., L. Han, P. Zhang, Q. Feng, and Y. Yin, "Convolution Coding and Amplitude Attenuation-Based Full Waveform Inversion," *IEEE Access*, Vol. 8, 2020, pp. 182996–183013.

[13] Lai, C.-H., and K. Kiasaleh, "Modified Viterbi Decoders for Joint Data Detection and Timing Recovery of Convolutionally Encoded PPM and OPPM Optical Signals," *IEEE Transactions on Communications,* Vol. 45, No. 1, January 1997, pp. 90–94.

[14] Tian, Y., Q. Zhang, Z. Ren, F. Wu, P. Hao, and J. Hu, "Multi-Scale Dilated Convolution Network Based Depth Estimation in Intelligent Transportation Systems," *IEEE Access*, Vol. 7, 2019, pp. 185179–185188.

[15] Kim, J. T., "Privacy and Security Issues for Healthcare System with Embedded RFID System on Internet of Things," *AdvSci Tech.*, Vol. 72, 2014 pp. 109–112.

[16] Grewal, M. S., and A. P. Andrews, "Kalman Filtering: Theory and Practice using MATLAB," 4th Ed., Hoboken, New Jersey: *John Wiley & Sons, Inc.*, 2015.

[17] Lu, Y., "Adaptive-Fuzzy Control Compensation Design for Direct Adaptive Fuzzy Control," *IEEE Transactions on Fuzzy Systems*, Vol. 26, No. 6, December 2018, pp. 3222–3231.

[18] Anderson, P. D., and M. A. Ingram, "The Performance of the Least Mean Squares Algorithm Combined with Spatial Smoothing," *IEEE Transactions on Signal Processing*, Vol. 45, No. 4, April 1997, pp. 1005–1012.

[19] Shi, D., B. Lam, W.-S. Gan, and S. Wen, "Optimal Leak Factor Selection for the Output-Constrained Leaky Filtered-Input Least Mean Square Algorithm," *IEEE Signal Processing Letters*, Vol. 26, No. 5, May 2019, pp. 670–674.

[20] Tsakiris, M. C., C. G. Lopes, and V. H. Nascimento, "An Array Recursive Least-Squares Algorithm with Generic Nonfading Regularization Matrix," *IEEE Signal Processing Letters*, Vol. 17, No. 12, December 2010, pp. 1001–1004.

[21] Mosquera, C., J. A. Gomez, and F. Perez-Gonzalez, "Filtered Error Adaptive IIR Algorithms and Their Application to Active Noise Control," *Proceedings of the 1998 IEEE International Conference on Acoustics, Speech and Signal Processing*, Vol. 3, 1998, pp. 1701–1704.

[22] Mishra, D. P., and S. K. Behera, "Multibit Coded Passive Hybrid Resonator Based RFID Transponder with Windowing Analysis," *IEEE Transactions on Instrumentation and Measurement*, Vol. 71, Art no. 8005907, 2022, pp. 1–7.

[23] Jonnalagadda, S. V. S. T., D. P. Mishra, and S. K. Behera, "Chipless RFID Sensors for Vital Signs Monitoring – A Comprehensive Review," *IEEE Microwave Magazine,* pp. 2–19, Online ISSN: 1557-9581.

[24] Mishra, D. P., T. K. Das, S. K. Behera, and N. C. Karmakar. "Modified Rectangular Resonator Based 15-Bit Chipless Radio Frequency Identification Transponder for Healthcare and Retail Applications," *International Journal of RF and Microwave Computer-Aided Engineering*, Vol. 32, No. 6, 2022, e23127.

[25] Mishra, D. P., and S. K. Behera, "Modified Rectangular Resonators Based Multi-Frequency Narrow-Band RFID Reader Antenna," *Microwave and Optical Technology Letters*, Vol. 64, No. 3, 2022, pp. 544–551.

[26] Mishra, D. P., A. Subrahmannian, and S. K. Behera, "Design of Multi-Bit Chipless RFID Tag Using Hybrid Resonator for Retail and Healthcare Applications," *Journal of Microwave Engineering & Technologies*, Vol. 8, No. 2, 2021, pp. 34–42.

CHAPTER 8

Antennas for RFID Reader Applications and Printing Techniques for Chipless RFID Tags

By making it easy to identify, track, and transmit data between things, RFID technology has transformed a number of sectors like retail, healthcare, and animal tracking. RFID tags and RFID readers are the two key elements in an RFID system. As opposed to conventional RFID tags, which communicate using ICs, chipless RFID tags take a novel approach by encoding data using specifically created materials or structures rather than application-specific ICs. Numerous applications in logistics, inventory control, identification, and other fields have now become possible for this autoidentification technology.

The function of antennas is crucial in chipless RFID technology. Antennas act as a gateway between the reader system and the tag, enabling the transmission of electromagnetic signals that convey significant information. Performance, read range, and adaptability of chipless RFID systems are greatly influenced by the design, properties, and placement of antennas. Additionally, innovations in printing methods have led to brand-new techniques for producing chipless RFID tags, offering scalable solutions that are affordable and adaptable to a variety of substrates. At the receiving end of the RFID system, the reader antenna is an essential part. A good reader indicates a wider tracking area and better reception. The various reader antenna types and the role they play in different RFID applications are covered in this chapter. Inkjet, thin-film, aerosol, flexography (flexo), screen, gravure offset, and other printing technologies are also examined as they pertain to mass manufacturing. This chapter explores a variety of antennas designed for RFID readers, analyzing their critical function in establishing reliable and efficient interactions with chipless RFID tags. Additionally, it examines the cutting-edge printing technologies used to create chipless RFID tags, highlighting the innovations that have expedited production procedures and expanded the range of possible applications.

8.1 Importance of Reader Antennas in RFID Reader Systems

The RFID reader antennas have much more importance in an RFID system. These are the components engaged in the interrogation of the tags/transponders. The extraction of data from the tag's signal is performed inside the signal processing unit by incorporating the demodulation techniques. In the case of a passive RFID tag, it is impossible to obtain a tag signal in the absence of a reader. In this scenario, the reader emits an interrogation signal first, and the tag reflects the signal in every direction, including the reader. Hence, this relationship can be compared with a master-slave scenario between the reader and the tags. In some cases, even the RFID readers can behave as a slave where the middleware software performs the work of the master. Middleware software application processes the data obtained from the RFID reader and transmits the command to the reader section. This relationship is presented in Figure 8.1.

RFID systems can be categorized into two types: passive-tag systems and active-tag systems. In the case of active tags, there is the provision of a battery, onboard amplification, and modulation of the signal, whereas passive tags do not have any battery and depend on the reader to obtain radiated energy. The energy is required in modulating and transmitting data by the passive tag. Hence, the active-tag systems are more independent in terms of reader dependency on the power supply. On the contrary, passive-tag systems require high radiation fields from the reader antennas to provide sufficient power to the tags for signal manipulation [1].

8.2 A Simple Reader Architecture

There are various types of RFID readers involved in the reading of chip-based and chipless RFID tags. In general, the basic architecture of any RFID reader consists of three components: (1) antenna, (2) RF unit, and (3) control unit [1]. The architectural layout of the RFID reader is shown in Figure 8.2.

The transmission and reception of RF signals are obtained by one or more antennas. The RF unit contains the circuits related to transmitter/receiver units. Hence, this can also be named as a transceiver unit. The output of the RF unit

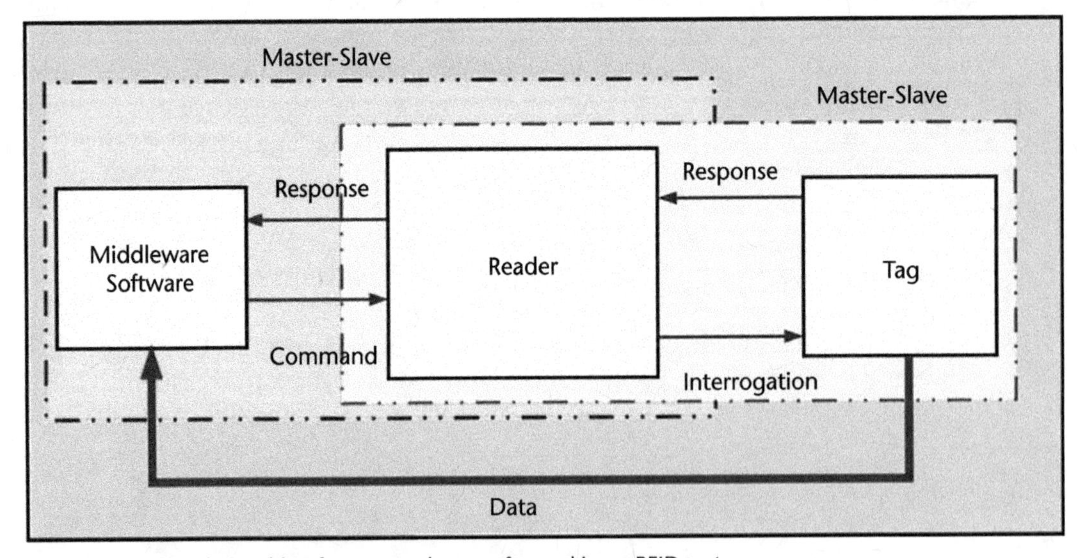

Figure 8.1 The relationship of a master-slave performed in an RFID system.

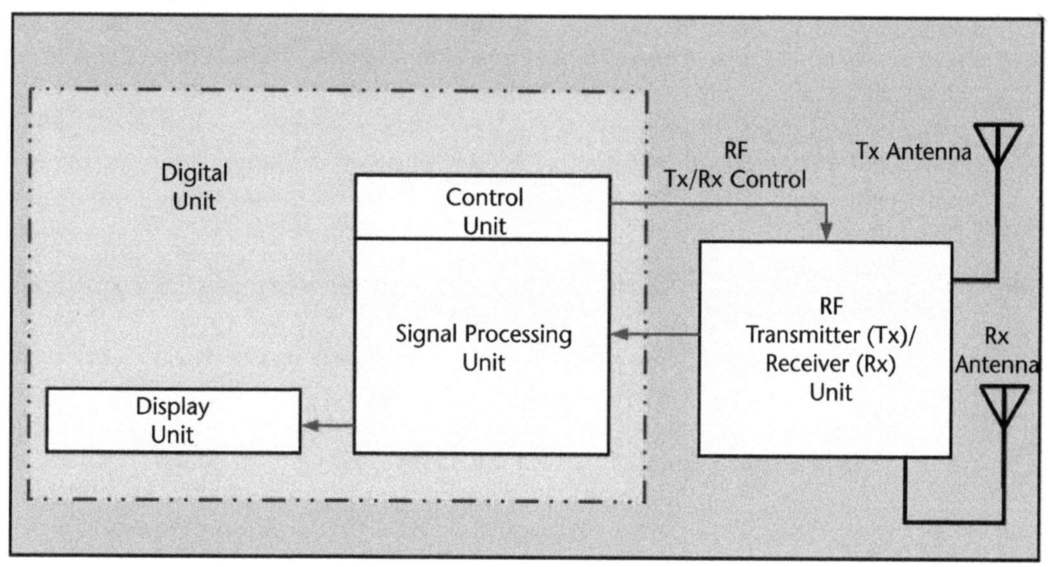

Figure 8.2 Architectural layout of RFID reader.

is fed into the control unit, which is connected to the signal processing unit. The data and signal processing are performed in this section. The control unit is usually employed in the digital unit with the aim of RF Tx/Rx unit control. This is the reason for customizing the RF unit to satisfy the required method of communication employed by the transponder.

Depending on the number of antennas used in measurement, the reader architecture can be monostatic and bistatic.

- The transmitting or receiving operations of the *monostatic reader* antenna are collocated, which refers to their placement as the same physical unit. Because it requires fewer antennas and components, the monostatic architecture is frequently utilized in applications where size or space restrictions are significant. Challenges in isolating transmitted and received signals can affect sensitivity and performance.
- The transmitting and receiving antennas are physically separated, which results in a more complicated setup in a bistatic reader design. This architecture is frequently used when a need for enhanced sensitivity or for better isolation between transmission and reception arises. It requires additional antenna hardware and careful alignment, potentially a larger setup.
- Readers of monostatic systems: Due to their small size and simplicity of integration, LP antennas might be useful in monostatic systems. CP antennas may be recommended if tag orientation is a concern in order to assure accurate reading regardless of tag orientation.
- Readers who use bistatic systems will profit from the improved separation of transmission and reception. With less interference between the two antennas, CP antennas can be especially helpful in this situation to enhance the signal quality in both broadcast and received messages.

The final decision between LP and CP antennas should consider the application environment, intended read range, varying tag orientation, and trade-offs between antenna complexity, integration, and performance. As per the types of RFID systems, in chipless systems, the absence of active modulation schemes on the tag restricts data extraction with the use of conventional communication methods. It is also important to highlight that encoding of data in chipless transponders occurs by using different techniques, which enhances the demand for extremely customized reader units in data extraction. It is crucial to design the mixer, demodulation, and LNA components in the context of reader systems when analyzing the data.

1. A mixer is a type of electrical component that mixes two or more input signals to create an output signal that contains the sum and difference frequencies of the input signals. A local oscillator (LO) signal is mixed with the incoming signals from the tag in the RFID reader systems to perform frequency translation, which changes the received signal to the appropriate frequency. It allows the conversion of high-frequency signals to lower frequencies that are easier to process and demodulate. A mixer's performance is assessed using factors such as conversion gain (the change in signal power from input to output), noise figure (how much the mixer adds to system noise overall), and linearity (how well the mixer handles various input power levels without distortion).
2. Demodulation is the process used in RFID reader systems to retrieve encoded data from backscattered or reflected signals from chipless RFID tags. Demodulation is essential for interpreting the data contained in chipless tags. It restores the modulated signal to its baseband representation, enabling the extraction and decoding of the encoded data. Important demodulation process parameters include demodulation accuracy (how accurately the original signal is recovered), noise tolerance (the capacity to withstand noise and interference without significantly affecting data quality), and modulation scheme compatibility (aligning the demodulation technique with the encoding scheme employed by chipless tags).
3. An LNA is a type of electrical amplifier that amplifies weak signals received by antennas with the least amount of noise feasible. LNAs are employed in RFID reader systems to strengthen the low strength signals coming from chipless tags or the receiving antenna. Due to distance and propagation losses, the signals received from chipless tags are frequently weak. In order to increase the overall system sensitivity, LNAs are essential for ensuring that the weak signals are amplified sufficiently before further processing. Gain (the amplification factor of an LNA), noise figure, and linearity—which measures how well an LNA retains signal quality while tolerating a range of input power levels—are some of the criteria that characterize the LNAs.

In some RFID and radar systems, the use of coherent detectors is crucial for recovering the phase information of backscattered signals reflected by the tags. For RFID applications, coherent detection makes it possible to extract the amplitude and phase information from the received signal. If phase shifts were used in the encoding process, the reader may decode the phase-encoded data and obtain the data from the reflected signals by extracting the phase information. Coherent detection, on the other hand, necessitates exact synchronization between the transmitted signal and the local oscillator and is more challenging than envelope detection (which merely considers the amplitude). As a result of this synchronization, the reference signal used for coherent detection is in line with the received signal. The chipless RFID readers can be made tag-specific, which is not useful in the presence of other tags. The reading process overview for various chipless tags can be of the time domain and frequency domain.

- *The time-domain reader antennas*: The encoding process of data is accomplished by utilizing the backscattered pulses from the passive tags. As these readers are time domain–based, the timing information of the received pulses is very essential in this case. The process includes the involvement of samplers of high speed as a signal processing element inside the digital unit [2–5].
- *The frequency-domain reader antennas*: As discussed in the previous case, here also, the encoding process depends on the reflected signals. But the frequency information is the important parameter. Hence, the reader section is focused on generating and receiving signals with a large bandwidth to cover the operating frequency of transponders [6–7]. As we have discussed previously, LP and CP antennas are the essential features in RFID reader systems depending on the applications; the following discussion is relevant in this regard.

8.3 LP and CP Antennas

The RFID reader antennas can be functional in two types depending upon the polarization, such as (1) LP reader antennas and (2) CP reader antennas. In the design of these antennas, the microstrip/planar antennas are substantial for various applications. The CP antennas require orthogonal fields, that is, having a phase difference between the field components [8]. Achieving the previous can be accomplished in a single patch antenna by two different kinds of feeding, such as (1) orthogonal feed network with a power-divider circuit (to generate phase difference) having a ground plane of larger dimension [8], or (2) single feed planar structures with different perturbations in the design [9]. The former method incurs more space whereas the latter is preferable for dimensionally small designs. In single-feed antennas, the feeding, along with the incorporated slots/slits/notches, can generate orthogonal modes, thus generating CP radiation [10]. A general idea of CP generation in different single-feed antennas used for RFID applications in various literature is given in Table 8.1. It is also essential to highlight that, in certain situations, the orientation of tags is fixed. Hence, the need for CP antennas is not very important in those situations. On the other hand, these LP antennas reduce transmitter power by 50% and also make the design simple [18–19]. A summary of various LP antennas used as RFID readers in different applications is presented in Table 8.2.

CP antennas can extract signals from chipless RFID tags regardless of their orientation. This is crucial when tags are randomly oriented or when the orientation changes dynamically. Whereas LP antennas can be optimized for efficient near-field

Table 8.1 Generation of CP in Various Single-Fed Microstrip Patch Antennas

Ref No.	*Antenna Type*	*Method of CP Generation*
[11]	Single layered, nearly square patch	Providing feed at a proper location along its diagonal line
[12]	Single layered, square MPA	Modification in the outer borders of square MPA by cutting notches
[13]	Single layered, square MPA	Incorporation of shorting posts
[14]	Single layered, square MPA	Chamfering corners
[15]	Single layered, square MPA	Modified square-MPA with stubs
[16]	Single layered, square MPA	A square MPA with a rectangular slot at the diagonal can be fed along the central axis
[17]	Single layered, square MPA	Addition of two pairs of slots having different dimensions onto the square patch and loading the coupled strips

Table 8.2 LP RFID Reader Antennas in Various Applications

Ref No.	*Antenna Type*	ε_r	f_r	*Gain*
[20]	Elliptical, single layered	4.55	900 MHz	5.5 dBi
[21]	Rectangular, multilayered	4.2	902 MHz and 2.4 GHz	1.5 dBi and 6 dBi
[22]	Circular, multilayered	3.5	923 MHz	1.4 dBi
[23]	Circular, single layered	2.45	2.43 GHz and 5.8 GHz	5.6 dBi and 7.5 dBi
[24]	Rectangular, single layered	4.4	5.8 GHz	5.1 dBi

operation, in chipless RFID systems where the distance between the reader and tag is short, LP antennas can provide reliable communication in close proximity. CP antennas are beneficial for accurate tag localization and tracking due to their ability to receive signals from different polarizations, allowing better estimation of tag positions. Whereas LP antennas might be less susceptible to certain types of interference to capture signals from various directions, these antennas can experience signal nulls when the tag orientation aligns with the polarization direction. CP antennas, with their circular polarization, experience fewer nulls, resulting in more reliable communication. In this regard, for compactness and wider bandwidth, fractal antennas are also used, which is discussed in the later sections.

8.4 Fractal Antennas

Self-replicating geometries are used to develop fractal antennas, which result in structures with complicated shapes. These antennas provide a number of benefits that can be helpful for chipless RFID systems.

1. *Compact size and miniaturization*: Fractal antennas can be made to have a small footprint and still operate effectively. Fractal antennas can assist in satisfying size restrictions in chipless RFID systems where integration into small devices or locations with limited space is essential. The self-replicating nature of fractal geometries allows for conformal designs that can be adapted to irregular surfaces. This can be useful when integrating antennas

into unconventional shapes or structures. Fractal antennas can be used to miniaturize the overall reader system, which can be beneficial for handheld or portable chipless RFID readers.

2. *Multiband and wideband*: Fractal antennas exhibit multiband and wideband characteristics. This means they can operate across multiple frequency bands or cover a wide range of frequencies, making them suitable for chipless RFID systems that might involve various frequency bands for tag interrogation.
3. *Frequency selectivity*: Fractal antennas can be designed to have frequency-selective properties, which is advantageous for chipless RFID systems that require specific frequency responses for decoding tag responses.
4. *Wide angle of coverage*: Some fractal antennas can offer wide-angle coverage, making them suitable for environments where tags may have varying orientations or positions relative to the reader.
5. *Reduced mutual coupling*: Fractal antennas can be designed to reduce mutual coupling effects when multiple antennas are used in close proximity. This can help maintain the isolation between antennas in an array, improving system performance.

It's important to note that fractal antennas also come with challenges like the complex geometry of fractal antennas, which can make their design, analysis, and optimization more intricate. Additionally, their performance can be sensitive to manufacturing tolerances and material properties. Therefore, careful design and testing are required to ensure that the fractal antenna meets the desired specifications in the chipless RFID reader system. Fractal shapes repeat a similar structure to obtain greater performance. When the fractal shapes are incorporated in antennas, they provide many features such as space-filling, periodicity, or self-similarity, which help in making the design compact and achieving better performance in terms of multiband and multimodal response [25–27]. In many cases, the improvement in antenna performance does not properly correlate with the attributes of the fractal geometry. Hence, the development of a novel fractal antenna requires the proper conceptualization and mathematical formulation of a suitable fractal shape utilizing the features of the fractal. There is a vast range of applications involving different fractal antennas to obtain better performances. In Section 8.4.1, some of the antenna applications related to Koch fractal geometry in RFID are highlighted.

8.4.1 Koch Fractal Antenna

Nowadays, RFID technology has a growing market in various sectors such as hospitals, warehouses, supermarkets, inventory control, healthcare, or access control [1]. The RFID system comprises of a reader/interrogator, tag, and middleware software. The tags are basically attached to the objects to be identified and are oriented in a random fashion. Therefore, circularly polarized reader antennas are essential for efficient tag reading [1]. In [28], a circularly polarized MPA with Koch fractal geometry was introduced for the UHF RFID system. The design consists of two asymmetrical Koch fractal geometry with four arrow-shaped slots. The fractal antenna resonates at 911 MHz with an impedance bandwidth (IBW) of 37.0 MHz (891.0 to 928.0 MHz) and axial ratio bandwidth (ARBW) of 8 MHz (907.0 to 915 MHz), respectively. Similarly, a compact square-shaped CP fractal antenna using Koch-fractal at the radiating edge was presented in [29] as an RFID reader at the ISM band. In [29], two antenna prototypes were fabricated, operating at two frequency bands at 2.43 GHz (antenna design 1) and 5.78 GHz (antenna design 2). The measurement results confirm 3 dB ARBW covering 2.432 to 2.439 GHz and 5.750 to 5.80 GHz for antenna designs 1 and 2, respectively. It was also evaluated that the lower-band antenna has a peak realized gain of 6.9 dBi at 2.43 GHz, whereas the upper-band antenna depicts a gain of 6.15 dBi at 5.78 GHz. The design in [30] is a coplanar waveguide (CPW) fed fractal monopole antenna developed for short-range RFID reading applications with a maximum reading distance of 1.32m. The antenna has an impedance bandwidth of 35.7% at the resonant frequency of 910 MHz and 3 dB ARBW of 4%, varying from 900 to 936 MHz.

8.5 Reader Antennas for Biomedical Applications

There is a heavy demand in the healthcare sector because of the increasing population, which consequently increases the number of patients. Hence, it enhances the issues associated with the medical care facilities. RFID can be a potential technology for identification, tracking, and monitoring. RFID can provide an opportunity in the healthcare sector for cost-saving, efficient operation, and safety through proper tagging of patients, assets, and inventory. One of the serious medical issues (i.e., sleep disorder) has been investigated, and the threat to life is highlighted. Critical diseases like the ones mentioned previously require continuous health monitoring and proper medical care. To provide ease in medication, wireless monitoring technology has been widely researched in recent times [23]. Various wireless technologies such as Bluetooth, Wi-Fi, ZigBee, and RFID are mostly utilized for the abovementioned purpose. The groups of researchers across the globe are mostly working in some of the unlicensed frequency bands, such as 900 MHz, 2.4 GHz, and 5.8 GHz, for the abovementioned purpose. To monitor and track all the vital data of sleep apnea patients by the implementation of a wireless device is also proposed in [23]. The major design criteria are focused on maximum data capturing capability, portability, and providing comfort to the patient. The various applications of RFID technology in the healthcare sector are highlighted in Figure 8.3. With the mentioned advantageous features in a wide range of applications (i.e., identification, tracking, monitoring, and providing drug compliance), RFID technology is a potential stakeholder in healthcare.

8.6 Antenna Arrays

Antenna arrays can offer spatial diversity, which enables the system to aggregate signals from many antennas to improve sensitivity and increase the read range. This is especially useful in chipless RFID systems where tags may be dispersed in wide areas and have different orientations. An antenna array can be utilized to produce directional radiation patterns, which enables the reader system to direct its energy

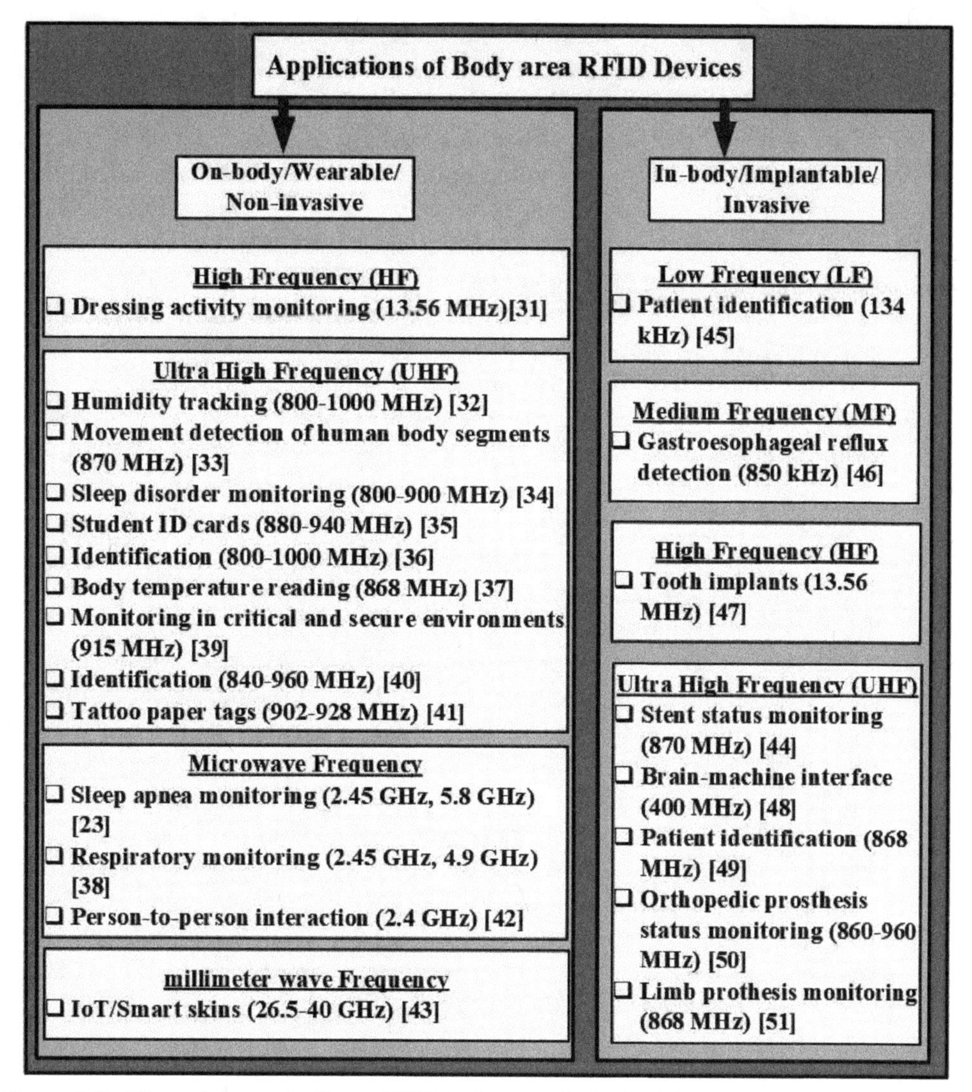

Figure 8.3 The various applications of RFID technology in the healthcare sector.

and reception in particular directions. This may aid in increasing selectivity and lowering noise or signal interference. The primary lobe of the radiation pattern can be directed in a certain direction using the beamforming technique using antenna arrays. This may be useful for focusing on particular tag locations or for following moving tags inside a coverage region. In order to decrease interference from particular directions and improve the signal-to-noise ratio, antenna arrays can be employed to spatially filter signals. This is especially crucial in settings with a lot of electromagnetic interference. Antenna arrays can help with the precise localization of chipless RFID tag placements. The system can determine the position of the tag with respect to the reader by examining the received signal levels from various antennas in the array. Signal fading and distortion in indoor situations can be brought on by multipath reflections. These effects can be reduced by using antenna arrays with the proper beamforming or diversity techniques, resulting in more accurate tag identification. The use of antenna arrays in chipless RFID reader systems can considerably improve the system's capacity to recognize, locate, and communicate with chipless RFID tags. To achieve the intended performance enhancements, antenna array design and implementation must pay close attention to variables such as array shape, spacing, beamforming algorithms, and calibration procedure.

The chipless RFID system makes use of the backscattered signal that consists of distinct spectral signatures. The read range of the chipless RFID system is confined to a few centimeters since the UWB signal transmitted has a regulation on the transmit power. The read range is a dependent function of the tag's RCS, the reader antenna performance, and the sensitivity of the reader. For a typical set of tags and a reader, the read range is limited due to the antenna gain, angular beamwidth for covering the reading zone, and the antenna matching condition [52]. In a chipless RFID system, the reader transmits a UWB signal of uniform amplitude, which powers the tag. The tag reflects the transmitted UWB signal with distinct spectral signatures (resonances/dissonances) to the reader. The identification of the tag is done at the reader side after the reflected signal is received by the reader antenna. The reader processes the received signal, and tag identification will be done accordingly. Since the UWB reader property is crucial for the smooth operation of chipless RFID systems, the reader antennas designed for deployment in these RFID systems should have high gain and wider bandwidth with less return loss. In the near field identification of the tag, it is more important to illuminate the entire read range uniformly, which is attained with the help of array synthesis and beam shaping. Designing reader antennas for chipless RFID applications, which have to provide high gain over ultrawideband, is a major challenge [52].

The single antenna elements (Figure 8.4) generally provide a wide beamwidth with low gain [9]. Moreover, in most of the applications, the requirements of high gain and a directive pattern for long-distance communication are necessary, which can be achieved by designing the antenna array. The arrangement of the antenna array can be done in a 1D linear, 2D, or nonuniform fashion. The excitation

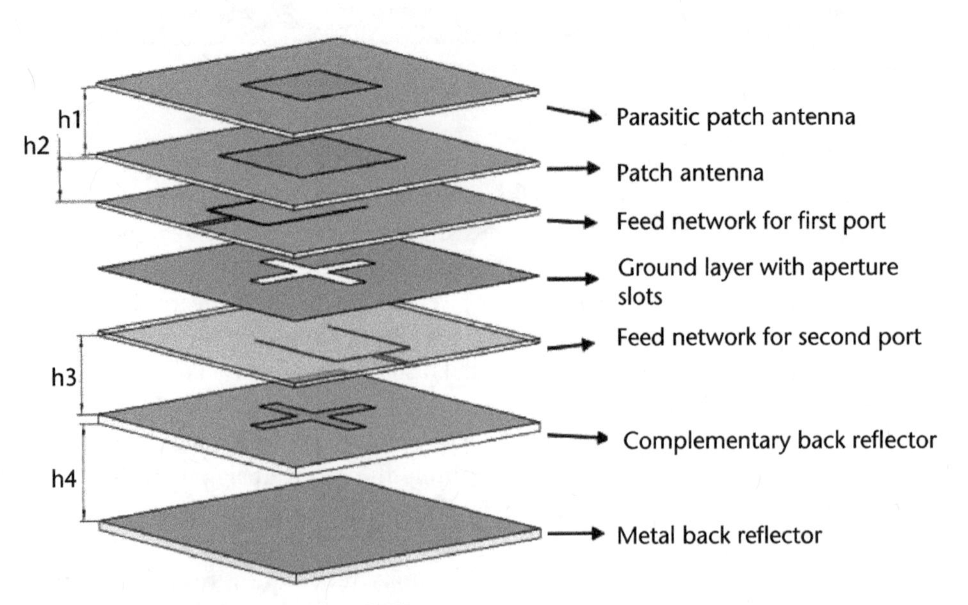

Figure 8.4 Single antenna structure [52].

coefficients and spacing of antenna elements are the two crucial factors that decide the radiation pattern of the antenna array. The effective radiation pattern is a multiplication of the radiation pattern of the single antenna element and the array factor. In [52], an array of modified dual polarized aperture coupled stacked microstrip patch antennas of high gain is proposed. To overcome the discussed challenges in antenna design for the detection of chipless RFID tags, an antenna array of 8 × 8 elements with high gain, high front-to-back ratio, dual polarized feature, and planar structure was successfully designed, fabricated, and measured. In the case that the interelement spacing is greater than or equal to λ, multiple maxima, which are called as grating lobes, will occur [53]. By utilizing the designed single element and implementing the array synthesis theory, an 8 × 8-element array with ultrawide bandwidth and a high gain was synthesized (Figure 8.5). The arrangement of this array antenna is symmetric with uniform excitation.

In applications like smart healthcare, household, and inventory management, reader antennas that are capable of steering radiation beams both electronically and dynamically will be an added advantage. Since modern chipless RFID tags have less sensitivity, this supports them to operate at less power levels and also eliminates the requirement of LOS. Due to the multipath effect produced in the read range, passive RFID tags can be activated. For interference management in wireless networks, employing reconfigurable antennas is a proven and viable option [54, 55]. Similar to that of reconfigurable antennas, the usage of a reconfigurable RFID reader antenna in a chipless RFID system improves the overall coverage area and aids in the movements of the tags that are attached to the targets.

Usually, RFID reader antennas exhibit high gain and front-to-back ratio in their radiation pattern. As a result of the factors mentioned earlier, passive RFID tags placed behind the antenna suffer from no illumination/power. The other important requirement of a reader antenna in the UHF band (902 to 928 MHz) is circular polarization with a good impedance match [56]. For longer read ranges and wider coverage areas, high gain is crucial. A polarization mismatch between the reader antenna (circularly polarized) and tag (linearly polarized) is found since most of the passive RFID tags are linearly polarized, and there is a path loss of 3 dB. In commercial applications, an RFID reader antenna with a maximum gain of 6 dBi or higher is preferred. Mutual interference between multiple readers and multipath fading effects can be minimized by using reader antennas with high gain. Also, an

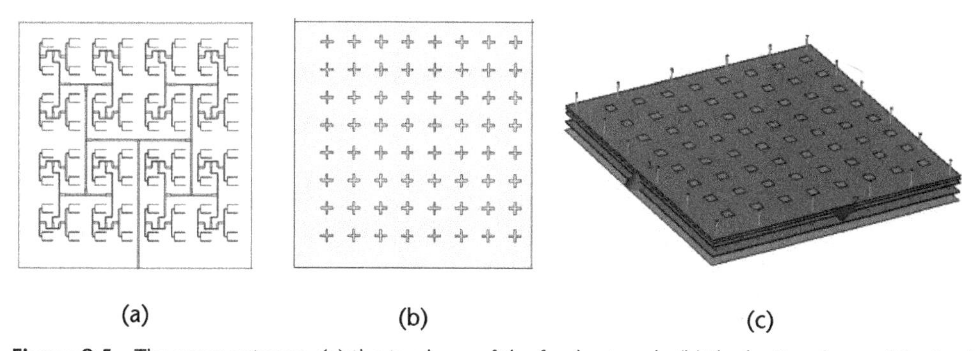

Figure 8.5 The array antenna: (a) the top layer of the feed network, (b) the bottom layer of the feed network (ground with cross-shaped slots), and (c) the array structure [52].

important point to be noted is that a high-gain antenna should be oriented in such a way that its directive beam pattern is along the line of tracking of tags. According to the stipulated regulations of the FCC (Federal Communications Commission), the maximum EIRP that is radiated by an RFID reader antenna must not exceed 36 dBm [57]. In other words, the effective power radiated from the reader antenna, which is the combination of input power and maximum antenna gain, should remain less than 36 dBm.

In [58], a new reconfigurable RFID reader antenna design that operates in the UHF band is demonstrated. The designed antenna array comprises four similar antenna elements (Figure 8.6, Figure 8.7, Figure 8.8). Some of the features of the reader antenna design include (1) good impedance match in the UHF band, (2) high gain with the electronically steering capability of reconfigurable beam pattern, which can track and identify tags in a dynamic environment with wider coverage for on-body applications, (3) circularly polarization for linearly polarized tags as well as for non-LOS applications of RFID, and (4) low cost due to the use of FR4 epoxy substrate instead of RT Duroid [59].

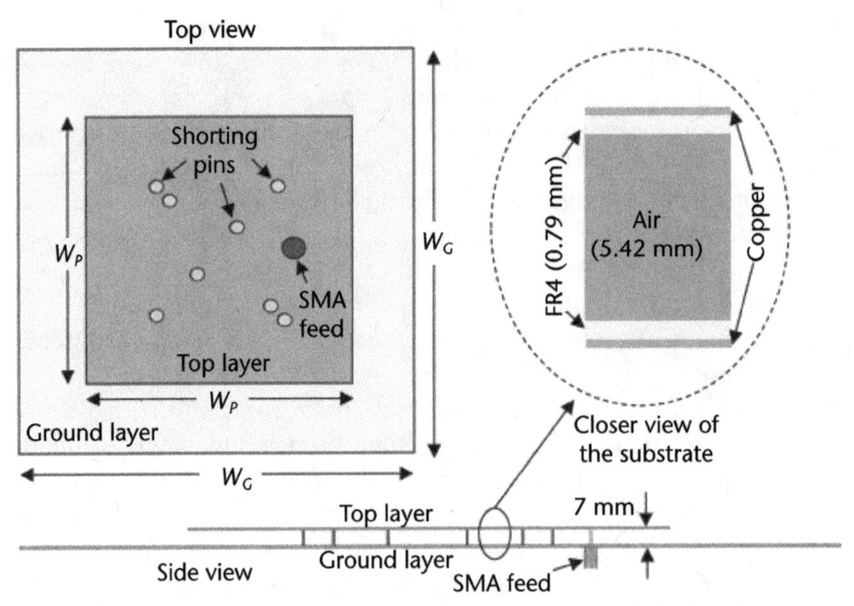

Figure 8.6 Top and side view of reconfigurable reader antenna array [58].

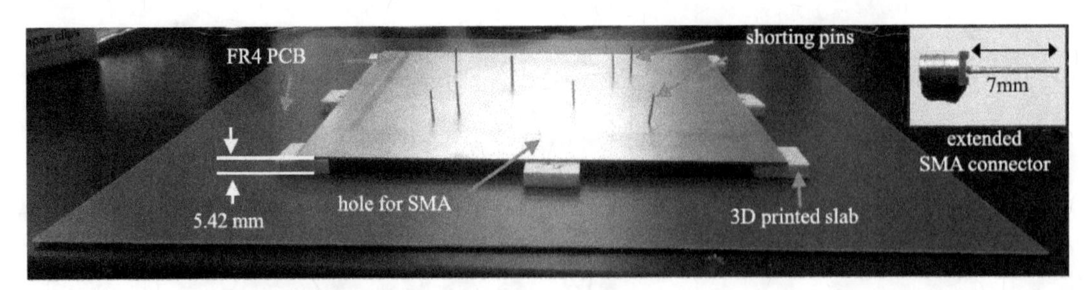

Figure 8.7 Single antenna during fabrication [58].

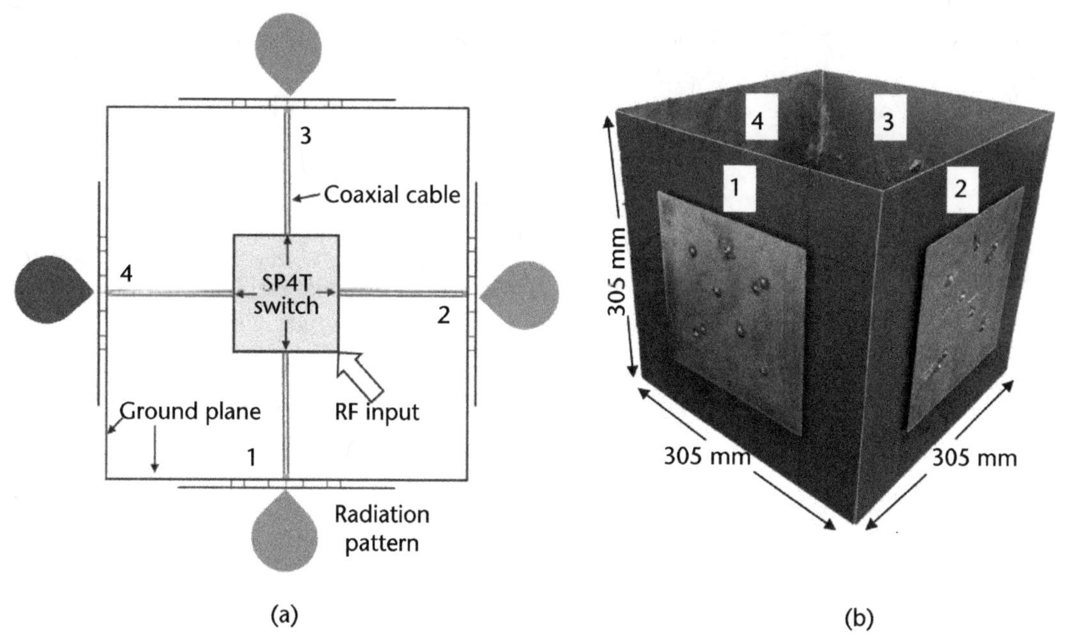

Figure 8.8 (a) Top view of the reconfigurable reader antenna system (antenna-4 is radiating, while the remaining three antennas are idle); (b) isometric view of the reconfigurable reader antenna system [58].

8.7 Limitations in RFID Applications

Out of both the technologies (i.e., the chip-based and chipless), the latter gained much research interest and marketplace owing to the low-cost and sustainability feature in different forms of usage. There are still limitations present in the case of chipless tags in certain areas, as presented in Sections 8.7.1 to 8.7.5.

8.7.1 The Read Range or Operating Range

- This is one important parameter in the design of an RFID system. If we compare the barcodes with RFID, the latter technology is more demanding with its non-line-of-sight (NLOS) characteristics, least interference by human beings, and longer read-range value [60].
- As discussed prior, the chip-based systems contain the transmitter in the onboard chip, which helps in the effective management of achieving larger read-range value. This value is quite large in comparison to the optical barcodes.
- But, the main challenge occurs while dealing with the chipless tags because of their passive nature and nonpresence of signal sources.
- Along with this, the design also has to tackle other possible sources of interference obstructing the path of obtaining a higher read-range, such as multipath transmission, or multiple reflections.

8.7.2 The Cost

- Generation of explicit EMS is an essential objective in the reading operation of a chipless tag. In this aspect, the reader section becomes overburdened due to the absence of an onboard chip. It results in a more complex reader architecture owing to the extra responsibility of signal processing.
- The cost of the reader unit is dependent on the reader components and its architecture. Hence, the cost is an essential limitation of the reader unit for chipless systems. For example, in RFID systems implementing the synthetic aperture radar method [60], the reader incurs more cost even though the passive tag has a very low budget.
- Therefore, the individual component cost can be a very important concern in the design of an RFID system.

8.7.3 The Orientation of Tag

- As discussed in Sections 8.7.1 and 8.7.2, along with cost and read range, the tag orientation also plays an essential role in the process of reading by chipless RFID readers. The reader antennas are playing a major role in this scenario.
- For example, in the case of chipless frequency signature-based RFID tags, the polarization of the transmitting (Tx) and receiving (Rx) antenna present on the tag should be matched to the reader antenna [61, 62] in order to have maximum transfer of radiation power. Similarly, for chipless retransmission-based RFID tags, the Tx and Rx antennas of the tag are placed in a cross-polarized fashion, which is also affected by the polarization of the reader antenna.
- Therefore, the passive tag orientation is always a vital parameter in RFID reader system design.

8.7.4 The Reading Rate

- The rate of reading or speed is another essential parameter that limits the RFID applications and is to be considered for the reader designs.
- The chipless RFID systems that are dependent on unique frequency signatures for encoding of data require a wide operating band [60, 63–65]. In this case, the generation of RF signal can be done by various methods such as usage of an oscillator with wide bandwidth to generate sweep signal, or generation of narrow ultrawideband pulses with short RF pulses [66–68].
- The abovementioned wide operating band signal generation also has to go through the major limitation of being more time-consuming in the former case and the need for high-speed ADC in the latter case. In any of the cases, ultimately, the process is more complex and time-consuming.

8.7.5 The Multiple-Tag Reading

- The backscattering process is automatic in the case of passive RFID tags due to the absence of any control over the reception and also retransmission of the RF signal.
- This creates the issue in the presence of multiple tags placed nearby, in which the collision occurs between the interrogating signal and the backscattered signals.
- The abovementioned problem again incurs more cost in terms of complexity to the reader section by incorporating the anticollision algorithm [69].

8.8 Chipless RFID Tag Printing and Various Parametric Effects on Printing

In recent days, RFID has become a prominent technology in the field of auto-ID and sensing. Owing to the increased demand and scope of the RFID technology, it is of primary concern to bring out cost-effective and economical solutions into the market in the field of this auto-ID technology. Also, in mass production, cost-effective solutions provide a clear edge over other identification technologies, and there will be a greater scope to replace those current auto-ID technologies. Features like NLOS, ease of fabrication, flexibility, and wider operating frequency range over greater temperature ranges are predominant as compared to the existing auto-ID technologies. As a matter of fact, the fabrication cost of a single chipless RFID tag has to be reduced below $0.01, so that a lot of tags can be utilized for tagging the most inexpensive items in the commercial arena.

Therefore, considering the cost, the objectives that were discussed for the chipless RFID tag printing, the factors that influence the real-time operation of chipless RFID, are bit density (or capacity), spectral signature in the desired frequency range, and EM characterization of the material in the working environment. EM characterization of material includes dielectric constant ε_r (or relative permittivity), loss tangent ($\tan \delta$), and substrate thickness (h). Ink conductivity is another factor that will affect the EM characteristics of the backscattered signal. The abovementioned factors will affect the proper working of chipless RFID technology. Chip-based tags that are capable of encoding multibits are utilized for identifying the costly products. But the presence of an ASIC chip increases the per tag cost. Producing high bit–density tags with compact size is a crucial factor in deciding the cost of the chipless RFID system. EM characterization of materials is also an important factor that has to be kept in mind while designing the chipless RFID system. Based on the EM characterization, the spectral signature of the tag will be determined. A flow diagram is presented in Figure 8.9 and is related to the effect of different parameters on chipless printing. The parameters in Figure 8.9 are profoundly discussed in the following paragraphs.

Ink Conductivity

To produce resonance at any frequency, some amount of conductive material has to be placed on the substrate of thickness (h). Different conductive materials can

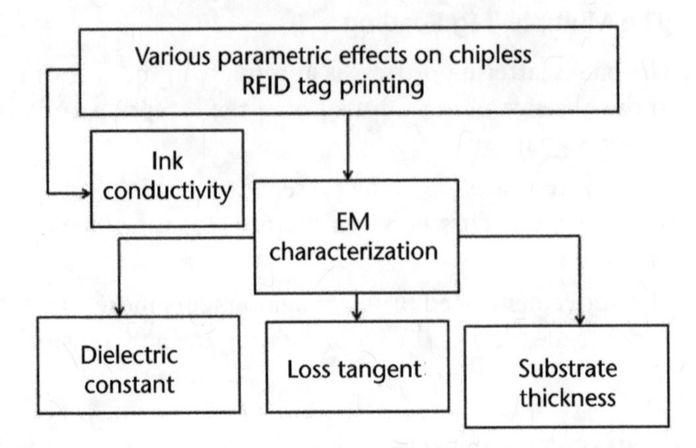

Figure 8.9 Representation of effects of various parameters on chipless RFID tag printing.

be used depending on the EM characterization. For the printing of tags using any conducting material, suitability of the conducting material, like conformability for wearable applications and frequency resonance at 2.4 GHz ISM band for healthcare applications, are some of the benchmarks that are to be addressed by the material used for printing chipless RFID tags. The conductive ink is printed in different shapes (also called resonators) on the substrate for producing the resonant frequencies. These conductive inks that are used for printing mainly comprise three ingredients: (1) silver (Ag) (45% to 65%), (2) a polymer (4% to 6%), and (3) a solvent (5.1% to 30%). Increasing the area of conductive ink on the substrate gradually improves the conductivity in addition to driving the cost to a maximum extent. There are other alternative printing techniques like screen, flexo, and gravure that are available in the market, which decrease the price of printing without compromising the ink conductivity. High-conductive inks are generally used by researchers to manufacture their tags, while in commercial applications, low-conductive inks are used. Researchers use copper conductive ink with a conductivity of 5.8×10^7 S/m while simulating their tag structures and in fabrication processes. There are many other conductive inks that are available in the market with conductivity less than the value mentioned. To clearly understand the effect of ink conductivity on the tags' response in terms of resonant bandwidth (BW), frequency (GHz), and notch depth (dBsm) at resonance are presented in [70]. The tag is simulated with different values of conductivity ranging from perfect electric conductor (PEC) to a metal with conductivity as low as 10 S/m. There will be an impact of ink conductivity on the resonant bandwidth (BW), frequency, and notch depth at the resonant frequency. It is to be observed that by improving the ink conductivity, this causes the BW to decrease in the conductivity range of 100 S/m to 10^4 S/m. Beyond 10^4 S/m, the BW remains constant, and there is no effect of increasing the ink conductivity. There is a drop of 1 GHz in BW in the range abovementioned. Similarly, there is a considerable effect on the resonant frequency when ink conductivity is increased. There is a rightward shift in the resonant frequency (f_r) when ink conductivity is increased. A notch depth of 40 dBsm is observed for PEC material, whereas it is decreased up to 10 dBsm for conductive material of conductivity 10^4 S/m and it further reduces if conductivity is decreased. It can be inferred that a low conductive material ($\sigma = 10^4$ S/m) is suitable for chipless RFID printing tags for the frequency domain.

Dielectric Constant

Dielectric constant plays an important role in determining the tag's response as it has a direct effect on resonant frequency. Different substrates are available in the software and by simulating the tag we can come to know the effect of permittivity on resonant frequency. Permittivity value differs from material to material and is crucial for detecting and sensing purposes in the field of healthcare and identification. Proper substrate characterization is important before RFID tag printing, and EM characterization has to be compared with the simulated ones for better real-time applicability. As low-cost substrates like paper and plastic materials have dielectric constant values in the range of 2 to 4, the simulation studies are carried out from 2 to 4 [70]. The resonant frequency decreases gradually while increasing relative permittivity; also, BW decreases gradually. Similarly, notch depth decreases while increasing relative permittivity from 2 to 4. Notch depth decreases from 28 dBsm to 18 dBsm in the range of 2 to 4. So, to decrease the cost of tag printing, low-cost substrates are to be selected accordingly, and aligned with the suitability for the applications desired. It is also inferred from the discussion that materials having low relative permittivity values are achieving good BW and notch depths at resonant frequencies.

Loss Tangent

Loss tangent explains the effect of losses within the dielectric materials. As losses hinder the proper functionality of the RFID tag, a study of loss tangent has to be done with respect to the tag's response in the real-time environment. Loss tangent ($\tan\delta$) degrades the tag performance. It tells about the quantity of current above the substrate variations relating to the changes in losses of the dielectric. Studies are carried out on a tag structure that is simulated by changing $\tan\delta$ values from 0.002 to 0.2 to study the effects of loss tangent on tag properties [70]. The BW improves from 0.4 GHz to 1 GHz while changing $\tan\delta$ values from 0.002 to 0.20. It is also an important point to note that resonant frequency in the range of 5.3 to 10.6 GHz is unaltered when the loss tangent is varied from 0.002 to 0.2. In the same way, notch depth is decreased substantially from 35 dB to 10 dB and further when the $\tan\delta$ value is varied from 0.002 to 0.2. It can be concluded that for attaining a considerable notch depth with smaller bandwidth, a lesser value of loss tangent is required. In an alternative way to get good notch depth, a dielectric material with a dielectric constant of 0.1 or higher is to be employed.

Substrate Thickness

The substrate thickness (h) of a tag plays an important role in the designing and fabrication of chipless RFID tags. Different materials have different thickness values. There is no hard rule that a particular thickness of substrate has to be used everywhere. Depending on the application and the area of employment, the tag thickness has to be maintained accordingly. For example, conformability can be achieved with low thick substrates that can be used in wearable applications. Varying substrate thickness implies a change in the EM characteristics of the tag response. It is to be noted that there is a slight variation in BW and resonant frequency when substrate thickness varies from 0.01 to 0.5 mm. Also, notch depth is reduced from 37 dBsm to 33 dBsm when substrate thickness varies from 0.01 to 0.1 mm. After 0.1 mm, notch depth is not varied until 0.3 mm and again decreased from 33 dBsm to 29 dBsm from 0.3 mm to 0.5 mm.

8.8.1 Chipless RFID Printing Technologies

It is also an important task to observe the printing processes and their outcomes in terms of the tags' performances in addition to the EM characterization. Printing processes also add up to the production cost of tags. It can be additive or subtractive (i.e., the undesired portions of the patterns on film materials are etched out by the photolithography process). Additive manufacturing includes binder jetting, material extrusion, power bed fusion, photo polymerization, sheet lamination, and directed energy deposition. Additive manufacturing can be used for 3D printing and brush painting. Since errors in printing affect the tags' performances, it is of greater importance to choose cost-effective, simple, and accurate printing processes for the fabrication of chipless RFID tags. For example, the electronic printing of chipless RFID can offer greater advantages like conformability (or flexibility), economics, robustness, and environmental friendliness. Different printing techniques have different advantages and disadvantages; hence, printing processes are to be chosen according to the type of desired application. The ink or conductive material that is printed on the substrate is generally a component of materials like silver or copper, which are used in a definite proportion. Recently, several conductive inks were produced that were used in the printing of semiconductor devices. It is very helpful to examine the rheological properties of the conductivity material used for printing. In addition to these rheological properties, the cohesive force of the ink on the substrate is a crucial factor in determining the printing capability. In Sections 8.8.1.1 to 8.8.1.12, different RFID tag printing processes are discussed.

8.8.1.1 Inkjet Printing

Features like digital processing and noncontact patterning can be achieved with this inkjet printing. The outstanding features of this printing technique are the nonrequirement of a pre-embedded master printing plate and substantial pressure on the sublayer. Inkjet printing can place conductive materials and semiconductor polymers that cannot be placed by traditional vacuum deposition techniques. Technical features of this printing technique are in line with the simple configuration, compact batch size, and mass production. It has received more attention due to its adaptability to many research areas that include mechanical and chemical engineering, and many more branches of science. Inkjet printing is much less time-consuming and economical when compared to other additive manufacturing techniques like 3D printing and brush painting. The other attractive feature of this inkjet printing technique is its ability to print complex patterns that are not fabricated manually. Sensors that are printed with the inkjet printing process are presented in Figure 8.10.

The ink that is to be deposited on the substrate can be soluble liquids, dispersions of nanoparticles, and mixtures (or blends). Inkjet printing can be broadly categorized into two types. They are (1) continuous ink jet printing and (2)

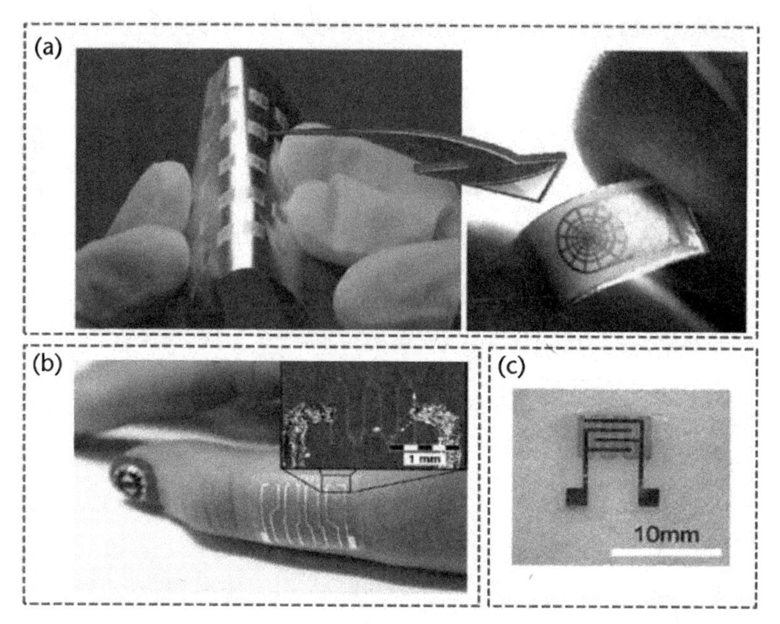

Figure 8.10 Inkjet printed: (a) proximity sensor, (b) temperature sensor array, and (c) temperature sensor attached to human skin [71].

drop-on-demand (DOD)-based inkjet printing. In the DOD-based method, the conductive ink is deposited on the substrate in a well-organized pattern. It only discharges the required drops that are to be deposited on the substrate for printing.

8.8.1.2 Paper Substrate–Based Inkjet Printing

Inkjet printing is compatible with many organic materials like liquid crystal polymer, paper, and wood. The high demand for inkjet printing in the industry is because of its quick prototyping and just-in-time manufacturing. It can quickly manufacture devices without photolithographic masking iterations. The motive behind using paper-based substrates is the benefits of paper materials at microwave frequencies with respect to electrical and dielectric performances in the UHF range up to 1 GHz. Print-on-slope is another powerful additive manufacturing technique by which small chips can be adhered onto thin substrates. An RFID chip can be manufactured that is based on a combination of silicon-on-flex and print-on-slope principles. The tag is flexible and conformable on any complex surface with any shape and size. So, the tag is a good solution for the on-body wearable applications.

8.8.1.3 Wood Substrate–Based Inkjet Printing

Tracking and identification of wooden materials for a long time is one of the exciting research fields. Externally attached coded labels may be lost on the wooden materials, which makes the identification more complex. Hence, embedded coding of labels is another alternative approach for wooden materials. Also, the coded labels are not affected by the environmental variations and provide accurate identification results. If the tags are embedded in the wood, they can prevail throughout the life cycle of wood from forest to finished product. This technology was already applied by several researchers for wood and plywood products. Wood is a powerful substrate for electronic applications. Longevity and greater strength-to-weight ratio are the additional features of wood-based substrates. Wood is a lossy substrate, and the loss is due to the moisture content present in the wooden material. Artificial plywood is an excellent alternative to mitigate the effects produced by the moisture content on wood-based substrates. Unlike natural plywood, artificial plywood doesn't absorb moisture. Thus, tags that are printed on or embedded in thin plywood-based substrates will be intact with the substrates and provide accurate measurements due to less interactions from the surrounding environment.

8.8.1.4 Brush Painting

It is a combination of brushing and sintering. Brush painting uses Ag- and Cu-based nanoparticle inks to effectively print UHF RFID tags on textile materials. It decreases the quantity of conductive ink used for printing as compared to other techniques. Since the ink is directly distributed on the brush with which tags are printed, the loss of conductive ink is much less, thereby improving the cost of the printing system. Brush painting involves two major processes, namely painting and sintering. From the observations made by the researchers, Cu-based ink is not suitable for the brush painting process. Also, with Ag ink, there are more variations in sintering parameters. The process of brush painting methods is explained in a step-by-step manner in Figure 8.11.

8.8.1.5 Doctor Blading Based on Graphene

The processing time in additive manufacturing can be minimized by choosing the proper materials for printing. Cu-based inks are seen as an alternative to Ag-based inks, but still, the cost of printing is high. Hence, the focus has shifted towards conductive inks like graphene and carbon nanomaterials. Graphene is a 2D material with features like zero band gap, good biocompatibility, and high charge mobility. Also, graphene-based nanomaterials are lightweight and compact, and the designs

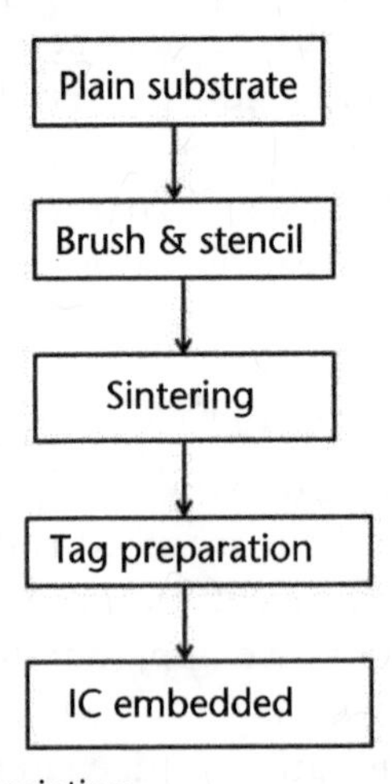

Figure 8.11 Process steps of brush painting.

based on these graphene-based designs are flexible, have a wider operating temperature range, and are economical when compared with other conductive inks. The features discussed earlier made graphene an excellent candidate for wireless identification technology. The graphene ink is printed on substrates like paper or cardboard; cardboard substrate is less expensive, more abundant, and ecofriendly. Cardboard-based substrates are used for two printing techniques: brush painting and doctor blading. Cardboard can be made from paper, fiber, or wood to fabricate RFID tags in the UHF range. A comparative study is conducted on Ag, Cu, and graphene-based conductive inks in terms of read range. It is to be noticed that Ag-based materials have achieved higher read ranges of 12m in the UHF range. In contrast, Cu-based materials have achieved a read range of 5m and graphene-based materials about 2m. With the usage of graphene-based nanoparticles as conductive inks, the cost of the printing process decreased drastically.

8.8.1.6 Thermal Printing

The working principle of thermal printing is based on thermal transfer print. The thermal print consists of a multilayered ribbon. A fine metal layer is attached to the ribbon with the help of a resin and acrylic glue is layered on the top surface of the thermal print. Printing is done by the application of a thermal print head to the desired area of heat-sensitive glue. This process results in the application of thin metal from a ribbon onto the substrate.

8.8.1.7 Screen Printing

With the reduction of printing costs for electronic devices, there is a huge demand created for R2R (roll-to-roll) processing, also known as web processing. It is the process of creating electronic devices on a roll of flexible plastic or metal foil. Suitable coatings, print, or other processors with a roll of flexible material can be used to create an output roll. Choosing this R2R fabricating technology will be a viable option for the fabrication of chipless RFID tags. It is a process in which a mask is used as a stencil, which will restrict some of the portions where the ink should be dropped on the material by applying physical pressure with the help of a roller. Three types of tag sensors can be manufactured by this technique. (1) *Gas sensors* which use carbon-based nanoparticle inks which are sensitive to gases and will detect the gases, (2) *humidity sensors* that can be used for detecting moisture content in smart packaging applications, and (3) *impedance sensors* that can be used as biosensors. Through this printing technique, by the deposition of dense layers of solid, high reliability and conductivity are achieved. The limitations of this fabrication technique include the usage of photolithography, which employs masks. A representation diagram of the screen-printing technique is presented in Figure 8.12. Electronic surveillance tags at 8.2 MHz with a relative permittivity of 3.2 and a film thickness of 25 Mils, and a conductivity of 40,000 S/cm of the ink are used in the fabrication process with a 300-stencil screen on polyethylene naphthalene (PEN) with a screen printer. During deposition, the PEN screen is kept a few Mils above the surface of the material. After deposition of the Ag ink on the screen, a rubber squeegee is brushed with a velocity across the plane of this screen. At this time, the ink will be passed from the screen to the surface of the substrate. When

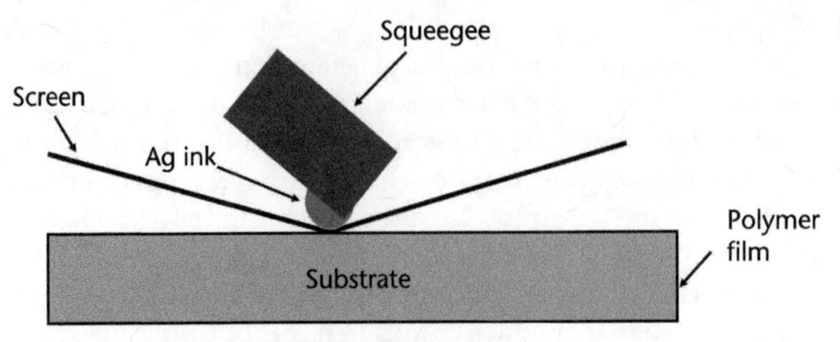

Figure 8.12 Representation of the screen-printing process.

the squeegee passes from one section to another, the screen detaches from the material and produces a continuous film. The technology can be similarly applied with a rotary-screen printer or high-speed R2R printers like Gravure, Offset, or Flexo. Since the abovementioned technique monitors the dimension and frequency of the tags it can be used to manufacture LC sensors.

8.8.1.8 3D Printing

Due to increased interest in on-body wireless sensing technologies, a huge amount of research is carried out in wearable applications. Also, textile materials or fabric patterning can be accomplished with this 3D printing technique. An nScrypt tabletop series 3D printer is used to manufacture RFID tags on textile substrate. This 3D printing technique provides a microprocessor-based approach to attain precision using dispensing pumps. A 3D axis rotation control system allows the deposition of ink on different planar and 3D surfaces with more accuracy. Materials like polyester and viscose are used in combination as the textile substrates for clothing based wearable applications. Since the textile-based tags are fabricated by screen printing of metallic ink and embroidery with metal-coated sewing thread, the tag designs have longer life spans. Also, they are very much suitable for stretchable antennas and textile-based wearable RFID tags. Copper oxide ink is used in 3D printing and brush painting to fabricate tags on pure cotton. Recently, a 3D-printed stretchable material and NinjaFlex® filament have been used to fabricate a wearable tag. The tag was printed with flexible Ag paste.

8.8.1.9 Gravure Printing

One-micrometer precision is the main motive in microelectronics printing methods. Attaining this goal is quite challenging but can be achieved by decreasing the pattern size from nanometers to micrometers. Within the available printing techniques, this gravure type of printing is the stand-out technique for achieving this 1-micrometer precision. This technique has high accuracy and speed. The benefits discussed earlier made the gravure printing technique suitable for tag manufacturing. The attractive material features like compressibility, smoothness, porosity, and wettability made this printing technique special. In this printing technique, the resolution of etching is restricted by the diameter of the laser beam. Generally, in standard gravure printing, a beam size of 40 micrometers is used, whereas, in an indirect laser system, a beam diameter of 10 to 20 micrometers is used. In the gravure printing technique, a thin ink layer was required for high conductivity, viscosity, and Ag content. A substantial layer of polymer binder that remained after curing makes the tag much stronger. This gravure printing based tag manufacturing technique is suitable for smart packaging and UHF RFID antennas.

8.8.1.10 Aerosol Jet Printing

A lot of research has been going on by using electrospinning and inkjet technique to print chipless RFID tags. The major hindrance to this technology is the ink viscosity. The traditional ink jet technique, the inks to be printed should have a viscosity in the range of 1 to 1,000 cP. The inks that are used in aerosol jet printing have a viscosity of up to 100 cP with low vapor pressure. The inks are printed using a pneumatic atomizer. Similarly, inks with viscosities of 1 to 5 cP can be printed with the ultrasonic atomizer. Multilayered device (transistors, capacitors, and photodetectors) manufacturing requires the deposition of inks on many layers stacked upon one another. Each layer has a separate ink viscosity. The printing techniques that were discussed can be used in combination to fabricate such components, thereby reducing the manufacturing complexity and cost. In this aerosol jet printing technique, a single direct write printing system, which has a high range of viscosities, is used. It uses a sheath of inert gas to tightly focus a beam of aerosol ink onto a material. This printing technique is computer aided design (CAD)-driven and uses a vector-based manufacturing technique. This technique offers more flexibility. Also, it is suitable for rapid prototyping of electronic components, the same as 3D printers. A detailed process of the aerosol jet printing technique is presented in Figure 8.13 [71]. From the previous discussions, we can arrive at a conclusion that this printing technique has the ability to print components on 2D and 3D materials with a wider viscosity range, and it allows fast and cost effective prototyping of microelectronic devices.

8.8.1.11 Thin Film Printing

The elasticity feature of thin film materials is used in various ways to realize flexible tags. In these conditions, the conductive layer is realized with conductive ink

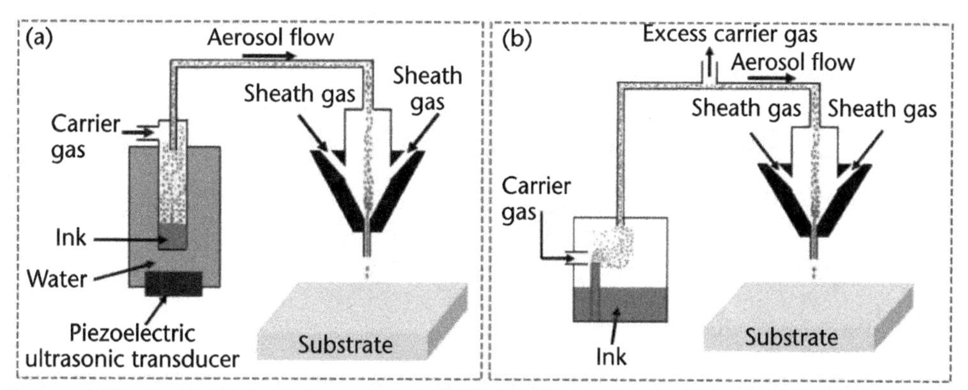

Figure 8.13 Aerosol jet printing process [71]. (a) Ultrasonic atomizer: high-frequency waves via piezoelectric transducer and (b) pneumatic atomizer: high-velocity gas through nozzle in ink reservoir.

or tape without metal that is extracted from the required thin film flexible material. Flexible RFID tags can be realized using polymer ceramic composite substrates. Transistor circuits made of printable thin films with disjointedness can be realized using TDR-based chipless RFID tags. A thin film transistor circuit (TFTC) is manufactured at a greater speed on flexible and inexpensive thin film materials like plastic. Also, organic TFTCs are an economical alternative for this thin film printing. In this type of organic TFTC printing, a bottom gate, bottom contact, OTFT-based additive manufacturing process are used.

8.8.1.12 Cellulose-Based Substrate Printing

In search of low-priced electronics, there is a huge rush for biopolymers. Using biopolymers for tag printing reduces the cost of printing. Cellulose fiber-based paper extracted from bark, leaves, or plant-based materials is an example of such type of biopolymer. Due to its sustainability and inexpensiveness, these cellulose-based substrates have the capacity to benefit the global economy. Combining optoelectronic and electronic features with the production techniques of the paper industry will open gates to enhance new utilities to traditional CBSs. These cellulose-based paper substrates can be used to print OTFT-based and electronic circuits. Researchers developed a field effect transistor (FET) structure with the help of a cellulose sheet used as the gate dielectric. The gate layer is made up of a conductive oxide that is deposited on one side, and a channel layer is on the other side, along with highly conductive source and drain terminals, as shown in Figure 8.14. The transistors fabricated with this technique attained a zero-threshold voltage and a high drain to source current modulation ratio. These results depict that this method has the potential to attain disposable electronics, smart labels, smart packaging, and healthcare applications [72–74].

Errors in the printing and effective solutions: In RFID tag design, EM parameters like relative permittivity (or dielectric constant) and loss tangent play a crucial role in determining the tags' response. Generally, the resonant frequency of the tag is shifted left or right if there is a discrepancy in the calculation of the dielectric constant. This error can be corrected by slightly varying the tags' dimensions. In the same way, the loss tangent of a paper substrate degrades the Q-factor of the resonators placed on the tag, thereby reducing the quantity of bits that are to be

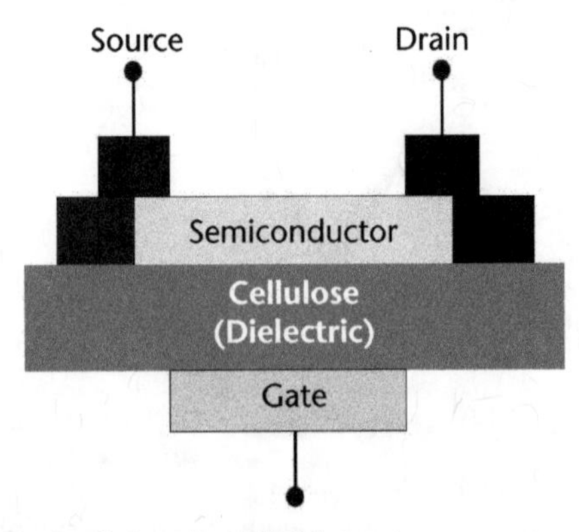

Figure 8.14 Representation of cellulose-based FET structure.

encoded on the substrate. The type of errors mentioned earlier can be rectified before the tag manufacturing process. But errors in the printing process are irreversible and impossible to correct once done. So, the utmost care has to be taken in the printing process to avoid the errors that will affect the tags' response. The following paragraphs discuss the possible errors in printing their effects and suitable solutions for them.

Errors in printing various shapes: It is observed that rectangular-shaped tags are fabricated more accurately when compared with circular-shaped tags using thermal printers. Due to the larger surface perimeters and roundedness of circular shapes, fabrication is prone to errors. U-slotted shaped tags attained perfectly fabricated results, whereas O-slotted shaped tags are prone to errors in printing at the submicron level. Hence, we can say that U-slotted tags have few errors in printing with thermal printer. To avoid these kinds of errors, it is better to make designs without rounded resonators. It is better to choose designs with rectangular-shaped or U-slotted designs for tag printing with a thermal printer [70].

8.9 Errors in Printing Vertical and Horizontal Edges

Errors in printing the horizontal and vertical edges are another source of printing errors. Due to the variation in the lengths of the vertical (V) and horizontal (H) edges, these discrepancies may arise in the output. These errors must be considered at the time of simulation to avoid discrepancies in the results of printing. The undesired effects of these discrepancies are shape variations like a circle to an oval, square to a rectangle, and variations in the *fr*, Q-factor of the resonating notch. From the studies that are carried out, it can be observed that with thermal printing on square tags, there are variations in the *V* edges, whereas the H edges are accurately printed. The rectangular edges are somewhat rounded into circular shapes using this thermal printer. It is to be noted that a square patch with width and length as L, as the input is printed with dimensions as width = L, and length = $L + \Delta L$. A variation in the length of the resonator is observed as a discrepancy in thermal printing. Furthermore, it is better to avoid this kind of printing error at the time of patch simulation rather than after manufacturing.

8.9.1 Impact of Aging on Tag Substrates and Conductive Inks

Most passive RFID tags are resistant to the aging effect. Paper-based substrates can be affected substantially (or gradually) by aging and extreme environmental conditions. There will be inevitable effects on the tags that are made of paper-based materials due to moisture. Moisture can deform the tags, and this leads to variations in the dielectric constant and loss tangent of the substrate material, thereby affecting the tags' performance in terms of resonant frequency and encoding capacity, respectively, due to variation in the parameters discussed previously. Deformation leads to a change in the thickness of the substrate. Moisture can alter the length of the resonators present on the tag. There is also a considerable effect of age on the ink conductivity if the tag is not protected properly. The aging effect on ink conductivity causes the encoded data of the tag to be changed. To avoid the impact of aging and other environmental factors on the tags' performance, an additional protective microwave-transparent layer made of plastic film has to be used.

8.10 Summary

The focus of this chapter was to examine the multiple benefits and drawbacks of printing methods in order to illustrate the significance of reader antennas in RFID systems. A reader architecture was explained with its classifications. Then, using an architectural explanation, the functionalities of LP and CP antennas were described. Following this, a table demonstrating CP generation methods using a number of relevant literatures was shown. In addition to being described, various fractal antennas' significance in RFID were also discussed. In this chapter, we concentrated on the useful contributions of Koch fractal antennas. Specific frequency ranges for various reader antennas used in biological applications were presented. The performance can be enhanced overall by using antenna arrays to increase features like directional gain and bandwidth dramatically. In Section 8.8, the limitations of RFID applications were briefly discussed. Ink conductivity, dielectric constant, loss tangent, and substrate thickness problems were discussed. Using real-world circumstances, various printing technologies, such as inkjet printing, paper and wood substrates, brush painting, and doctor blading, were explained. Furthermore, two techniques that might work well for creating RFID tags are 3D printing and gravure printing, as were described. For particular RFID printing, this chapter also emphasized aerosol and thin film printing.

References

[1] Karmakar, N. C., *Handbook of Smart Antennas for RFID Systems,* John Wiley & Sons, 2011.

[2] Sasaki, N., K. Kimoto, W. Moriyama, and T. Kikkawa, "A Single-Chip Ultra-Wideband Receiver with Silicon Integrated Antennas for Inter-Chip Wireless Interconnection," *IEEE Journal of Solid-State Circuits,* Vol. 44, No. 2, 2009, pp. 382–393.

[3] Pavlina, J., N. Kozlovski, B. Santos, and D. Malocha, "Saw RFID Spread Spectrum OFC and the Technology," in *2009 IEEE International Conference on RFID, IEEE*, 2009, pp. 110–116.

[4] Plessky, V. P., and L. M. Reindl, "Review on Saw RFID Tags," *IEEE Transactions on Ultrasonics, Ferroelectrics, and Frequency Control,* Vol. 57, No. 3, 2010, pp. 654–668.

[5] Zito, D., D. Pepe, M. Mincica, F. Zito, A. Tognetti, A. Lanatà, and D. De Rossi, "Soc CMOS uwb Pulse Radar Sensor for Contactless Respiratory Rate Monitoring," *IEEE Transactions on Biomedical Circuits and Systems,* Vol. 5, No. 6, 2011, pp. 503–510.

[6] Yogev, A., and S. Dukler, "Radio Frequency Data Carrier and Method and System for Reading Data Stored in the Data Carrier," February 14, 2006, U.S. Patent 6,997,388.

[7] Preradovic, S., I. Balbin, N. C. Karmakar, and G. F. Swiegers, "Multiresonator-Based Chipless RFID System for Low-Cost Item Tracking," *IEEE Transactions on Microwave Theory and Techniques,* Vol. 57, No. 5, 2009, pp. 1411–1419.

[8] Reddy, T., "Review of Compact Broadband Circularly Polarized Microstrip Antennas," IETE Tech Rev, Vol. 22, No. 4, 2005 pp. 249–58.

[9] Balanis, C.A., *Antenna Theory: Analysis and Design*, Wiley, 2009.

[10] Stutzman, W.L. and G. A. Thiele, *Antenna Theory and Design*, John Wiley & Sons, 2012.

[11] Sharma, P and K. Gupta, "Analysis and Optimized Design of Single Feed Circularly Polarized Microstrip Antennas," *IEEE Trans Antennas Propag.*, Vol. 31, No. 6, 1983, pp. 949–55.

[12] Wong K.-L. and J.-Y. Wu, "Single-Feed Small Circularly Polarized Square Microstrip Antenna," Electron. Lett., Vol. 33, No. 22, 1997, pp. 1833–4.

[13] Huang C. Y., J. Y. Wu, and K. L. Wong, "Broadband Circularly Polarized Square Microstrip Antenna Using Chip-Resistor Loading," *IEE Proceedings of Microwaves, Antennas Propag.*, Vol. 146, No. 1, 1999, pp. 94–96.

[14] Qing X., et al., "Compact Asymmetric-Slit Microstrip Antennas for Circular Polarization," *IEEE Trans Antennas Propag.*, Vol. 59, No. 1, 2011, pp. 285–288.

[15] Xie, L., Y. Li, and Y. Zheng, "A Wide Axial-Ratio Beamwidth Circularly Polarized Microstrip Antenna," *2016 IEEE International Conference on Ubiquitous Wireless Broadband (ICUWB), IEEE*, 2016, pp. 1–4.

[16] Kumar, S., B. K. Kanaujia, M. K. Khandelwal, and A. Gautam, "Single-Feed Superstrate Loaded Circularly Polarized Microstrip Antenna for Wireless Applications," *Wireless Pers. Commun.*, Vol. 92, No. 4, 2017, pp. 1333–46.

[17] Wang, M.-S., X.-Q. Zhu, Y.-X. Guo, and W. Wu, "Compact Circularly Polarized Patch Antenna with Wide Axial-Ratio Beamwidth," *IEEE Antennas Wireless Propag. Lett.*, Vol. 17, No. 4, 2018, pp. 714–718.

[18] Drori, R., "Selecting RFID Antennas," White Paper, MTI Wireless Edge Ltd.

[19] Catarinucci, L., S. Tedesco, and L. Tarricone, "Customized Ultra-High Frequency Radio Frequency Identification Tags and Reader Antennas Enabling Reliable Mobile Robot Navigation," *IEEE Sensors Journal*, Vol. 13, No. 2, 2013, pp. 783–791.

[20] Zaman, M., R. Azim, N. Misran, M. Asillam, and T. Islam "Development of a Semielliptical Partial Ground Plane Antenna for RFID and GSM-900," *Int. J. Antennas Propag.* 2014.

[21] Hsu, H.-T. and T.-J. Huang, "A 12 Dual-Band Antenna Array for Radio-Frequency Identification (RFID) Handheld Reader Applications," *IEEE Trans. Antennas. Propag.*, Vol. 62, No. 10, 2014, pp. 5260–5267.

[22] Wang, Y., A. Pretorius, and A. Abbosh, "Low-Profile Antenna with Elevated Toroid-Shaped Radiation for the On-Road Reader of RFID-Enabled Vehicle Registration Plate," *IEEE Trans. Antennas Propag.*, Vol. 64, No. 4, 2016, pp. 1520–1525.

[23] Karmakar, N. C., Y. Yang, and A. Rahim, "Case Study: Microwave Sleep Apnea Monitoring," in *Microwave Sleep Apnea Monitoring*, Springer, 2018, pp. 295–302.

[24] Tanmaya, K. D., B. Dwivedy, and S. K. Behera, "Design of a Meandered Line Microstrip Antenna with a Slotted Ground Plane for RFID Applications," *AEÜ - Int. J. of Electron. and Commun.*, Vol. 118, 2020, p. 15313.

[25] Vinoy, K. J., "Fractal Shaped Antenna Elements for Wide- and Multi-Band Wireless Applications," PhD dissertation, Pennsylvania State University, 2002.

[26] Miller, E., and T. Sarkar, *Frontiers in Electromagnetics*, D. H. Werner and R. Mittra (eds.), Piscataway, NJ: IEEE Press, 2000.

[27] Choukiker, Y. K., "Investigations on Some Compact Wideband Fractal Antennas," PhD dissertation, National Institute of Technology, Rourkela, India, 2013.

[28] Farswan, A., A. K. Gautam, B. K. Kanaujia, and K. Rambabu, "Design of Koch Fractal Circularly Polarized Antenna for Handheld UFF RFID Reader Applications," *IEEE Transactions on Antennas and Propagation*, Vol. 64, No. 2, 2015, pp. 771–775.

[29] Das, T. K., B. Dwivedy, D. Behera, S. K. Behera, and N. C. Karmakar, "Design and Modelling of a Compact Circularly Polarized Antenna for RDDD Applications," *AEU-International Journal of Electronics and Communications*, Vol. 123, 2020, p. 153313.

[30] Raviteja, C., C. Varadhan, M. Kanagasabai, A. K. Sarma, and S. Velan, "A Fractal-Based Circularly Polarized UHF RFID Reader Antenna," *IEEE Antennas and Wireless Propagation Letters*, Vol. 13, 2014, pp. 499–502.

[31] Preradovic, S., I. Balbin, N. C. Karmakar, and G. Swiegers, "Chipless Frequency Signature Based RFID Transponders," in *2008 European Conference on Wireless Technology, IEEE*, 2008, pp. 302–305.

[32] Kalansuriya, P., N. C. Karmakar, and E. Viterbo, "On the Detection of Frequency-Spectra-Based Chipless RFID Using UWB Impulsed Interrogation," *IEEE Transactions on Microwave Theory and Techniques*, Vol. 60, No. 12, 2012, pp. 4187–4197.

[33] Kalansuriya, P., and N. Karmaka "UWB-IR Based Detection for Frequency-Spectra Based Chipless RFID," in *2012 IEEE/MTT-S International Microwave Symposium Digest, IEEE*, 2012, pp. 1–3.

[34] Kiourti, A., "RFID Antennas for Body-Area Applications: From Wearables to Implants," *IEEE Antennas and Propagation Magazine*, Vol. 60, No. 5, 2018, pp. 14–25.

[35] Darwish, A., and A. E. Hassanien, "Wearable and Implantable Wireless Sensor Network Solutions for Healthcare Monitoring," Sensors, Vol. 11, No. 6, 2011, pp. 5561–5595.

[36] Kiourti, A., K. A. Psathas, and K. S. Nikita, "Implantable and Ingestible Medical Devices with Wireless Telemetry Functionalities: A review of Current Status and Challenges," *Bioelectromagnetics*, Vol. 35, No. 1, 2014, pp. 1–15.

[37] Stoppa, M., and A. Chiolerio, "Wearable Electronics and Smart Textiles: A Critical Review," *Sensors*, Vol. 14, No. 7, 2014, pp. 11957–11992.

[38] Kiourti, A., and J. L. Volakis, "High-Geometrical-Accuracy Embroidery Process for Textile Antennas with Fine Details," *IEEE Antennas and Wireless Propagation Letters*, Vol. 14, 2014, pp. 1474–1477.

[39] Want, R., "An Introduction to RFID Technology," *IEEE Pervasive Computing*, Vol. 5, No. 1, 2006, pp. 25–33.

[40] Marrocco, G., "The Art of UHF RFID Antenna Design: Impedance-Matching and Size-Reduction Techniques," *IEEE Antennas and Propagation Magazine*, Vol. 50, No. 1, 2008, pp. 66–79.

[41] Matic, A., P. Mehta, J. M. Rehg, V. Osmani, and O. Mayora, "Monitoring Dressing Activity Failures Through RFID and Video," *Methods of Information in Medicine,* Vol. 51, No. 1, 2012, pp. 45–54.

[42] Akbari, M., L. Sydänheimo, Y. Rahmat-Sami, J. Virkki, and L. Ukkonen, "Implementation and Performance Evaluation of Graphene-Based Passive UHF RFID Textile Tags," in *2016 URSI International Symposium on Electromagnetic Theory (EMTS), IEEE*, 2016, pp. 447–449.

[43] Amendola, S., L. Bianchi, and G. Marrocco, "Movement Detection of Human Body Segments: Passive Radio-Frequency Identification and Machine-Learning Technologies," *IEEE Antennas and Propagation Magazine*, Vol. 57, No. 3, 2015, pp. 23–37.

[44] Occhiuzzi, C., and G. Marrocco, "The RFID Technology for Neurosciences: Feasibility of Limbs' Monitoring in Sleep Diseases," *IEEE Transactions on Information Technology in Biomedicine*, Vol. 14, No. 1, 2009, pp. 37–43.

[45] Tsai, M.-C., C.-W. Chiu, H.-C. Wang, and T.-F. Wu, "Inductively Coupled Loop Antenna Design for UHF RFID On-Body Applications," *Progress In Electromagnetics Research*, No. 143, 2013, pp. 315–330.

[46] Kellomäki, T., J. Virkki, S. Merilampi, and L. Ukkonen, "Towards Washable Wearable Antennas: A Comparison of Coating Materials for Screen-Printed Textile-Based UHF RFID Tags," *International Journal of Antennas and Propagation*, 2012.

[47] Juels, A., "RFID Security and Privacy: A Research Survey," *IEEE Journal on Selected Areas in Communications*, Vol. 24, No. 2, 2006, pp. 381–394.

[48] Singh, A., and V. M. Lubecke, "Respiratory Monitoring and Clutter Rejection Using a CW Doppler Radar with Passive RF Tags," *IEEE Sensors Journal*, Vol. 12, No. 3, 2011, pp. 558–565.

[49] Ziai, M. A., and Batchelor, J. C., "Temporary On-Skin Passive UHF RFID Transfer Tag," *IEEE Transactions on Antennas and Propagation*, Vol. 59, No. 10, 2011, pp. 3565–3571.

[50] Santiago, A. G., J. R. Costa, and C. A. Fernandes, "Broadband UHF RFID Passive Tag Antenna for Near-Body Applications," *IEEE Antennas and Wireless Propagation Letters*, Vol. 12, 2013, pp. 136–139.

[51] Sanchez-Romaguera, V., M. A. Ziai, D. Oyeka, S. Barbosa, J. S. Wheeler, J. C. Batchelor, E. A. Parker, and S. G. Yeates, "Towards Inkjet-Printed Low Cost, Passive UHF RFID Skin Mounted Tattoo Paper Tags Based On Silver Nanoparticle Inks," *Journal of Materials Chemistry* C, Vol. 1, No. 39, 2013, pp. 6395–6402.

[52] Babaeian, F., and N. C. Karmakar, "A High Gain Dual Polarized Ultra-Wideband Array of Antenna for Chipless RFID Applications," *IEEE Access*, Vol. 6, 2018, pp. 73702–73712.

[53] Zomorrodi, M., and N. C. Karmakar, "On the Application of the EM-Imaging for Chipless RFID Tags," *Wireless Power Transf.*, Vol. 2, September 2015, pp. 86–96.

[54] Gulati, N., and K. R. Dandekar, "Learning State Selection for Reconfigurable Antennas: A Multi-Armed Bandit Approach," *IEEE Trans. Antennas Propag.*, Vol. 62, No. 3, March 2014, pp. 1027–1038.

[55] Piazza, D., N. J. Kirsch, A. Forenza, R. W. Heath, Jr., and K. R. Dandekar, "Design and Evaluation of a Reconfigurable Antenna Array for MIMO Systems," *IEEE Trans. Antennas Propag.*, Vol. 56, No. 3, March 2008, pp. 869–881.

[56] Dobkin, D. M., "Reader Antennas," in *The RF RFID: Passive UHF RFID in Practice*, Chapter 6, D. M. Dobkin (ed.), Burlington, VT: Newnes, 2008, pp. 241–303.

[57] FCC Electronic Code of Federal Regulations. Accessed: April 8, 2020.

[58] Tajin, M. A. S., and K. R. Dandekar, "Pattern Reconfigurable UHF RFID Reader Antenna Array," *IEEE Access*, Vol. 8, 2020, pp. 187365–187372.

[59] RT/duroid 5870 /5880 High Frequency Laminates, https://rogerscorp.com/-/media/project/rogerscorp/documents/advanced-electronics-solutions/english/data-sheets/rt-duroid-5870---5880-data-sheet.pdf. Accessed: May 13, 2020.

[60] Koswatta, R. V., "Readers for Frequency Signature-Based Chipless RFID Tags," PhD thesis, Monash University, 2013.

[61] Pettus, M. G., "RFID System Utilizing Parametric Reflective Technology," December 2, 2008, U.S. Patent 7,460,014.

[62] Vena, A., E. Perret, and S. Tedjini, "High-Capacity Chipless RFID Tag Insensitive to the Polarization," *IEEE Transactions on Antennas and Propagation*, Vol. 60, No. 10, 2012, pp. 4509–4515.

[63] Vena, A., E. Perret, and S. Tedjini, "Chipless RFID Tag Using Hybrid Coding Technique," *IEEE Transactions on Microwave Theory and Techniques*, Vol. 59, No. 12, 2011, pp. 3356–3364.

[64] Islam, M. A., and N. C. Karmakar, "A Novel Compact Printable Dual-Polarized Chipless RFID System," *IEEE Transactions on Microwave Theory and Techniques*, Vol. 60, No. 7, 2012, pp. 2142–2151.

[65] Preradovic, S., and N. C. Karmakar, "Design of Short Range Chipless RFID Reader Prototype," in *2009 International Conference on Intelligent Sensors, Sensor Networks, and Information Processing (ISSNIP), IEEE*, 2009, pp. 307–312.

[66] Hu, S., Y. Zhou, C. L. Law, and W. Dou, "Study of a Uniplanar Monopole Antenna for Passive Chipless UWB-RFID Localization System," *IEEE Transactions on Antennas and Propagation*, Vol. 58, No. 2, 2009, pp. 271–278.

[67] Gupta, S., B. Nikfal, and C. Caloz, "Chipless RFID System Based on Group Delay Engineered Dispersive Delay Structures," *IEEE Antennas and Wireless Propagation Letters*, Vol. 10, 2011, pp. 1366–1368.

[68] Kalansuriya, P., and N. Karmakar, "Time Domain Analysis of a Backscattering Frequency Signature Based Chipless RFID Tag," in *Asia-Pacific Microwave Conference 2011, IEEE*, 2011, pp. 183–186.

[69] Hartmann, C., P. Hartmann, P. Brown, J. Bellamy, L. Claiborne, and W. Bonner, "Anti-Collision Methods for Global Saw RFID Tag Systems," *IEEE Ultrasonics Symposium*, IEEE, Vol. 2, 2004, pp. 805–808.

[70] Islam, M. A., and N. C. Karmakar, "Real-World Implementation Challenges of a Novel Dual-Polarized Compact Printable Chipless RFID Tag," *IEEE Transactions on Microwave Theory and Techniques*, Vol. 63, No. 12, December 2015, pp. 4581–4591.

[71] Shah, M. A., D.-G. Lee, B.-Y. Lee, and S. Hur, "Classifications and Applications of Inkjet Printing Technology: A Review," *IEEE Access*, Vol. 9, 2021, pp. 140079–140102.

[72] Casula, G. A., G. Montisci, and G. Mazzarella, "A Wideband PET Inkjet-Printed Antenna for UHF RFID," *IEEE Antennas and Wireless Propagation Letters*, Vol. 12, 2013, pp. 1400–1403.

[73] Putaala, J., et al., "Capability Assessment of Inkjet Printing for Reliable RFID Applications," *IEEE Transactions on Device and Materials Reliability*, Vol. 17, No. 2, June 2017, pp. 281–290.

[74] Fortunato, E., N. Correia, P. Barquinha, L. Pereira, G. Goncalves, and R. Martins, "High-Performance Flexible Hybrid Field-Effect Transistors Based on Cellulose Fiber Paper," *IEEE Electron Device Letters*, Vol. 29, No. 9, September 2008, pp. 988–990.

CHAPTER 9

Challenges and Future Research Directions

9.1 Introduction

Radio frequency identification technology has revolutionized various industries by enabling efficient and automated data capture and identification processes. Traditional RFID systems utilize microchips or integrated circuits embedded in the tags to store and transmit information. However, the evolution of RFID technology has given rise to a promising alternative known as chipless RFID.

Chipless RFID refers to a class of RFID technology that operates without the use of integrated circuits or microchips in the tags. Instead, these tags rely on innovative techniques to encode and transmit information, making them more cost-effective, lightweight, and versatile than their chip-based counterparts. The absence of microchips not only reduces production costs but also simplifies the manufacturing process, allowing for potential mass production on a larger scale. The concept of chipless RFID is captivating researchers, engineers, and industries alike, as it opens up a plethora of applications and possibilities across diverse domains. From logistics and supply chain management to healthcare, agriculture, smart cities, and beyond, chipless RFID holds immense potential to transform how data is captured, tracked, and analyzed. In this context, this chapter delves into the challenges faced by chipless RFID technology in realizing its full potential in various real-world applications. Understanding these challenges is vital to overcoming them and paving the way for successful implementation in practical scenarios. We explore the limitations in data capacity, read range, reading speed, and potential interference that currently hinder the seamless integration of chipless RFID in different industries.

Furthermore, this chapter explores the exciting future research directions that can propel chipless RFID technology to new heights. By addressing the identified challenges, researchers and engineers can focus on improving tag performance, enhancing data analytics, increasing read ranges, and developing more energy-efficient solutions. These research directions will unlock innovative applications, optimize existing systems, and open up new possibilities for chipless RFID adoption in various sectors. The importance of data security, user privacy, and regulatory compliance cannot be overlooked, especially as chipless RFID becomes increasingly prevalent in applications that involve sensitive information. We also explore the critical aspect of integrating chipless RFID seamlessly with existing infrastructure and systems, ensuring compatibility and efficient data exchange. As we embark on this journey of exploring the challenges and potential of chipless RFID technology, it becomes evident that it has the power to reshape industries, drive technological advancements, and create more efficient, secure, and interconnected systems. This chapter seeks to shed light on the opportunities that lie ahead for researchers, engineers, and industries to harness the full potential of chipless RFID and unlock the next generation of RFID technology.

9.2 Implementational Challenges

Chipless RFID technology offers promising advantages for various industrial applications, including inventory management, asset tracking, supply chain management, and anticounterfeiting. Unlike traditional RFID, chipless RFID tags do not contain integrated circuits, making them more cost-effective and versatile. However, the implementation of chipless RFID in industrial settings is not without its challenges. Chipless RFID tags generally have shorter read ranges compared to traditional RFID tags. Ensuring reliable read ranges in various environments and applications is essential for the technology to be viable for different use cases. The absence of universal standards for chipless RFID technology can lead to compatibility issues and interoperability challenges. Establishing industry-wide standards is essential for seamless integration and cooperation between different chipless RFID systems. Chipless RFID systems must incorporate robust encryption and authentication measures to protect sensitive information from unauthorized access and potential cyberthreats. Chipless RFID technology must adhere to environmental regulations, especially concerning the use of materials and waste management. Ensuring compliance with industry-specific regulations is vital for market acceptance. Chipless RFID faces competition from other identification and tracking technologies, such as traditional RFID, barcode systems, and other IoT devices. Demonstrating the unique advantages of chipless RFID and its cost-effectiveness is vital to gain a competitive edge. This section examines the key challenges faced by industries when deploying chipless RFID systems, including technical, operational, regulatory, and economic hurdles. Additionally, potential solutions and best practices to overcome these challenges are discussed, aiming to facilitate the successful integration of chipless RFID in the industrial environment. Details of related studies were well described in Chapters 2, 3, 4, and 5. Here, we will only discuss the key issues while implementing the chipless RFID [1–4].

1. The first thing we observe from the chipless RFID is that the capacity is less as compared to the chipped one. Nowadays, the chip is turning out to be less costly as a bulk when produced. In such a situation, chipless RFID has to grow up to an advanced level in terms of data capacity and EM detection efficiency.
2. The problems associated with the deployment are a big issue as the range of chipless RFID is limited to meters; a little change in the outer environ-

ment can lead to a loss in frequency signature, especially a change in EM characteristics of the ambience.

3. Technical challenges like read range and sensitivity: Limited read range of chipless RFID tags compared to traditional RFID is of concern for mass production. Environmental factors affect the tag readability and signal strength as no filters/amplifiers/repeaters are used.
4. Data accuracy and integrity: Ensuring accurate data capture and transmission from chipless RFID tags and mitigating errors caused by tag collisions and signal interference are crucial during real-time applications.
5. Tag manufacturing and printing: Consistency and reliability of chipless RFID tag manufacturing processes, printing, and encoding techniques for high-volume applications is essential before its mass production.
6. Scalability and integration with existing systems: Implementing chipless RFID systems across large industrial environments and integrating with the existing enterprise resource planning (ERP) and logistics systems is a very important aspect. Integrating chipless RFID technology with existing infrastructure and systems can be complex and costly. Compatibility with legacy systems is vital to avoid disruption and facilitate a smooth transition.
7. Real-time data processing: Handling and analyzing vast amounts of data collected from chipless RFID tags need efficient data storage, retrieval, and analytics infrastructure.
8. Robustness and reliability: Ensuring chipless RFID tags' durability and performance in demanding industrial conditions and mitigating risks of tag damage, tampering, and loss during operations should be taken care of. Ensuring the durability and reliability of chipless RFID tags is crucial for their successful deployment in various industries.
9. Regulatory and security challenges: Here, the challenge regards data privacy and data security during real-time applications. When we use chipless tags, the pattern is fixed, and the signal is backscattered without any coding or processing ability at the reader end. Hacking/duplicating may be a severe issue in this domain. Therefore, implementing encryption and authentication mechanisms to safeguard data is very essential.
10. Industry standards and interoperability: Navigating complex standards and protocols for chipless RFID technology and ensuring compatibility and interoperability with other RFID systems and technologies is essential.
11. Economic and cost-related challenges like initial investment and return on investment (ROI) where assessing the initial costs associated with deploying chipless RFID systems and evaluating the ROI and long-term cost benefits are analyzed. Total cost of ownership (TCO): Understanding the complete cost implications of chipless RFID adoption and identifying cost-saving opportunities in the overall supply chain and operational processes are studied. The primary challenge is the cost of chipless RFID tags compared to traditional RFID tags and barcodes. Chipless RFID technology may be less expensive to produce, impacting its scalability for mass deployment, including receiver design. Reducing production costs and achieving economies of scale are crucial for wider market adoption.

12. Industry-specific challenges like manufacturing and supply chain management are a big issue in both medium and large industries. Overcoming challenges in tracking raw materials, work-in-progress, and finished goods, chipless RFID with existing manufacturing and logistics workflows is very much helpful. In retail and inventory management, challenges like enhancing inventory visibility and accuracy in warehouse environments need to be focused. Addressing the challenges of tag readability in densely packed inventory is also an important factor for business employment.
13. In healthcare and pharmaceuticals, chipless RFID holds an important aspect for the new generations. (1) Ensuring compliance with regulatory requirements in sensitive healthcare applications and (2) managing unique identification and tracking of pharmaceutical products are the most general issues to be addressed while ensuring the chipless RFID technology and auto-ID at less cost. Case studies and best practices by examining successful chipless RFID implementations in various issues/cases need to be explored for better utilization of the technology, thereby highlighting best practices and lessons learned from real-world deployments. The in-body and on-body chipless RFID implementation still required a huge advancement when compared to chip-based RFID systems [1, 5, 6].
14. Mathematical modeling of resonators in chipless RFID: We often ignore the analysis part in the chipless domain, but it is very important to study the reflection and scattering behavior of the resonators with mathematical modeling to properly quantify the radar cross-section. Though the empirical mathematical aspects may vary depending upon the resonator shape and material used in the tag, the analysis is useful before implementing the chipless RFID in the real world. This will greatly reduce the errors after real-world implementation.
15. Receiver design: As no chip is used in the chipless tag, the backscattered signal strength is low. In such a situation, a receiver design is the most important part of a chipless RFID system.
16. Performance in challenging environments: Some environments, such as those with metals or liquids, can cause interference and degrade chipless RFID tag performance. Developing solutions to handle challenging environments is crucial for diverse applications.
17. Education and awareness: Many potential users and decision-makers in the market may not be familiar with chipless RFID technology and its capabilities. Educating stakeholders and creating awareness about the benefits and applications of chipless RFID is essential for its adoption. As chipless RFID is a relatively newer technology, there might be a scarcity of experts and professionals with in-depth knowledge. Building a skilled workforce and fostering research and development efforts are essential to address this challenge.

Addressing these challenges will require collaborative efforts from technology developers, industry stakeholders, regulatory bodies, and end-users. By overcoming these obstacles, chipless RFID can find broader acceptance and play a significant role in revolutionizing various industries with its unique benefits and capabilities.

9.3 Future Research

The advancement of chipless RFID sensor design is very rapid with the latest progress in technology in this field. With the increasing requirements of sensing technologies, especially in the healthcare sector, there is ample room for enhancement of the quality of chipless sensor design. The following points highlight some probable future investigations required in this direction:

1. Fully printable chipless RFID tag sensors on fabrics are greatly demanded to replace existing technologies in the healthcare domain [5, 6].
2. The selection of a flexible sensing material is desirable for the comfort of the patient.
3. The integration of sensing technology with IoT provides a novel path to chipless RFID sensing technology shortly as it enables continuous heartbeat monitoring remotely.
4. Implementation of more efficient windowing techniques along with noise reduction algorithms is essential for the proper detection of the heartbeat signal.
5. Designing and fabricating suitable reader antennas for the proposed tags.
6. To design a chipless RFID scheme with more encoding bits and a smaller footprint than currently available tags.

Advancements in chipless RFID technology continue to open up new possibilities and applications across various industries. Future research in this field aims to address existing limitations, enhance performance, explore new materials and fabrication techniques, and develop innovative applications. In this comprehensive overview, we'll discuss several key areas for future research in chipless RFID.

1. *Enhancing read range and sensitivity*: Improving the read range and sensitivity of chipless RFID tags remains a critical area of research. In this regard, Chapters 2 and 3 provided considerable information [7–9]. Researchers can explore novel antenna designs, materials, and signal processing techniques to extend the reading distance of chipless tags while maintaining high signal integrity. Advanced signal modulation schemes, beamforming, and array antenna configurations could help achieve longer read ranges and enhanced performance in challenging environments.
2. *Multiband and wideband chipless RFID*: Current chipless RFID systems typically operate within specific frequency bands. Future research can focus on developing multiband and wideband chipless RFID solutions, enabling tag reading across a broader range of frequencies. This approach could enhance compatibility with existing RFID systems and provide greater flexibility in diverse applications.
3. *Anticollision and tag density handling*: As chipless RFID tags become more prevalent in the market, the issue of anticollision and tag density management becomes increasingly important. Future research can focus on developing advanced algorithms and protocols for efficient anticollision schemes, allowing simultaneous reading of multiple tags without interfer-

ence. Such advancements would boost scalability and streamline inventory management and asset tracking applications [7].

4. *Miniaturization and integration*: Miniaturization of chipless RFID tags is an area of interest [10–13], particularly for applications requiring tags to be embedded in smaller objects or used in constrained spaces. Researchers can explore microfabrication techniques, nanotechnology, and alternative materials to create ultra-small, flexible, and conformal chipless tags suitable for various IoT and wearable applications.
5. *Energy harvesting and battery-less tags*: One of the challenges of chipless RFID is its reliance on external RF energy for tag operation. Future research can focus on energy-harvesting solutions to power chipless tags, making them more self-sustaining and potentially battery-less. Integrating energy harvesting mechanisms, such as solar, kinetic, or RF harvesting, into chipless tags could extend their applicability and improve their ecofriendliness.
6. *Robustness and reliability*: Enhancing the robustness and reliability of chipless RFID tags is essential for real-world deployment. Researchers can investigate the impact of environmental factors, such as temperature, humidity, and mechanical stress, on tag performance. Designing tags with self-healing capabilities or redundancy mechanisms could improve their resilience in harsh operating conditions.
7. *Security and privacy*: Security and privacy are crucial concerns in RFID technology. Future research can focus on developing robust encryption, authentication, and access control mechanisms to protect sensitive data from unauthorized access and potential attacks. Implementing privacy-preserving protocols and ensuring compliance with data protection regulations will be critical for widespread adoption in various industries.
8. *Flexible and printed electronics*: Flexible and printed electronics have the potential to revolutionize chipless RFID technology by enabling cost-effective, large-scale manufacturing of tags on flexible substrates. Future research can explore advancements in printable materials, conductive inks, and roll-to-roll manufacturing processes to develop low-cost, high-performance chipless tags.
9. *On-chip sensing and data processing*: Integrating sensing capabilities and on-chip data processing in chipless RFID tags can add value to various applications. Future research can explore the integration of sensors for environmental monitoring, temperature sensing, humidity detection, and more. This would enable chipless RFID tags to function not only as identifiers but also as data loggers and real-time monitoring devices.
10. *Biodegradable and ecofriendly tags*: In an era of environmental consciousness, researchers can investigate the development of biodegradable and ecofriendly chipless RFID tags. Using sustainable materials and designing tags that degrade over time in environmentally friendly ways could have a significant impact on reducing electronic waste and promoting green RFID solutions.
11. *Thermal and chemical sensing*: Expanding chipless RFID tags to include thermal and chemical sensing capabilities would open up new applications

in healthcare, food safety, and environmental monitoring. Research can focus on exploring suitable sensing materials and mechanisms that respond to changes in temperature or chemical presence.

12. *Artificial intelligence and machine learning*: Integrating AI and machine learning techniques with chipless RFID systems could enhance tag read accuracy, anticollision algorithms, and data analysis. Future research can explore AI-powered solutions for intelligent tag identification, localization, and predictive maintenance in various industrial applications.
13. *Biometric and secure authentication*: Investigating the integration of biometric authentication into chipless RFID tags could revolutionize secure access control and identity verification systems. Research in this area could explore biometric recognition methods such as fingerprint, iris, or facial recognition embedded within the chipless tag for enhanced security.
14. *Harmonization of standards*: As chipless RFID technology advances, harmonizing industry standards becomes crucial. Future research can focus on collaborations between regulatory bodies, standardization organizations, and industry stakeholders to establish unified guidelines and protocols for chipless RFID systems. This will ensure seamless interoperability and promote wider adoption.
15. *Cross-disciplinary collaboration*: Future research in chipless RFID should encourage cross-disciplinary collaboration between researchers from different fields, such as electrical engineering, materials science, nanotechnology, computer science, and data analytics. Collaborative efforts can foster innovation and lead to breakthroughs in chipless RFID technology.

The future of chipless RFID technology holds immense potential for transformative advancements in various industries. By addressing the challenges outlined above and conducting research in these areas, we can unlock new possibilities, create innovative applications, and propel chipless RFID technology into broader market adoption. The combination of cutting-edge research, industry partnerships, and regulatory support will play a vital role in shaping the future of chipless RFID.

9.4 Limitations

Even though it is possible to design compact chipless RFID sensors with high bit capacity, some limitations are to be met for practical utilization. The introduction of sensors on tags is not simple, especially for commercial tags. Only a few commercial tag versions permit the mounting of onboard sensors. Shortly, we observe the following challenges during the beginning of the implementation of chipless RFID:

1. The less flexible nature of the substrates makes the tag designs less comfortable for the user, whereas a flexible substrate could be more comfortable and simpler to manage.
2. The tag sensors may not be integrated into fabrics and thus do not provide a wearable solution for the monitoring of life signal parameters.

3. The fabrication procedure may lead to the wastage of material and less accurate results. Low-cost printing technologies like inkjet printing, thermal printing, or screen printing are well-suited for efficient tag manufacturing.
4. The sensing material that varies its properties concerning the surrounding environment is not utilized for specific use. Instead, the RCS generated by the copper resonators has to be relied upon for finding out a change in the signal [13–15].
5. The noise reduction algorithms are not fully efficient in the presence of noise.

Chipless RFID technology, while promising and versatile, comes with certain limitations that can impact its practical implementation in real-world domains. In this comprehensive review, we will discuss the limitations of chipless RFID technology across several domains.

Retail and inventory management:

- *Read range and sensitivity*: One of the significant limitations of chipless RFID in retail and inventory management is its relatively shorter read range compared to traditional RFID. In large warehouses or stores, where fast and long-range tag reading is essential, this limitation can pose challenges in achieving efficient inventory tracking and management.
- *Tag density and anticollision*: As the number of items in a retail environment increases, the likelihood of tag collision rises. The anticollision algorithms in chipless RFID systems may struggle to handle a high tag density efficiently, leading to potential data inaccuracies and slower tag identification, hindering real-time inventory updates.
- *Integration with existing systems*: Retrofitting chipless RFID technology into existing retail and inventory management systems can be complex and costly. Ensuring compatibility with legacy systems, such as barcode scanners and RFID readers, may require additional investments and customization.
- *Cost*: The cost of implementing chipless RFID technology, including tags, readers, and backend systems, may be higher compared to traditional barcode systems. For small and medium-sized retailers, this cost factor can present a barrier to entry.

Manufacturing and supply chain:

- *Scalability and production cost*: Scaling up chipless RFID tag production to meet the demands of the manufacturing and supply chain industry can be challenging. The complexity of chipless tag designs and fabrication processes may lead to higher production costs, potentially limiting their widespread adoption.
- *Environmental factors*: Manufacturing and supply chain environments often involve metal and other materials that can interfere with chipless RFID tag signals, leading to reduced read range and accuracy. Addressing these environmental challenges is crucial to ensure reliable tag performance.

- *Asset tracking in harsh conditions*: In certain manufacturing processes or supply chain activities involving extreme temperatures, chemicals, or mechanical stress, chipless RFID tags may face durability and reliability issues. Research into ruggedized chipless tags and protection mechanisms is essential for overcoming these limitations.
- *Data security and privacy*: In the supply chain, data security and privacy are paramount, especially when dealing with sensitive information about products, logistics, and customers. Chipless RFID systems need robust encryption and authentication mechanisms to safeguard against data breaches.

Healthcare and medical applications:

- *Size and form factor*: In healthcare, the size and form factor of chipless RFID tags can be critical, especially when dealing with medical devices or patient identification. Achieving miniaturization while maintaining reliable performance remains a challenge.
- *Biocompatibility*: Some chipless RFID tag materials may not be biocompatible, limiting their direct use in implantable medical devices or applications requiring close contact with the human body. Research into biocompatible materials for chipless tags is crucial for healthcare applications [16].
- *Real-time tracking*: In critical medical applications, such as tracking surgical instruments or monitoring patient vital signs, real-time tracking is vital. Ensuring high-speed and accurate tag identification for time-sensitive medical procedures is a challenge that requires optimization of chipless RFID systems.
- *Regulatory compliance*: The healthcare industry is subject to strict regulations concerning patient data privacy and medical device certifications. Chipless RFID solutions must comply with these regulations to gain acceptance and adoption.

Logistics and transportation:

- *Read performance in variable environments*: In logistics and transportation, chipless RFID tags may encounter varying environmental conditions, such as temperature fluctuations, humidity, and exposure to harsh weather. Ensuring consistent read performance under these conditions is a challenge.
- *Vehicle tracking and speed*: Real-time vehicle tracking is essential in logistics and transportation for optimizing routes and delivery schedules. Achieving accurate vehicle tracking and high-speed tag identification in moving vehicles is a technological challenge.
- *Interoperability*: In the logistics industry, different companies and entities may use various RFID systems, including chipless and traditional RFID. Ensuring interoperability and seamless data exchange between these systems is essential for efficient supply chain management.
- *Cost and scalability*: Large-scale deployment of chipless RFID technology across an extensive logistics network can incur significant costs. Ensuring

scalability and cost-effectiveness is essential for logistics companies seeking to adopt chipless RFID for tracking and asset management.

Agriculture and food supply chain:

- *Harsh environmental conditions*: In agricultural and food supply chain applications, chipless RFID tags may encounter harsh conditions such as exposure to moisture, dirt, and pesticides. Developing ruggedized tags capable of withstanding these conditions is crucial for successful implementation [17].
- *Real-time monitoring*: Real-time monitoring of perishable goods, such as fruits and vegetables, is vital for ensuring food quality and safety. Chipless RFID technology must support real-time data updates for time-sensitive food supply chain operations.
- *Scalability and cost*: The agriculture and food industry often deals with a large volume of products. The cost of implementing chipless RFID on a massive scale for tracking and traceability may be a limiting factor for widespread adoption [18, 19].
- *Compliance with food safety standards*: The agriculture and food supply chain industries are heavily regulated to ensure food safety. Chipless RFID solutions must meet these standards and provide accurate traceability data to comply with regulations.

Environmental monitoring and IoT:

- *Sensing and data logging*: Integrating sensing capabilities into chipless RFID tags for environmental monitoring and IoT applications requires addressing power consumption and data logging challenges. Power-efficient sensor integration and data storage are essential for extended autonomous operation.
- *Network connectivity*: Chipless RFID tags operating in IoT environments need seamless network connectivity to relay data to cloud-based systems. Research into efficient communication protocols and integration with IoT platforms is necessary.
- *Localization and positioning*: In environmental monitoring and IoT applications, localization and positioning of chipless RFID tags play a significant role. Advancements in localization algorithms and technologies are needed to enable accurate spatial tracking and mapping.
- *Interference and noise mitigation*: In IoT deployments, multiple devices and networks can generate interference and noise, affecting the performance of chipless RFID tags. Techniques to mitigate interference and improve signal robustness are critical for reliable IoT implementations [20].

Security and access control:

- *Authentication and authorization*: Chipless RFID systems must ensure secure authentication and authorization of users to prevent unauthorized access. Robust access control mechanisms and encryption protocols are essential for secure applications such as building access or secure document management [21–23].

- *Counterfeiting and antitampering*: Chipless RFID tags may face counterfeiting and tampering risks in applications such as product authentication or secure packaging. Research into advanced anticounterfeiting features and tamper-evident mechanisms is crucial for preventing fraud.
- *Signal spoofing and jamming*: Ensuring the security and integrity of chipless RFID systems against signal spoofing and jamming attacks is essential. Implementing antijamming techniques and secure communication protocols is vital for safeguarding sensitive data.

Smart cities and infrastructure:

- *Massive tag deployment*: Smart city applications often require a massive deployment of chipless RFID tags across diverse infrastructures. Addressing the logistical challenges and ensuring efficient tag management becomes crucial for successful implementation.
- *Data analytics and integration*: Smart cities generate vast amounts of data from various sensors and devices, including chipless RFID tags. Integrating and analyzing this data to enable data-driven decision-making and improve city services is a significant research area.
- *Privacy and ethical considerations*: Smart city deployments must address privacy and ethical concerns related to data collection, surveillance, and citizen tracking. Ensuring data anonymization, informed consent, and adherence to privacy regulations are crucial for public acceptance.
- *Sustainability and green solutions*: As smart cities focus on sustainability and energy efficiency, chipless RFID technology must align with these principles. Developing green RFID solutions with ecofriendly materials and reduced energy consumption will be a priority.

Defense and security:

- *Localization and tracking*: In defense and security applications, chipless RFID tags may be used for asset tracking and personnel identification. Achieving precise localization and tracking capabilities in complex environments is critical for mission success.
- *Secure communication*: Security is paramount in defense and military applications. Developing chipless RFID systems with secure communication channels and antijamming features is essential for safeguarding sensitive data.
- *Stealth and covert operations*: Chipless RFID tags used in covert operations must possess characteristics like low radar cross section and reduced electromagnetic emissions to avoid detection and maintain mission secrecy.
- *Reliability in harsh conditions*: In military operations, chipless RFID tags may be subjected to extreme conditions, including high temperatures, shock, and vibrations. Ensuring tag reliability and ruggedness in harsh environments is crucial.

Education and campus management:

- *Student tracking and attendance*: In educational institutions, chipless RFID tags may be used for student tracking and attendance monitoring. Ensuring reliable and efficient attendance recording, especially in crowded environments, is a research challenge.
- *Integration with campus infrastructure*: Implementing chipless RFID technology in educational campuses requires integration with existing infrastructure such as access control systems, library management, and student ID cards. Ensuring seamless integration is vital for efficient campus management.
- *Privacy and data protection*: Balancing data collection for campus management and student privacy is essential. Designing chipless RFID systems with privacy-preserving features and adherence to data protection regulations is crucial for gaining acceptance in educational institutions.
- *Cost-effectiveness*: Educational institutions often operate on tight budgets. Ensuring the cost-effectiveness of chipless RFID solutions for various campus management applications will be critical for adoption.

Entertainment and events:

- *Real-time access control*: In entertainment events, chipless RFID tags may be used for access control and ticketing. Ensuring real-time access validation and smooth entry for large crowds is essential for enhancing the attendee experience.
- *Event analytics and crowd management*: Chipless RFID tags can provide valuable data for event organizers to analyze attendee behavior, crowd flow, and venue utilization. Developing analytics tools and visualization techniques for event management is an area of interest.
- *Integration with event infrastructure*: Successful deployment of chipless RFID technology in entertainment events requires seamless integration with ticketing systems, payment gateways, and venue access points.
- *RFID wristbands and wearables*: RFID wristbands and wearables are popular in entertainment events. Researchers can explore innovative designs and materials to enhance the comfort, aesthetics, and functionality of chipless RFID wearables.

Environment and wildlife conservation:

- *Remote tracking and monitoring*: Chipless RFID tags can be used for remote tracking and monitoring of wildlife and environmental conditions. Achieving long-range tracking capabilities in remote locations is essential for conservation applications.
- *Energy-efficient sensing*: Environmental monitoring often involves long-term data collection. Developing chipless RFID tags with energy-efficient sensing capabilities and low-power modes is crucial for extended autonomous operation.

- *Data analysis and predictive modeling*: The vast amount of data generated from chipless RFID tags in conservation applications require advanced data analytics and predictive modeling to inform conservation strategies effectively.
- *Bioinformatics and biodiversity studies*: Integrating chipless RFID technology with bioinformatics and biodiversity studies can provide valuable insights into species behavior, migration patterns, and ecosystem dynamics.

Automotive and traffic management:

- *Vehicle identification and tolling*: Chipless RFID tags can be used for vehicle identification and electronic toll collection. Ensuring accurate and high-speed tag reading at toll booths and checkpoints is essential for efficient traffic management.
- *On-vehicle integration*: Researchers can explore integrating chipless RFID technology within vehicles for identification, authentication, and secure communication with smart infrastructure.
- *Traffic analytics and smart traffic management*: Chipless RFID tags can be used for traffic analytics to optimize traffic flow and congestion management. Research into smart traffic management algorithms and infrastructure is essential for smart city implementations.
- *Integration with smart infrastructure*: Integrating chipless RFID technology with existing smart city infrastructure, such as traffic signals and road sensors, requires interoperability and seamless data exchange.

Banking and payment systems:

- *Contactless payments*: Chipless RFID technology can be used for contactless payments in banking and financial systems. Ensuring secure, fast, and reliable payment transactions is crucial for user acceptance.
- *Security and fraud prevention*: Implementing chipless RFID in payment systems requires robust security features and fraud prevention measures to protect against unauthorized access and data breaches.
- *Integration with mobile payment solutions*: Integrating chipless RFID technology with mobile payment platforms and digital wallets requires standardization and compatibility with various payment systems.
- *User acceptance and trust*: Building user trust and acceptance for chipless RFID-based payment systems is essential for successful adoption.

Hospitality and tourism:

- *Guest experience*: Chipless RFID technology can enhance guest experience in hotels and tourism destinations by enabling seamless access to rooms, amenities, and attractions. Ensuring a smooth and frictionless guest experience is crucial for customer satisfaction.

- *Privacy and personalization*: Balancing data collection for personalization with guest privacy is essential. Designing chipless RFID systems with privacy-preserving features while providing personalized services is a research challenge.
- *Integration with hospitality infrastructure*: Successful deployment of chipless RFID technology in hospitality and tourism requires integration with property management systems, access control systems, and guest engagement platforms.
- *Real-time analytics and service optimization*: Leveraging chipless RFID data for real-time analytics can optimize hospitality services, resource allocation, and customer engagement.

Real estate and facilities management:

- *Smart building integration*: Chipless RFID technology can be used for access control and occupancy tracking in smart buildings. Integrating chipless RFID with other smart building technologies is essential for efficient facilities management.
- *Asset and inventory tracking*: Chipless RFID tags can help in tracking assets and inventory within large facilities and buildings. Achieving reliable and accurate asset tracking is vital for optimizing resource utilization.
- *Energy efficiency and sustainability*: Chipless RFID technology can contribute to energy efficiency and sustainability initiatives in buildings. Research into energy-harvesting chipless tags and integration with smart building systems can enhance sustainability efforts.
- *Data security and privacy*: Ensuring data security and privacy in real estate and facilities management applications is crucial, especially when dealing with sensitive information about building occupants and assets.

Education and learning systems:

- *Student tracking and engagement*: In educational institutions, chipless RFID tags can be used to track student attendance and engagement. Ensuring student privacy while promoting engagement and interaction is essential.
- *Library and resource management*: Chipless RFID tags can enhance library management by enabling efficient book tracking and inventory management. Research into smart library systems and data analytics for resource utilization is vital.
- *Secure access control*: Implementing chipless RFID technology for secure access control in educational buildings requires robust authentication mechanisms and user verification.
- *Integration with e-learning platforms*: Integrating chipless RFID technology with e-learning platforms and virtual classroom systems can enhance the student experience and facilitate personalized learning.

Waste management and recycling:

- *Asset and waste tracking*: Chipless RFID tags can be used for asset tracking and waste management in recycling centers and waste facilities. Ensuring efficient waste sorting and recycling requires accurate tag identification.
- *Environmental sensing*: Integrating sensing capabilities into chipless RFID tags for environmental monitoring can help assess waste disposal practices and environmental impact.
- *Smart bin technology*: Chipless RFID can be integrated with smart bins to enable automated waste collection and optimize waste management routes.
- *Cost and scalability*: The cost-effectiveness and scalability of chipless RFID solutions for waste management are essential factors to consider.

Pharmaceutical and healthcare supply chain:

- *Counterfeiting prevention*: Chipless RFID technology can help prevent counterfeit drugs by enabling product authentication and traceability in the pharmaceutical supply chain.
- *Temperature and environmental monitoring*: Ensuring temperature monitoring and environmental conditions during pharmaceutical transportation and storage is crucial for drug efficacy and patient safety [24, 25].
- *Regulatory compliance*: The pharmaceutical industry is heavily regulated. Chipless RFID solutions must comply with regulatory standards and requirements for product labeling and tracking.
- *Data security and privacy*: Protecting patient data and sensitive pharmaceutical information is paramount. Robust data security and privacy features are essential for chipless RFID systems used in healthcare supply chains.

Sports and entertainment venues:

- *Ticketing and access control*: Chipless RFID tags can be used for contactless ticketing and access control in sports and entertainment venues. Ensuring fast and reliable tag identification at entry points is crucial for seamless event management.
- *Enhancing fan experience*: Chipless RFID technology can enhance the fan experience in sports events by enabling cashless payments, personalized services, and interactive engagement.
- *Data analytics and fan engagement*: Analyzing chipless RFID data can provide valuable insights into fan behavior, preferences, and engagement, enabling targeted marketing and fan engagement strategies.
- *Security and safety*: Chipless RFID systems in sports venues must prioritize security and safety, especially concerning crowd management and emergency response.

Water management and conservation:

- *Environmental monitoring*: Chipless RFID tags can be used for remote water quality monitoring and environmental data collection. Developing tags with sensors for water parameters and implementing efficient data transmission mechanisms is essential.
- *Water usage tracking*: Chipless RFID technology can be integrated into water meters for tracking water usage and promoting water conservation practices.
- *Smart irrigation systems*: Chipless RFID tags with environmental sensors can contribute to the development of smart irrigation systems, optimizing water usage in agriculture and landscaping.
- *Data integration and analysis*: Research into data integration platforms and analytics for water management is crucial for optimizing resource allocation and improving water conservation efforts.

Energy and utility management:

- *Integration with smart grids*: Chipless RFID technology can be integrated into smart grid systems for efficient energy management and monitoring. Ensuring interoperability and data exchange with smart meters and utility infrastructure is essential.
- *Energy consumption monitoring*: Chipless RFID tags with energy sensing capabilities can enable energy consumption monitoring in buildings and industries, facilitating energy conservation efforts.
- *Cost-effectiveness*: The cost-effectiveness of chipless RFID solutions for energy and utility management will be a critical factor for widespread adoption.
- *Data privacy and security*: Protecting consumer data and ensuring secure communication between chipless RFID tags and utility systems is crucial for energy and utility management applications.

Law enforcement and public safety:

- *Identification and authentication*: Chipless RFID tags can be used for identification and authentication in law enforcement applications. Ensuring secure and accurate identification is crucial for public safety.
- *Evidence tracking and chain of custody*: Chipless RFID technology can aid in evidence tracking and maintaining the chain of custody in law enforcement investigations.
- *Integration with surveillance systems*: Integrating chipless RFID technology with surveillance systems and public safety infrastructure requires robust data integration and interoperability.
- *Privacy and ethical considerations*: Law enforcement applications must balance data collection for public safety with individual privacy and ethical considerations.

Humanitarian aid and disaster response:

- *Rapid deployment and scalability*: In disaster response scenarios, chipless RFID technology must enable rapid deployment and scalability to facilitate efficient aid distribution and tracking.
- *Remote tracking and identification*: Chipless RFID tags can be used for remote tracking and identification of affected individuals and disaster relief supplies in hard-to-reach areas.
- *Data analytics for aid allocation*: Analyzing chipless RFID data can inform aid allocation decisions and optimize resource distribution in humanitarian operations.
- *Resilience and durability*: Chipless RFID tags used in disaster response scenarios must be resilient and capable of withstanding harsh environmental conditions.

Space exploration and satellite applications:

- *Miniaturization and space constraints*: In space exploration and satellite applications, chipless RFID tags must be miniaturized to fit within the constraints of spacecraft and satellite platforms.
- *Reliability in space environment*: Ensuring the reliability and durability of chipless RFID tags in the extreme conditions of space, including radiation exposure and vacuum, is crucial.
- *Localization and satellite tracking*: Chipless RFID tags can contribute to satellite tracking and localization in space missions.
- *Data communication and telemetry*: Chipless RFID technology can be used for data communication and telemetry in satellite systems, requiring research into efficient communication protocols.

Oil and gas industry:

- *Asset tracking and condition monitoring*: Chipless RFID tags can be used for asset tracking and condition monitoring in the oil and gas industry. Ensuring reliable performance in remote and hazardous environments is essential.
- *Safety and compliance*: Chipless RFID systems used in the oil and gas industry must adhere to safety standards and regulations, considering the potential risks associated with the industry.
- *Interference and harsh environments*: The oil and gas industry involve complex and harsh environments, including electromagnetic interference and extreme temperatures. Research into interference mitigation techniques and ruggedized tags is necessary.
- *Energy efficiency and sustainability*: Developing energy-efficient and eco-friendly chipless RFID solutions aligning with sustainability initiatives in the oil and gas industry is crucial.

Mining and resource extraction:

- *Asset tracking and inventory management*: Chipless RFID tags can aid in asset tracking and inventory management in the mining and resource extraction industry. Ensuring reliable tracking of equipment, tools, and resources in harsh environments is essential.
- *Safety and compliance*: Chipless RFID technology can contribute to safety and compliance in mining operations. Ensuring worker safety, equipment maintenance, and compliance with regulatory requirements are critical.
- *Durability and reliability*: In mining and resource extraction applications, chipless RFID tags must be durable and reliable to withstand rough conditions and mechanical stress.
- *Real-time monitoring*: Real-time tracking and monitoring of mining operations can be achieved with chipless RFID systems, enabling timely decision-making and optimization of mining processes.

Aerospace and aviation:

- *Aircraft part identification*: Chipless RFID technology can be used for aircraft part identification and maintenance tracking. Ensuring accurate identification and tracking of parts is crucial for aircraft safety.
- *Security and antitampering measures*: Chipless RFID tags used in aerospace and aviation must possess antitampering features to ensure the integrity of critical components.
- *Read performance at high altitudes*: Ensuring reliable tag reading at high altitudes, such as during flight or space missions, is a research challenge.
- *Data management and analysis*: An important consideration is to manage and analyze chipless RFID data in the aerospace industry to inform maintenance schedules and optimize operations.

Art and cultural heritage preservation:

- *Artwork and artifact identification*: Chipless RFID tags can be used for noninvasive identification and tracking of artworks and cultural artifacts. Ensuring minimal impact on valuable objects is crucial for preservation.
- *Museum asset management*: Chipless RFID technology can aid in museum asset management and collection tracking, contributing to improved inventory and conservation efforts.
- *Data security and preservation*: Protecting sensitive data related to cultural heritage and artwork is essential. Ensuring secure data storage and access control is crucial for preserving historical and artistic information.
- *Integration with conservation efforts*: Research into integrating chipless RFID technology with conservation efforts, such as environmental monitoring and artifact preservation, is valuable.

Automotive and vehicle identification:

- *Vehicle registration and identification*: Chipless RFID technology can be used for vehicle registration and identification, streamlining toll collection and traffic management.
- *Integration with vehicle systems*: Integrating chipless RFID technology with vehicle systems and smart infrastructure can enhance vehicle-to-infrastructure communication and enable smart city applications.
- *Data security and privacy*: Protecting vehicle data and user privacy in automotive chipless RFID systems is essential.
- *Reliability in harsh automotive conditions*: Ensuring the reliability and durability of chipless RFID tags in automotive environments, including exposure to heat, vibrations, and road conditions, is a consideration.

Environmental remediation and contaminant detection:

- *Contaminant sensing*: Chipless RFID tags can be used for contaminant sensing and environmental remediation applications. Developing sensitive and selective sensing mechanisms is essential.
- *Real-time monitoring*: Real-time monitoring of environmental contaminants requires chipless RFID tags with low power consumption and efficient data transmission.
- *Data analytics and environmental analysis*: Chipless RFID data can contribute to environmental analysis and decision-making for remediation efforts.
- *Deployment in remote areas*: In environmental remediation applications, chipless RFID tags may be deployed in remote and hard-to-reach areas. Ensuring the reliability of communication and data retrieval from such areas is critical.

Construction and building management:

- *Asset tracking and inventory management*: Chipless RFID tags can aid in asset tracking and inventory management in construction projects. Ensuring efficient tracking of materials and equipment is essential for project efficiency.
- *Building access and security*: Chipless RFID technology can be used for secure building access and personnel identification in construction sites and building complexes.
- *Integration with BIM systems*: Integrating chipless RFID technology with building information modeling (BIM) systems can enhance project management and resource allocation.
- *Durability and ruggedness*: Chipless RFID tags used in construction must be durable and rugged to withstand construction site conditions.

Financial services and banking:

- *Contactless payments*: Chipless RFID technology can be used for contactless payments in financial services. Ensuring secure and reliable payment transactions is critical for customer trust.
- *Integration with mobile wallets*: Integrating chipless RFID technology with mobile payment platforms and digital wallets requires standardization and compatibility.
- *User authentication*: Chipless RFID systems used in financial services must prioritize user authentication and security features.
- *Data security and privacy*: Protecting customer financial data and ensuring secure communication between chipless RFID tags and financial systems is paramount.

Forest management and wildlife conservation:

- *Wildlife tracking*: Chipless RFID tags can be used for wildlife tracking and monitoring in forest management and conservation efforts.
- *Antipoaching and security measures*: Integrating chipless RFID technology with antipoaching efforts and security measures can aid in wildlife protection.
- *Forest inventory and resource management*: Chipless RFID tags can contribute to forest inventory and resource management, enabling efficient tracking of timber and forest products.
- *Durability and environmental impact*: Ensuring the durability of chipless RFID tags in forest environments and minimizing their environmental impact is important for conservation efforts.

Precision agriculture and farming:

- *Crop monitoring and sensing*: Chipless RFID tags with environmental sensors can aid in precision agriculture by monitoring crop conditions and environmental parameters.
- *Livestock management*: Chipless RFID tags can be used for livestock tracking and health monitoring in precision farming applications.
- *Data integration and analysis*: Research into data integration platforms and analytics for precision agriculture is crucial for optimizing resource utilization and crop yield.
- *Energy efficiency*: Developing energy-efficient chipless RFID tags and systems for precision agriculture can contribute to sustainable farming practices.

Robotics and automation:

- *Tag localization and object recognition*: Chipless RFID tags can be used for tag localization and object recognition in robotics and automation applications.

- *Integration with robotic systems*: Integrating chipless RFID technology with robotic systems requires seamless data exchange and efficient communication protocols.
- *Anticollision and navigation*: Ensuring efficient anticollision algorithms and navigation capabilities for robotic systems using chipless RFID technology is vital.
- *Reliability and robustness*: Chipless RFID tags used in robotics and automation must be reliable and robust to withstand the dynamic and unpredictable nature of robotic environments.

Textile and fashion industry:

- *Product tracking and authentication*: Chipless RFID tags can aid in product tracking and authentication in the textile and fashion industry, addressing issues like counterfeit products.
- *Supply chain transparency*: Chipless RFID technology can contribute to supply chain transparency in the textile and fashion industry, enabling traceability of raw materials and manufacturing processes.
- *Data security and privacy*: Ensuring secure communication and data privacy in textile and fashion applications using chipless RFID is crucial.
- *Integration with retail systems*: Integrating chipless RFID technology with retail systems, such as inventory management and point-of-sale terminals, is essential for seamless operations.

Renewable energy and grid management:

- *Integration with renewable energy sources*: Chipless RFID technology can be integrated with renewable energy sources for efficient grid management and monitoring.
- *Smart grid integration*: Research into smart grid integration of chipless RFID technology can optimize renewable energy utilization and grid stability.
- *Energy storage and battery management*: Chipless RFID tags with energy sensing capabilities can aid in energy storage and battery management for renewable energy systems.
- *Data analytics for energy optimization*: Analyzing chipless RFID data can provide insights into energy consumption patterns and inform energy optimization strategies.

Oceanography and marine science:

- *Environmental sensing*: Chipless RFID tags with environmental sensors can be used for oceanography and marine science applications, monitoring water quality and environmental conditions.
- *Marine life tracking*: Chipless RFID technology can contribute to marine life tracking and migration studies.

- *Data transmission and telemetry*: Ensuring efficient data transmission and telemetry in underwater environments using chipless RFID is a research challenge.
- *Durability and marine-friendly materials*: Chipless RFID tags used in oceanography must be durable and constructed from materials that do not harm marine life.

Data center and IT asset management:

- *IT asset tracking*: Chipless RFID tags can be used for IT asset tracking and management in data centers, ensuring efficient inventory control and equipment monitoring.
- *Security and access control*: Chipless RFID systems used in data centers must prioritize security and access control for sensitive information and equipment.
- *Energy efficiency*: Developing energy-efficient chipless RFID solutions for data center asset management can contribute to overall energy savings.
- *Integration with IT infrastructure*: Integrating chipless RFID technology with data center management systems and IT infrastructure requires seamless data exchange and interoperability.

Mining and resource extraction:

- *Asset tracking and inventory management*: Chipless RFID tags can aid in asset tracking and inventory management in mining and resource extraction industries, ensuring efficient tracking of equipment and resources.
- *Safety and compliance*: Chipless RFID technology can contribute to safety and compliance in mining operations, ensuring worker safety, equipment maintenance, and compliance with regulatory requirements.
- *Durability and ruggedness*: In mining and resource extraction applications, chipless RFID tags must be durable and rugged to withstand harsh conditions and mechanical stress.
- *Real-time monitoring*: Real-time tracking and monitoring of mining operations can be achieved with chipless RFID systems, enabling timely decision-making and optimization of mining processes.

This section highlighted the limitations of chipless RFID technology across diverse domains in the real world. While chipless RFID holds immense potential for various applications, addressing these limitations through innovative research and development is crucial for achieving widespread adoption and realizing the full benefits of this technology in practical scenarios. As technology continues to advance, addressing these limitations will pave the way for a more efficient, secure, and interconnected future in which chipless RFID plays a vital role.

9.5 Summary

Recapitulation of the key challenges in implementing chipless RFID in industry and emphasizing the importance of collaboration, innovation, and ongoing improvements for successful integration will make easy access to chipless RFID. By addressing the challenges discussed in the previous sections and understanding their implications, industries can effectively leverage chipless RFID technology to enhance operational efficiency, increase visibility, and unlock new possibilities for the future of industrial applications. This chapter covered a wide range of issues related to chipless RFID technology and its applications in various industries and domains. Chipless RFID refers to a type of radio frequency identification technology that does not require an integrated circuit or microchip in the tag, making it more cost-effective and versatile than traditional RFID tags. The technology has numerous potential applications across industries, including logistics and supply chain management, healthcare, agriculture, smart cities, automotive, defense, and more. Chipless RFID offers unique advantages in each domain, such as low-cost mass production, long-range reading capabilities, and resistance to harsh environments. However, chipless RFID faces several challenges and limitations despite its potential. These include limitations in data capacity, read range, and reading speed, as well as potential interference from surrounding materials. Research and development efforts are needed to overcome these limitations and improve the technology's performance and capabilities. In logistics and supply chain management, chipless RFID can revolutionize inventory tracking and reduce counterfeiting. Healthcare applications can benefit from chipless RFID for patient monitoring and drug authentication. Agriculture can use the technology for precision farming and livestock tracking, while smart cities can optimize services and infrastructure using chipless RFID. In automotive and aerospace industries, chipless RFID can aid in vehicle identification and tracking, as well as aircraft part authentication. Defense and security applications can benefit from chipless RFID for asset tracking and secure communication. Challenges arise in data analytics, integration with existing systems, privacy concerns, and cost-effectiveness. Efforts are required to ensure data security, user privacy, and regulatory compliance in various applications. Despite these challenges, there are numerous opportunities for future research and development in chipless RFID technology. Researchers can focus on improving tag performance, enhancing data analytics, increasing read ranges, and developing more energy-efficient solutions. Overall, chipless RFID holds immense potential in transforming industries and driving technological advancements. Addressing the limitations and challenges will be crucial for realizing the full benefits of this technology in the real world. With ongoing research and innovation, chipless RFID has the potential to revolutionize various sectors, making them more efficient, secure, and interconnected.

References

[1] Islam, M. A., and N. C. Karmakar, "Real-World Implementation Challenges of a Novel Dual-Polarized Compact Printable Chipless RFID Tag," *IEEE Transactions on Microwave*

Theory and Techniques, Vol. 63, No. 12, December 2015, pp. 4581–4591, doi: 10.1109/TMTT.2015.2495285.

[2] Mishra, D. P., and S. K. Behera, "Resonator Based Chipless RFID: A Frequency Domain Comprehensive Review," *IEEE Transactions on Instrumentation and Measurement*, Vol. 72, Art no. 5500716, 2023, pp. 1–16, doi 10.1109/TIM.2022.3225027.

[3] Suresh, K., V. Jeoti, S. Soeung, M. Drieberg, M. Goh, and M. Z. Aslam, "A Comparative Survey on Silicon Based and Surface Acoustic Wave (SAW)-Based RFID Tags: Potentials, Challenges, and Future Directions," *IEEE Access*, Vol. 8, 2020, pp. 91624–91647, doi: 10.1109/ACCESS.2020.2976533.

[4] Behera, S. K., "Chipless RFID Sensors for Wearable Applications: A Review," *IEEE Sensors Journal*, Vol. 22, No. 2, January 15, 2022, pp. 1105–1120, doi: 10.1109/JSEN.2021.3126487.

[5] Behera, S. K., and N. C. Karmakar, "Chipless RFID Printing Technologies: A State of the Art," *IEEE Microwave Magazine*, Vol. 22, No. 6, June 2021, pp. 64–81, doi: 10.1109/MMM.2021.3064099.

[6] Subrahmannian, A., and S. K. Behera, "Chipless RFID Sensors for IoT-Based Healthcare Applications: A Review of State of the Art," *IEEE Transactions on Instrumentation and Measurement*, Vol. 71, Art no. 8003920, 2022, pp. 1–20, doi: 10.1109/TIM.2022.3180422.

[7] Mishra, D. P., and S. K. Behera, "Multibit Coded Passive Hybrid Resonator Based RFID Transponder with Windowing Analysis," *IEEE Transactions on Instrumentation and Measurement*, Vol. 71, Art no. 8005907, 2022, pp. 1–7, doi: 10.1109/TIM.2022.3201538.

[8] Sethy, P., and S. K. Behera, "Chipless RFID Sensors for Bioimplants: A Comprehensive Review," *IEEE Microwave Magazine*, Vol. 24, No. 7, July 2023, pp. 41–60, doi: 10.1109/MMM.2023.3265465.

[9] Mishra, D. P., T. K. Das, and S. K. Behera, "Design of a 3-Bit Chipless RFID Tag Using Circular Split-Ring Resonators for Retail and Healthcare Applications," *2020 National Conference on Communications (NCC),* Kharagpur, India, 2020, pp. 1–4, doi: 10.1109/NCC48643.2020.9056018.

[10] Mishra, D. P., and S. K. Behera, "A Novel Technique for Dimensional Space Reduction in Passive RFID Transponders," *2021 2nd International Conference on Range Technology (ICORT),* Chandipur, Balasore, India, 2021, pp. 1–4, doi: 10.1109/ICORT52730.2021.9581971.

[11] Mishra, D. P., T. Kumar Das, P. Sethy, and S. K. Behera, "Design of a Multi-Bit Chipless RFID Tag Using Square Split-Ring Resonators," *2019 IEEE Indian Conference on Antennas and Propogation (InCAP),* Ahmedabad, India, 2019, pp. 1–4, doi: 10.1109/InCAP47789.2019.9134626.

[12] Sahu, P. P., D. Prasad Mishra, T. K. Das, and S. K. Behera, "Design of a Chipless RFID Tag for 2.4 GHz and 5.8 GHz ISM Band Applications," *2020 IEEE International Students' Conference on Electrical, Electronics and Computer Science (SCEECS)*, Bhopal, India, 2020, pp. 1–4, doi: 10.1109/SCEECS48394.2020.108.

[13] Mishra, D. P., and S. K. Behera, "Analysis of Chipless RFID Tags Based on Circular-SRR and Koch Snowflake Fractal for Space-Reduction and Bit-Coding Improvement," *2021 IEEE Indian Conference on Antennas and Propagation (InCAP),* Jaipur, Rajasthan, India, 2021, pp. 621–624, doi: 10.1109/InCAP52216.2021.9726478.

[14] Mishra, D. P., I. Goyal, and S. K. Behera, "A Comparative Study of Two Different Octagonal Structure-Based Split Ring Resonators," *2022 IEEE Microwaves, Antennas, and Propagation Conference (MAPCON),* Bangalore, India, 2022, pp. 1007–1012, doi: 10.1109/MAPCON56011.2022.10047484.

[15] Mishra, D. P., I. Goyal, and S. K. Behera, "A Technique to Improve RCS in Passive Chipless RFID Tags by Incorporating Array Structure," *2022 IEEE Microwaves, Antennas, and Propagation Conference (MAPCON),* Bangalore, India, 2022, pp. 1478–1483, doi: 10.1109/MAPCON56011.2022.10047155.

[16] Subrahmannian, A., D. P. Mishra, and S. K. Behera, "Multi-Bit Passive RFID Tag Design using Concentric Rectangular Strip Resonators for 2.4 GHz and 5.8 GHz ISM Band," *2022*

IEEE Microwaves, Antennas, and Propagation Conference (MAPCON), Bangalore, India, 2022, pp. 1434–1438, doi: 10.1109/MAPCON56011.2022.10047522.

[17] Miao, F., Y. Han, and B. Tao, "ZnO/MoS2/rGO Nanocomposite Non-Contact Passive and Chip-Less LC Humidity Sensor," *IEEE Sensors Journal*, Vol. 22, No. 14, July 15, 2022, pp. 13891–13897, doi: 10.1109/JSEN.2022.3169152.

[18] Shi, G., X. Shen, Y. He, and H. Ren, "Passive Wireless Detection for Ammonia Based on 2.4 GHz Square Carbon Nanotube-Loaded Chipless RFID-Inspired Tag," *IEEE Transactions on Instrumentation and Measurement,* Vol. 72, Art no. 9510812,2023, pp. 1–12, doi: 10.1109/TIM.2023.3300133.

[19] Amin, E. M., M. S. Bhuiyan, N. C. Karmakar, and B. Winther-Jensen, "Development of a Low Cost Printable Chipless RFID Humidity Sensor," *IEEE Sensors Journal*, Vol. 14, No. 1, January 2014, pp. 140–149, doi: 10.1109/JSEN.2013.2278560.

[20] Alves, A. A. C., D. H. Spadoti, and L. L. Bravo-Roger, "Optically Controlled Multiresonator for Passive Chipless Tag," *IEEE Microwave and Wireless Components Letters*, Vol. 28, No. 6, June 2018, pp. 467–469, doi: 10.1109/LMWC.2018.2824726.

[21] Arjomandi, L. M., G. Khadka, Z. Xiong, and N. C. Karmakar, "Document Verification: A Cloud-Based Computing Pattern Recognition Approach to Chipless RFID," *IEEE Access*, Vol. 6, 2018, pp. 78007–78015, doi: 10.1109/ACCESS.2018.2884651.

[22] Herrojo, C., J. Mata-Contreras, A. Núñez, F. Paredes, E. Ramon, and F. Martín, "Near-Field Chipless-RFID System with High Data Capacity for Security and Authentication Applications," *IEEE Transactions on Microwave Theory and Techniques*, Vol. 65, No. 12, December 2017, pp. 5298–5308, doi: 10.1109/TMTT.2017.2768029.

[23] Khadka, G., B. Ray, N. C. Karmakar, and J. Choi, "Physical-Layer Detection and Security of Printed Chipless RFID Tag for Internet of Things Applications," *IEEE Internet of Things Journal*, Vol. 9, No. 17, September 1, 2022, pp. 15714–15724, doi: 10.1109/JIOT.2022.3151364.

[24] Feng, Y., L. Xie, Q. Chen, and L.-R. Zheng, "Low-Cost Printed Chipless RFID Humidity Sensor Tag for Intelligent Packaging," *IEEE Sensors Journal*, Vol. 15, No. 6, June 2015, pp. 3201–3208, doi: 10.1109/JSEN.2014.2385154.

[25] Vena, A., L. Sydänheimo, M. M. Tentzeris, and L. Ukkonen, "A Fully Inkjet-Printed Wireless and Chipless Sensor for CO2 and Temperature Detection," *IEEE Sensors Journal*, Vol. 15, No. 1, January 2015, pp. 89–99, doi: 10.1109/JSEN.2014.2336838.

CHAPTER 10

Conclusion

10.1 Book Summary

We explored the fascinating realm of chipless RFID printing technologies in this book, covering a wide range of topics. Each chapter provided useful insights and in-depth knowledge, taking us on a journey through the accomplishments, problems, and future prospects of chipless RFID technology. As we get to the end of this book, consider the important findings and contributions of each chapter.

10.2 Chapter 1

In the first chapter, we laid the groundwork for our investigation by introducing chipless RFID printing technologies. We covered the fundamentals, benefits, and prospective applications of this innovative technology. The chapter was a key guide to comprehending the underlying concepts that form the foundation of future chapters. We began our journey through the intriguing world of RFID technology, investigating both chip-based and chipless RFID systems, as well as numerous tracking technologies such as smart cards, biometric tracking, fingerprinting, OCR, and barcoding. This comprehensive introduction has established the framework for understanding the basic principles and applications of RFID in a variety of sectors and scenarios. As we conclude this chapter, reflection on the key points are as follows:

Key insights:

- *Introduction to RFID*: Exploration by understanding the basics of RFID, a technology that utilizes radio waves for automatic data capture and identification of objects or individuals, is commended.
- *Chip-based versus chipless RFID*: We distinguished between chip-based and chipless RFID systems, with a note that chip-based RFID tags rely on integrated circuits, and chipless RFID tags use pattern-based techniques for identification.
- *Tracking technologies*: The chapter introduced various tracking technologies such as smart cards, biometric tracking, fingerprinting, OCR, and barcode, highlighting their specific applications and functionalities.

- *RFID devices and barcoding*: We explored the working principles of RFID devices and their comparison to traditional barcoding systems, emphasizing the advantages of RFID in terms of read range and data capacity.
- *Introduction to chipless RFID*: The concept of chipless RFID intrigued us, as we learned that these tags do not require an integrated circuit, allowing for reduced cost and increased flexibility in manufacturing.
- *RCS of a tag*: We gained insight into the RCS of a chipless RFID tag with calculation.
- *Chipless RFID printing technology*: The chapter introduced chipless RFID printing technology, which enables the cost-effective and scalable production of RFID tags using various printing techniques.
- *Tag printing process*: An overview of the tag printing process highlighted the importance of material selection, characterization, and precise fabrication for optimal tag performance. Chipless RFID offers an exciting alternative, revolutionizing the field of RFID with its cost-effectiveness, flexibility, and potential for integration into a wide range of industries.

Finally, we offered an outline of the following chapters, providing an overview of the in-depth investigation of many elements of chipless RFID printing technology. The advent of chipless RFID printing technology opens up new opportunities for innovation in a variety of fields, including supply chain management, logistics, healthcare, retail, and others. The ability to print RFID tags using scalable printing processes opens up new options for large-scale deployment, promoting RFID technology adoption in situations where chip-based solutions were previously difficult to implement.

10.3 Chapter 2

The second chapter provided a thorough assessment of present research and breakthroughs in chipless RFID. We aggregated and examined a wide range of scholarly publications to get a complete grasp of the field's present state-of-the-art. This chapter provided important context for the findings presented in the next chapters. We began our investigation with an introduction to RFID, laying a good framework for our subsequent investigation. We investigated conventional RFID systems, learning about their operation and uses in many industries. The distinction between active and passive RFID sensors clarifies their varied properties and power requirements. The chapter discussed the introduction of fully printable smart sensing materials, which enable flexible and low-cost RFID tag fabrication. The concept of multiple parameter sensing further captivated us, as we discovered the potential to integrate diverse sensing capabilities into a single RFID tag, transforming the landscape of real-time data capture and monitoring. Reader antennas' critical significance in RFID applications, guaranteeing seamless tag reading and data distribution were explored.

Key insights:

- *Conventional RFID*: Our exploration of conventional RFID systems revealed their significance in tracking and identification, serving as the backbone of numerous industries.
- *Active and passive RFID sensors*: The distinction between active and passive RFID sensors highlighted their respective applications and power requirements, influencing their suitability for different scenarios.
- *Passive resonators in chipless RFID*: The revelation of passive resonators in chipless RFID marked a significant advancement, offering cost-effective and scalable RFID tags with enhanced flexibility.
- *Fully printable smart sensing materials*: The emergence of fully printable smart sensing materials introduced a new paradigm of RFID tag fabrication, enabling cost-effective and flexible production.
- *Multiple parameter sensing*: The concept of multiple parameter sensing opened doors to enhanced data capture capabilities, paving the way for innovative applications in diverse domains.
- *Reader antennas for RFID applications*: Our exploration of reader antennas underscored their critical role in efficient RFID tag reading and data exchange.
- *Applications*: Witnessing the diverse applications of RFID technology showcased its transformative impact across industries, redefining operational efficiency and data management.
- *Motivation*: Our underlying motivation to explore chipless RFID printing technology and smart sensing materials inspired us to unveil their potential and real-world applications.
- *Objectives*: With clear objectives in mind, our journey through the subsequent chapters will focus on deepening our understanding of chipless RFID and exploring its transformative impact on multiple parameter sensing and real-world applications.

The insights gained from this chapter have profound implications for the future of RFID technology and smart sensing materials. The advent of chipless RFID printing technology, passive resonators, and fully printable smart sensing materials opens new avenues for innovation, scalability, and integration in various industries. The ability to capture multiple parameters within a single RFID tag presents opportunities for transformative applications in healthcare, environmental monitoring, logistics, and beyond. The subsequent chapters continued to unravel the intricacies of chipless RFID and smart sensing materials, delving deeper into cutting-edge advancements and exploring real-world implementations. We invite readers to join us on this enlightening journey as we explore the future of smart sensing and identification systems fueled by the remarkable capabilities of chipless RFID and fully printable smart materials.

10.4 Chapter 3

Chapter 3 focused on smart materials and their crucial role in chipless RFID printing technologies. We explored various types of smart materials, such as conductive inks, nanomaterials, and functional polymers, which enable the creation of printable chipless RFID tags. The chapter highlighted the significance of material selection and its impact on the performance of RFID tags.

Key insights:

- *Smart materials*: We explored the world of smart materials, which possess the extraordinary ability to sense and respond to environmental changes, making them ideal candidates for RFID sensing applications.
- *Classification of smart materials for RF sensing*: The classification of smart materials provided valuable insights into their specific sensing functionalities, with each category catering to distinct environmental parameters.
- *Temperature sensing materials*: We delved into materials that excel in temperature sensing, enabling precise monitoring of thermal changes in diverse environments.
- *Humidity sensing materials*: Our exploration of humidity sensing materials showcased their significance in applications requiring accurate humidity measurements.
- *pH sensing materials*: The concept of pH sensing materials captivated us as we recognized their potential to enable pH monitoring in various fields, such as agriculture, healthcare, and industrial processes.
- *Gas sensing materials*: We witnessed the transformative impact of gas sensing materials, enabling the detection of specific gases for environmental monitoring and safety applications.
- *Strain and crack sensing materials*: The exploration of strain and crack sensing materials highlighted their role in structural health monitoring and damage detection in critical infrastructures.
- *Graphene*: Our exploration of graphene showcased its exceptional properties and potential applications in RFID sensing and beyond.

Smart materials possess the extraordinary ability to sense and respond to various environmental parameters, enabling a new era of RFID sensing applications. The diverse classifications of smart materials offer endless possibilities for real-world implementations in industries such as healthcare, agriculture, environmental monitoring, and beyond.

10.5 Chapter 4

In Chapter 4, we delved into the critical process of characterization of smart materials for printing chipless RFID tags. We discussed various characterization techniques, such as electrical measurements, spectroscopy, and microscopy, to evaluate material properties and optimize tag performance. This chapter underscored the importance of precise material characterization for successful chipless RFID tag fabrication. From X-ray diffraction to Raman scattering spectroscopy, secondary ion mass spectrometer, transmission electron microscopy, scanning electron microscope, atomic force microscopy, UV–visible spectrophotometers, electrical conductivity measurement, and basic microwave-material interaction aspects, we delved into each technique's capabilities and applications.

Key insights:

- *Introduction*: The chapter's introduction emphasized the importance of the characterization of smart materials, enabling us to harness their unique properties and optimize their performance for RFID sensing applications.
- *Characterization of sensing materials*: We gained valuable insights into the significance of characterizing smart materials, which aids in understanding their structural, optical, electrical, and microwave properties.
- *X-ray diffraction*: The exploration of X-ray diffraction showcased its ability to analyze the crystal structure of materials, shedding light on their atomic arrangement and phase identification.
- *Transmission electron microscopy*: We marveled at the capabilities of transmission electron microscopy, which enables high-resolution imaging and analysis of nanoscale structures and defects.
- *Scanning electron microscope*: The significance of scanning electron microscope was highlighted, offering 3D imaging and surface analysis of materials at high magnification.

The diverse range of characterization techniques presented in this chapter highlighted their indispensability in comprehending smart materials' unique characteristics. Each method contributes to the in-depth analysis of structural, optical, electrical, and microwave properties, guiding researchers and engineers in optimizing these materials for chipless RFID applications. The comprehensive analysis aids in identifying the most suitable materials for specific sensing parameters and real-world applications. The advanced techniques discussed there empower researchers and engineers to comprehend the intricate details of smart materials and guide their integration into chipless RFID tags. The information gleaned from these characterization techniques facilitates the optimization of materials for specific sensing applications, contributing to the advancement of RFID technology in diverse industries.

10.6 Chapter 5

The fifth chapter focused on the intriguing potential of passive chipless RFID tags in biomedical applications. The transformational influence of chipless RFID technology in the healthcare arena was highlighted in this chapter. We investigated design ideas and numerous geometries, focusing on chipless sensors based on MSRR and rectangular resonator. The chapter also included simulated and measured RCS data, as well as current distribution patterns, which provided useful insights into the performance of these tags.

Key insights:

- *Introduction*: The chapter's introduction highlighted the pivotal role of passive printable chipless RFID tags in biomedical applications, enabling nonintrusive and cost-effective data capture for medical sensing and monitoring.
- *Design principle of passive resonator-based tags*: We gained a comprehensive understanding of the design principles behind passive resonator–based tags, exploring their unique characteristics and functionalities.
- *Chipless sensor*: We delved into the concept of MSRR based chipless sensors, recognizing their potential for high-performance biomedical applications. The exploration of rectangular resonator–based chipless sensors unveiled their capabilities in biomedical sensing, offering distinct advantages for specific applications.
- *Simulated and measured RCS*: The comparison of simulated and measured RCS data provided valuable insights into the accuracy and reliability of the tags' response patterns.
- *Current distribution patterns*: We investigated the current distribution patterns within the tags, understanding their influence on the tags' performance and resonance frequencies.
- *Results and discussions*: The chapter presented detailed results and discussions, elucidating the findings from the experiments and simulations, leading to a comprehensive understanding of the tags' behaviors.

The design principles, geometries, and resonator-based sensors discussed in this chapter demonstrated the versatility of these tags, catering to various biomedical parameters with precision and accuracy. The underlying physics revealed by current distribution patterns further contributes to the optimization and refinement of these tags.

10.7 Chapter 6

Chapter 6 focused on WBAN applications of passive tags. The chapter underscored the significance of chipless RFID technology in advancing personalized healthcare solutions. The chapter began with an introduction emphasizing the importance of RFID technology in the context of body area networks, setting the way for our investigation of chipless sensors based on modified spiral resonator, high bit density based on MSRR, and modified CSRR.

Key insights:

- *Introduction*: The chapter's introduction highlighted the pivotal role of passive printable chipless RFID tags in body area networks, enabling nonintrusive, efficient, and seamless data capture for healthcare monitoring and applications.

- *Modified spiral resonator–based chipless sensor*: We gained insights into the unique design and working principles of modified spiral resonator–based chipless sensors, showcasing their potential in WBAN applications.
- *High bit density chipless sensor based on MSRR*: The exploration of high bit density chipless sensors based on MSRR revealed their capability to accommodate a large amount of data in a compact format, ideal for medical sensing and monitoring.
- *Modified CSRR–based chipless sensor*: We delved into the concept of modified CSRR-based chipless sensors, presenting detailed analysis through simulated and measured RCS data and current distribution patterns.
- *Comparison of literature*: The comparison of the chapter's findings with existing literature provided valuable insights into the advancements and contributions of the chipless sensors in the context of BAN applications.
- *Results and discussions*: The chapter presented in-depth results and discussions, offering a comprehensive understanding of the chipless sensor's behavior and performance in BAN scenarios.

These ground-breaking passive tags provide a nonintrusive, low-power, high-density solution for medical sensing and monitoring, paving the way for more efficient and tailored healthcare services. The comparison with current literature verifies the chipless sensor's originality and contributions in the domain of BAN applications. The simulated and measured RCS data, as well as the current distribution patterns, provide vital insights into the behavior of the tags, assisting researchers and engineers in refining their design and performance.

10.8 Chapter 7

In Chapter 7, resonators in chipless RFID Technologies and various signal processing techniques were explored. We discussed the design and optimization of resonators for enhancing tag sensitivity and reliability. Signal processing techniques for data extraction and tag authentication for robust chipless RFID systems were also investigated. We investigated the sensor comparison, categorizing the numerous chipless sensors and analyzing their diverse uses. The chapter moved on to investigate signal processing techniques in chipless RFID, such as the convolutional encoder, Viterbi algorithm, windowing techniques, wavelet transform, and so on. We thoroughly examined the performance and efficacy of various strategies through data and conversations.

Key insights:

- *Comparison of chipped and chipless sensors*: We gained valuable insights into the differences between chipped and chipless sensors, understanding their unique characteristics, benefits, and limitations.
- *Categories of chipless sensors*: The exploration of various categories of chipless sensors showcased their versatility in addressing diverse sensing parameters and applications.

- *Applications of chipless sensors*: The diverse applications of chipless sensors in industries such as healthcare, retail, logistics, and environmental monitoring were discussed.
- *Signal processing techniques in chipless RFID*: Our journey into signal processing techniques for chipless RFID unveiled their critical role in enhancing data accuracy and reliability in RFID sensing.
- *Convolutional encoder*: The exploration of the convolutional encoder highlighted its capability to improve data transmission and error correction in chipless RFID systems.
- *Viterbi algorithm*: We delved into the Viterbi algorithm, which enables efficient decoding and retrieval of data from chipless RFID tags, optimizing signal processing.
- *Windowing techniques*: The significance of windowing techniques was underscored, offering a means to segment and process chipless RFID signals with improved accuracy.
- *Wavelet transform*: We explored the transformative potential of wavelet transform, enabling multiresolution analysis and denoising RFID signals.
- *Noise reduction algorithms*: The chapter presented an array of noise reduction algorithms, catering to various types of noise to enhance the accuracy of RFID data.

Chapter 7 was influenced by the transformative potential of resonators in printing technologies and signal processing techniques for chipless RFID. These technologies play a key role in advancing the capabilities of both chipped and chipless sensors, enhancing their precision, sensitivity, and reliability. The categorization of chipless sensors showcases their versatility, catering to various industries and sensing parameters with ingenuity. Signal processing techniques emerged as a crucial component in the success of chipless RFID systems. The results and discussions provided a comprehensive analysis of their effectiveness, guiding researchers and engineers in selecting appropriate techniques for specific applications.

10.9 Chapter 8

Chapter 8 focused on antennas for RFID reader applications and printing techniques for chipless RFID tags. We discussed the design and optimization of RFID reader antennas for efficient tag reading. Moreover, we explored various printing techniques, such as inkjet and flexographic printing, to achieve cost-effective and scalable chipless RFID tag production.

Key insights:

- *Importance of reader antennas in RFID reader system*: The chapter began with an elucidation of the crucial role that reader antennas play in RFID reader systems, enabling efficient and reliable communication between RFID readers and tags.

- *A simple reader architecture*: We gained insights into the basic architecture of an RFID reader, understanding the components and their functions in the communication process.
- *LP and CP antennas*: The exploration of linearly polarized and circularly polarized antennas provided valuable information about their orientations and applications in RFID systems.
- *Reader antennas for biomedical applications*: The significance of reader antennas in biomedical applications was highlighted, presenting their use in medical sensing and monitoring scenarios.
- *Antenna arrays*: We explored the concept of antenna arrays, showcasing their benefits in enhancing signal strength, directionality, and coverage in RFID systems.
- *Limitations in RFID applications*: The chapter presented an analysis of the limitations faced in RFID applications, shedding light on challenges that need to be addressed for further advancements.
- *Fabrication techniques and classifications*: The exploration of various fabrication techniques for chipless RFID tags provided insights into the diverse methods used to create these tags.

The chapter provided a comprehensive overview of the role of reader antennas in RFID systems, enabling seamless communication between RFID readers and tags. The exploration of LP and CP antennas showcased their unique orientations and applications in RFID systems, catering to various requirements and scenarios. Fractal antennas, with their compact designs and enhanced performance, emerged as a fascinating area of study, opening up new possibilities for efficient RFID communication. The application of reader antennas in biomedical scenarios holds immense potential for medical sensing and monitoring, offering noninvasive and real-time data capture. The use of antenna arrays enhances the coverage and directionality of RFID systems, ensuring robust performance in diverse environments. However, the chapter also highlighted the limitations faced in RFID applications, guiding researchers and engineers in addressing these challenges to further enhance the capabilities of RFID technology.

10.10 Chapter 9

In Chapter 9, the challenges in chipless RFID technologies and future research directions were discussed. We also discussed the implementational challenges faced in the adoption and deployment of chipless RFID systems. The hurdles related to material selection, tag readability, and signal processing are studied. Moreover, we presented a visionary outlook on potential research areas, including security enhancements and advanced manufacturing techniques. The chapter highlighted several potential areas of exploration, such as enhancing data encoding techniques, optimizing reader sensitivity, extending reading range, developing energy-efficient tags, and advancing manufacturing processes. Acknowledging the limitations faced in the current state of chipless RFID, we recognized areas that require attention for overcoming hurdles in commercial adoption. These limitations encompassed tag detection accuracy, tag density, signal-to-noise ratio, read rate, and scalability. The complexities arising from hardware limitations, signal interference, standardization, and security concerns call for a concerted effort from researchers, engineers, and industry stakeholders to overcome these hurdles. Addressing these challenges will pave the way for the seamless integration of chipless RFID systems in diverse applications, revolutionizing industries such as retail, logistics, healthcare, agriculture, and beyond. As researchers and engineers collaborate to overcome implementational challenges and address limitations, we can anticipate significant advancements in the field. By investing in innovative research, exploring novel techniques, and embracing interdisciplinary approaches, we can elevate chipless RFID to new heights, fostering its integration into various sectors and enriching human experiences.

10.11 Concluding Remarks

We have taken an interesting voyage through the world of chipless RFID technology in this book, studying its different sides, uses, obstacles, and future research possibilities. Our journey began with an introduction to chip-based and chipless RFID, RFID devices, and printing technologies, laying the foundation for an illuminating journey into this groundbreaking field. We have unlocked the potential of this cutting-edge technology by first comprehending the principles and problems of tag printing, then anticipating future advances and overcoming limits.

Our journey through the chapters has demonstrated the versatility and importance of chipless RFID in a variety of applications, including healthcare, logistics, retail, and environmental monitoring. We learned about the relevance of antennas in RFID reader systems, the role of smart materials in RFID sensing, and the potential of signal processing techniques in enhancing RFID data accuracy. The problems and constraints revealed are opportunities for researchers, engineers, and industry stakeholders to join forces and design a future in which chipless RFID technology is effortlessly incorporated into our daily lives, driving innovation and improving human experiences. Chipless RFID technology's future prospects are limitless, encouraging us to start on a voyage of creativity and discovery.

As we come to the end of this book on chipless RFID printing technologies, we celebrate the incredible advances made in this sector as well as the limitless possibilities that lay ahead. Each chapter has added to our grasp of this transformative technology, from the fundamental principles to the cutting-edge applications. RFID without chips has the potential to change a variety of industries, including healthcare, logistics, and consumer electronics. The voyage does not end here; rather, it ushers in a new era of chipless RFID technology. It is sustained by the passion, curiosity, and dedication of researchers and inventors all across the world. Let us accept the difficulties, take the possibilities, and continue to push the boundaries of knowledge and innovation as we move forward. Let us begin on a journey of discovery together, influencing the future of chipless RFID technology and driving positive global change.

APPENDIX A

Case Studies in Chipless RFID Applications

In this appendix, we delve into real-world case studies that illustrate the diverse applications and benefits of chipless RFID technology in various industries. Each case study highlights how chipless RFID has been implemented to address specific challenges and deliver tangible results.

Case Study 1: Supply Chain Management and Logistics

Problem Statement: A global logistics company was facing issues related to inventory management and tracking of goods in their warehouses. Traditional barcoding systems were proving to be inefficient and prone to errors. They needed a more robust and automated solution to streamline their operations.

Solution: The company adopted chipless RFID technology for inventory tracking. They implemented fully printable chipless RFID tags that were cost-effective and allowed them to tag each item in their inventory easily. This significantly improved the accuracy and speed of inventory management.

Results: With chipless RFID, the company achieved real-time tracking of inventory, reducing errors, and enabling faster order processing. They reported a significant decrease in lost or misplaced items and an increase in overall operational efficiency. The cost savings from the implementation were substantial, making it a successful case of chipless RFID adoption in logistics.

Case Study 2: Healthcare and Patient Monitoring

Problem Statement: A large hospital was seeking a solution to enhance patient monitoring and streamline the process of tracking patients' vital signs and medications. The existing manual processes were time-consuming and had the potential for errors.

Solution: The hospital integrated chipless RFID technology into patient wristbands. These wristbands contained chipless RFID tags that could capture patient information, vital signs, and medication data. RFID readers placed throughout the hospital automatically collected this information, creating a real-time patient monitoring system.

Results: The implementation of chipless RFID in patient monitoring significantly improved the hospital's ability to provide timely and accurate care.

Nurses and doctors could access real-time patient data, leading to quicker responses to critical situations. Medication errors were reduced, and the hospital reported an overall improvement in patient care and safety.

Case Study 3: Agriculture and Precision Farming

Problem Statement: A large agricultural cooperative was looking for ways to optimize their farming processes and enhance crop yield. They needed a system to monitor and manage crop conditions, soil quality, and livestock.

Solution: The cooperative introduced chipless RFID technology with environmental sensors in their fields. These tags could monitor soil moisture, temperature, and other environmental parameters. Livestock were tagged with RFID for tracking and health monitoring. Data from these tags was integrated into a centralized system that provided real-time insights into crop conditions and livestock health.

Results: The use of chipless RFID in precision farming allowed the cooperative to make data-driven decisions. They optimized irrigation, fertilizer application, and livestock management. As a result, crop yields increased, and livestock health improved. This case study demonstrates the potential of chipless RFID in revolutionizing agriculture and food production.

APPENDIX B

Future Research and Development in Chipless RFID Technology

In this appendix, we explore potential areas for future research and development in chipless RFID technology. As the technology continues to evolve, researchers and innovators have the opportunity to address existing challenges and unlock new possibilities. The following areas represent promising directions for advancing chipless RFID:

Enhanced data encoding techniques: Developing more efficient and secure data encoding methods for chipless RFID tags will be crucial. Advanced encoding techniques can improve data capacity, security, and retrieval speed.

Optimizing reader sensitivity: Research into improving RFID reader sensitivity will extend the reading range and enhance the reliability of data capture. This is particularly important in applications that require long-range tracking.

Energy-efficient tags: Creating energy-efficient chipless RFID tags is essential, especially in scenarios where long-lasting battery life is crucial. Energy harvesting and power management strategies are areas for further exploration.

Advanced manufacturing processes: Innovations in manufacturing techniques for chipless RFID tags can lead to cost-effective and scalable production. The development of more precise and versatile printing methods is an exciting avenue for research.

Enhancing security features: In applications where data security is paramount, such as financial services and healthcare, research in enhancing security features for chipless RFID systems is vital. Encryption, authentication, and anticounterfeiting measures need continuous improvement.

Overcoming hardware limitations: Hardware limitations, such as read range and tag density, are ongoing challenges in chipless RFID. Researchers can explore novel approaches to address these limitations and unlock new applications.

Interference mitigation: Signal interference in crowded environments can impact chipless RFID performance. Future research can focus on interference mitigation techniques to ensure reliable data capture.

Standardization: Standardization of chipless RFID technology will promote interoperability and widespread adoption. Researchers and industry stakeholders should collaborate to establish common standards.

Scalability: Developing scalable solutions for large-scale deployments of chipless RFID technology is essential. This includes addressing issues related to tag density and data management.

Multisensing tags: Expanding the capabilities of chipless RFID tags to include multisensing functionality for environmental monitoring, healthcare, and other applications can open new avenues for research and development.

Chipless RFID technology continues to evolve, offering a multitude of opportunities for researchers and innovators. By addressing these research areas, we can advance the technology's capabilities and drive its integration into various industries, enhancing efficiency, security, and connectivity in the world of IoT and beyond.

APPENDIX C

Implementing Chipless RFID in Retail Inventory Management

In this appendix, we will explore the practical implementation of chipless RFID technology in retail inventory management. We will provide a step-by-step guide for retailers looking to adopt chipless RFID for more efficient and accurate inventory tracking.

Retailers often face challenges in inventory management, including inaccurate stock counts, time-consuming manual processes, and the risk of stockouts or overstock. Chipless RFID offers a solution to these challenges by providing real-time, accurate inventory tracking. In this chapter, we will outline the steps to implement chipless RFID in a retail environment.

Step 1: System Design and Planning

Before implementing chipless RFID, it's essential to plan the system. Retailers should consider the store layout, types of products, and existing infrastructure. Key steps in system design include:

- Determine the RFID tag format suitable for products;
- Select the RFID reader and antenna setup;
- Plan the placement of RFID readers throughout the store;
- Choose a data management system for RFID data.

Step 2: Tagging Merchandise

Once the system is planned, the next step is to tag merchandise with chipless RFID tags. The tagging process includes:

- Selecting appropriate RFID tags for various product types;
- Attaching or embedding RFID tags on products or packaging;
- Encoding each tag with product information, such as SKU and price.

Step 3: Installation of RFID Readers

Install RFID readers and antennas in strategic locations within the store. These readers capture RFID tag data and send it to the central database. Consider factors like reader placement and coverage area for optimal performance.

Step 4: Software Integration

Integrate RFID software with the retailer's inventory management system. This step ensures that RFID data is seamlessly integrated into the existing software infrastructure. The integration should enable real-time updates and alerts.

Step 5: Staff Training

Train store staff to work with the new chipless RFID system. This includes understanding how to handle tagged merchandise, operate RFID readers, and utilize the RFID data for inventory management.

Step 6: Inventory Audits and Monitoring

Implement regular inventory audits and monitoring to verify the accuracy and reliability of the chipless RFID system. This step helps identify and address any issues or discrepancies.

Step 7: Real-Time Inventory Management

With the chipless RFID system in place, retailers can perform real-time inventory management. Benefits include:

- Accurate stock counts;
- Reduced stockouts and overstock;
- Improved inventory turnover;
- Efficient restocking processes.

Implementing chipless RFID technology in retail inventory management streamlines processes, reduces errors, and enhances overall efficiency. By following these steps, retailers can harness the benefits of chipless RFID for their operations.

APPENDIX D

Chipless RFID in Healthcare: Patient Tracking and Medication Management

This appendix focuses on the practical use of chipless RFID technology in healthcare, specifically for patient tracking and medication management. We will explore how healthcare facilities can benefit from implementing chipless RFID systems.

D.1 Introduction

Healthcare facilities, including hospitals and clinics, are increasingly adopting chipless RFID technology to enhance patient tracking and medication management. The practical implementation of chipless RFID in healthcare offers numerous advantages, including improved patient care, medication accuracy, and operational efficiency.

D.2 Patient Tracking

Problem Statement: Healthcare facilities often face challenges in accurately tracking patients, especially in busy and high-traffic areas such as emergency rooms. Manual tracking methods can lead to errors and delays in patient care.

Solution: Implement chipless RFID-based patient tracking. Patients wear RFID wristbands or tags that contain unique identifiers. RFID readers placed throughout the facility automatically detect and log the patient's location and movement.

Benefits:

- Real-time tracking of patient movements;
- Faster response times in emergencies;
- Improved patient flow and reduced wait times;
- Enhanced security, preventing unauthorized access.

D.3 Medication Management

Problem Statement: Medication errors, such as administering the wrong medication or incorrect dosages, are a critical concern in healthcare. Traditional manual methods of medication management are error-prone.

Solution: Employ chipless RFID for medication management. RFID tags are attached to medication containers, and RFID readers are placed in medication storage areas and at medication administration points. When a healthcare professional administers medication, they scan the RFID tag on the container, ensuring accurate medication dispensing.

Benefits:

- Reduced medication errors;
- Enhanced medication inventory management;
- Automated tracking of medication expiration dates;
- Improved patient safety and quality of care.

D.4 Practical Implementation Steps

This appendix outlined the practical steps healthcare facilities can take to implement chipless RFID for patient tracking and medication management, including system design, RFID tag selection, reader placement, and staff training.

The adoption of chipless RFID technology in healthcare offers practical solutions to critical challenges in patient tracking and medication management. By following best practices and implementing the technology effectively, healthcare facilities can provide safer and more efficient patient care.

APPENDIX E

Chipless RFID in Environmental Monitoring

In this appendix, we explore the practical applications of chipless RFID technology in environmental monitoring. We will focus on how chipless RFID can be used for various environmental parameters, including air quality, water quality, and climate monitoring. Environmental monitoring is essential for assessing and managing various environmental parameters that impact our planet's health. Chipless RFID technology can play a vital role in collecting and transmitting environmental data in real time. This chapter provides insights into practical implementations of chipless RFID in environmental monitoring.

E.1 Air Quality Monitoring

Problem Statement: Poor air quality is a significant concern in many urban areas. Monitoring air quality is crucial for public health and environmental assessment. Traditional monitoring methods involve fixed monitoring stations, which may not provide comprehensive coverage.

Solution: Implement chipless RFID–based mobile air quality sensors. These sensors can be attached to vehicles or drones, allowing them to collect air quality data as they move through urban areas. RFID readers positioned strategically throughout the city can capture and transmit real-time air quality data.

Benefits:

- Real-time air quality assessment;
- Comprehensive coverage of urban areas;
- Early detection of air pollution hotspots;
- Data-driven air quality improvement measures.

E.2 Water Quality Monitoring

Problem Statement: Ensuring the quality of water sources is crucial for public health and environmental conservation. Traditional water quality monitoring methods involve manual sampling and laboratory analysis, leading to delays in detecting contaminants.

Solution: Deploy chipless RFID-based water quality sensors. These sensors can be placed in bodies of water, such as rivers and lakes. They monitor parameters like water temperature, pH levels, and the presence of contaminants. RFID readers collect data in real-time and transmit it to central monitoring systems.

Benefits:

- Continuous and real-time water quality monitoring;
- Early detection of contamination events;
- Improved management of water resources;
- Enhanced conservation efforts.

E.3 Climate Monitoring

Problem Statement: Understanding climate changes and patterns is crucial for addressing climate-related challenges, such as extreme weather events and rising temperatures. Traditional climate monitoring methods involve fixed weather stations, limiting coverage.

Solution: Implement chipless RFID–based mobile climate sensors. These sensors can be placed on weather balloons, drones, or even satellites to collect climate data at various altitudes and locations.

These appendixes provided additional context and insights into the practical applications and potential future directions of chipless RFID technology, as discussed in the previous chapters of the book.

Glossary

Active RFID Radio frequency identification technology that uses an integrated power source to transmit data to a reader. Active RFID tags are self-powered and often used for tracking high-value assets over long ranges.

Antenna arrays Arrays of multiple antennas used to enhance signal strength, directionality, and coverage in RFID systems.

Barcode A method of representing data in a visual, machine-readable form, often consisting of a series of parallel lines and numbers. Barcodes are commonly used in retail for product identification and inventory management.

Biomedical applications The use of chipless RFID tags in healthcare settings, enabling nonintrusive and cost-effective data capture for medical sensing and monitoring.

Biometric tracking The process of identifying individuals based on unique physical or behavioral characteristics, such as fingerprints, retinal scans, or facial recognition. Biometric tracking is used for authentication and access control.

Chip-based RFID Radio frequency identification technology that relies on integrated circuits (chips) to store and transmit data. Chip-based RFID tags are durable and commonly used for supply chain management.

Chipless RFID RFID technology that does not use integrated circuits for data storage and relies on pattern-based techniques for identification. Chipless RFID tags are often more cost-effective and flexible than chip-based tags.

Chipless sensors RFID tags that do not rely on integrated circuits for data storage and use resonator-based methods for identification.

Convolutional encoder A signal processing technique used in chipless RFID systems to improve data transmission and error correction. Convolutional encoders add redundancy to data for error recovery.

Data encoding techniques Methods used to encode data on chipless RFID tags, allowing for efficient data storage and retrieval.

Electrical conductivity measurement A characterization technique used to measure the ability of a material to conduct electricity. Electrical conductivity is important in the development of conductive inks and smart materials for RFID tags.

Energy-efficient tags RFID tags designed to consume minimal energy, extending their operational life, and reducing the need for frequent battery replacement.

Flexographic printing A printing technique used for producing chipless RFID tags that is cost-effective and scalable. It involves a flexible relief plate and is often used for high-volume printing.

Fully printable smart sensing materials Materials that enable the fabrication of RFID tags through printing processes, making them cost-effective and flexible. These materials can sense and respond to environmental changes, making them ideal for RFID sensing.

Graphene A unique material with exceptional properties, including high electrical conductivity and mechanical strength. Graphene is explored for its potential applications in RFID sensing and as a component of conductive inks.

Hardware limitations Constraints in RFID technology related to the physical components and devices used in RFID systems.

High-bit density A feature of chipless RFID tags based on MSRR that allows them to store a large amount of data in a compact format. High bit density tags are valuable for applications requiring extensive data storage.

Inkjet printing A printing technique used for producing chipless RFID tags, known for its precision and accuracy. Inkjet printers create images by propelling droplets of ink onto a substrate, making them suitable for printing intricate patterns.

Linearly polarized antennas Antennas with a specific orientation that allows them to transmit and receive signals in a linear manner. Linear polarization is often used in RFID reader antennas.

Logistics The management and coordination of the flow of goods, information, and resources from the point of origin to the point of consumption. RFID technology is widely used in logistics for tracking and managing inventory.

Manufacturing processes Techniques and methods for the production of RFID tags, including inkjet and flexographic printing.

Material characterization The process of evaluating and understanding the properties of materials used in chipless RFID tag fabrication. Material characterization techniques include electrical measurements, spectroscopy, and microscopy.

Noise reduction algorithms Algorithms used to reduce or eliminate noise in RFID signals, improving data accuracy. Noise reduction is essential for reliable data extraction from RFID tags.

Optical character recognition Technology used to convert different types of documents, such as scanned paper documents, PDF files, or images captured by a digital camera, into editable and searchable data. OCR technology is used for document digitization and data retrieval.

Passive resonators in chipless RFID Passive resonators used in chipless RFID tags for cost-effective and scalable RFID solutions. These resonators do not require integrated circuits and are essential components of chipless RFID technology.

Passive RFID Radio frequency identification technology that doesn't have an integrated power source and relies on energy from the RFID reader for operation. Passive RFID tags are commonly used for inventory tracking and access control.

Printing techniques Various methods, including inkjet and flexographic printing, used to produce chipless RFID tags. Printing techniques are crucial for cost-effective and scalable tag production.

Privacy concerns Concerns related to the security and privacy of data stored and transmitted by RFID technology. Privacy concerns include the protection of personal information and the prevention of unauthorized access to data.

RCS A measurement of how detectable an object is by radar. RFID tags have specific RCS patterns that can be used for identification.

RCS patterns Patterns of RCS specific to chipless RFID tags, which can be used for tag identification and differentiation.

Read range The maximum distance at which an RFID reader can successfully detect and read data from an RFID tag.

RFID A technology that uses radio waves for automatic data capture and identification of objects or individuals. RFID technology is used in various applications, including supply chain management, access control, and healthcare.

Scalability The ability of chipless RFID technology to be deployed and adapted to various applications and industries, from small-scale to large-scale implementations.

Scanning electron microscope A characterization technique used to analyze materials at high magnification. Scanning electron microscopes provide detailed images of material surfaces and structures.

Security concerns Concerns related to the protection of RFID data from unauthorized access and potential cyberthreats.

Security enhancements Ongoing improvements and measures to enhance the security of chipless RFID technology. Security enhancements aim to protect RFID data from unauthorized access and cyberthreats.

Signal interference Interference and noise that affect RFID signals and can impact the accuracy of data capture.

Signal processing techniques Methods used to enhance data accuracy and reliability in RFID sensing, including convolutional encoder and Viterbi algorithm. Signal processing techniques are essential for extracting data from RFID signals accurately.

Smart materials Materials with the ability to sense and respond to environmental changes, enabling a new era of RFID sensing applications. Smart materials are used in various industries for real-time data capture and monitoring.

Standardization The process of establishing industry-wide standards for RFID technology to ensure compatibility and interoperability among different devices and systems.

Strain and crack sensing materials Materials used in RFID tags to monitor structural health and detect damage. These materials are essential for infrastructure and asset monitoring.

Supply chain management The management of the flow of goods and services, including the movement and storage of raw materials, work-in-progress inventory, and finished goods, from point of origin to point of consumption. RFID technology is widely used in supply chain management to improve efficiency and traceability.

Tag density The number of RFID tags that can be used within a specific area or volume, which can affect data accuracy and readability.

Tag detection accuracy The precision of RFID tag detection in RFID systems. High tag detection accuracy is critical for applications where data accuracy is essential.

Tag printing The process of producing RFID tags, with a focus on material selection, characterization, and fabrication. Tag printing techniques play a crucial role in the production of chipless RFID tags.

Tag readability The ability of RFID tags to be accurately and reliably read by RFID readers. Tag readability is essential for data capture and monitoring.

Temperature sensing materials Materials used in RFID tags for monitoring temperature changes in various environments. Temperature sensing materials are valuable for applications such as healthcare and environmental monitoring.

Ultrahigh-frequency RFID A specific frequency band used in RFID technology for communication between RFID readers and tags. UHF RFID is known for its longer read range.

UV–visible spectrophotometers A characterization technique used to measure the absorption and transmission of ultraviolet (UV) and visible light by materials. UV-visible spectrophotometers are used to evaluate material properties and characteristics.

Viterbi algorithm A signal processing technique used for efficient decoding and retrieval of data from RFID tags. The Viterbi algorithm is critical for accurate data extraction from RFID signals.

Wavelet transform A signal processing technique used in chipless RFID.

Windowing techniques Methods for segmenting and processing RFID signals with improved accuracy and precision.

List of Symbols

Γ	Reflection coefficient
ε_0	Permittivity of free space
ε_r	Relative permittivity
ε_{eff}	Effective permittivity
η_0	Free-space intrinsic impedance
λ	Wavelength
λ_0	Free-space wavelength
λ_g	Guided wavelength
μ_0	Permeability of free space
μ_r	Relative permeability
ρ	Impedance ratio
σ	Radar cross section (RCS)
σ_{tag}	RCS of tag
φ	Phase difference
ω	Angular frequency
Ω	Ohm (unit of resistance)
A	Cross sectional area
c	Speed of light in free space
C	Capacitance
D	Directivity
f_c	Center frequency
$G_{norm\ tag}$	Normalized gain of tag
G_{reader}	Gain of the reader antenna
G_{tag}	Gain of the antenna attached to tag
k_0	Free-space wave number
L	Inductance
P_{in}	Input power
P_{LF}	Polarization loss factor
Q	Quality factor
R	Resistance
R_a	Real part of impedance
R_{max}	Maximum reading distance
S^{tag}	S-parameter of tag

$\tan\delta$	Loss tangent
Z	Impedance
Za	Impedance of antenna attached to tag
Zc	Impedance of chip

About the Authors

Santanu Kumar Behera received his BSc (Eng.) degree in electronics and telecommunication engineering from the Veer Surendra Sai University of Technology (VSSUT) Burla, India, in 1990, and his ME degree in electronics and telecommunication engineering and PhD degree (Eng.) from Jadavpur University Kolkata, West Bengal, India in 2001 and 2008, respectively. He completed his postdoctoral research from Monash University Clayton, Victoria, Australia under the Endeavour Executive fellowship program, Ministry of Education, Government of Australia in 2018.

From May 1992 to June 2007, he worked as a senior lecturer in the Department of Electronics and Telecommunication Engineering, Government Polytechnic Dhenkanal. From August 2017 to November 2017, he was with the Department of Telecommunications, Asian Institute of Technology Bangkok as a visiting associate professor duly seconded by Ministry of Education, Government of India. From July 2007 to date, he is with the Department of Electronics and Communication Engineering, National Institute of Technology Rourkela, India. Currently, he is working as a professor in the same institute.

Dr. Behera's research interests include electromagnetic theory, planar antennas, dielectric resonator antennas, fractal antennas, reconfigurable antennas, and radio frequency identification systems. He has published more than 100 referred journal and conference papers, and three book chapters. He is a member of the editorial board on a number of journals and associate editor-in-chief of the *ACES Journal*.

Durga Prasad Mishra currently serves as an assistant professor in the Department of Electronics and Communication Engineering at LNMIIT (LNM Institute of Information Technology), located in Jaipur, Rajasthan, India. His journey in academia has been distinguished by several significant roles and notable contributions. His prowess in teaching and his valuable contributions to the field of education led to his appointment as an assistant professor at National Institute of Technology (NIT), Rourkela in 2022 on contract. In addition, he has held the esteemed position of research associate at Indian Institute of Technology (IIT), (ISM) Dhanbad, India. He has also dedicated a substantial duration of his career as an assistant professor at the prestigious Gandhi Institute for Technology Autonomous Institute in Bhubaneswar, Odisha.

Born in 1989 in Gopinathpur Sasan, Tigiria, Cuttack, Odisha, India, Dr. Mishra hails from a family with a rich tradition of educators, and his wife, Anisha Mohapatra, is also an educator. A pivotal moment in his life, the birth of his son Sriram Mishra, ignited a newfound passion for writing, resulting in the commencement of a book. Dr. Mishra embarked on his academic journey with the pursuit of dual bachelor's degrees, majoring in electronics and telecommunication engineering from the Biju Patnaik University of Technology (BPUT), Odisha, and mathematics and physical sciences from ANU AP. He continued his quest for knowledge by earning a master's degree in electronics and communication engineering from BPUT, Odisha. His academic journey reached its pinnacle with the attainment of a PhD from NIT, Rourkela, where he specialized in the field of design of chipless RFID transponders for retail and healthcare applications. Dr. Mishra's academic contributions are well-documented through his publications in esteemed journals, conferences, and book chapters, published by reputable publishers such as IEEE (conferences/magazines/transactions), Wiley Journals, CRC Press U.S., and Nova Science U.S. His areas of research expertise encompass chipless radio frequency identification, reader antennas, electromagnetics, and radio signal processing.

Index

U

V

W

X

Y

Z

Artech House
Electromagnetic Analysis Library

Christos Christodoulou, Series Editor
Vince Rodriguez, Series Editor

Advanced Computational Electromagnetic Methods and Applications, Wenhua Yu, Wenxing Li, Atef Elsherbeni, Yahya Rahmat-Samii, editors

Advanced FDTD Methods: Parallelization, Acceleration, and Engineering Applications, Wenhua Yu, et al.

Advances in FDTD Computational Electrodynamics: Photonics and Nanotechnology, Allen Taflove, editor

Analysis Methods for Electromagnetic Wave Problems, Volume 1, Eikichi Yamashita, editor

Analysis Methods for Electromagnetic Wave Problems, Volume 2, Eikichi Yamashita, editor

Analytical and Computational Methods in Electromagnetics, Ramesh Garg

Analytical Modeling in Applied Electromagnetics, Sergei Tretyakov

Anechoic Range Design for Electromagnetic Measurements, Vince Rodriguez

Applications of Neural Networks in Electromagnetics, Christos Christodoulou and Michael Georgiopoulos

CFDTD: Conformal Finite-Difference Time-Domain Maxwell's Equations Solver, Software and User's Guide, Wenhua Yu and Raj Mittra

Chipless RFID Printing Technologies, Santanu Kumar Behera and Durga Prasad Mishra

Computational Electrodynamics: The Finite-Difference Time-Domain Method, Second Edition, Allen Taflove and Susan C. Hagness

Computational Electromagnetics, Prabha Umashankar and Allen Taflove

Electromagnetic Fields in Multilayered Structures: Theory and Applications, Arun Bhattacharyya

Electromagnetics for Engineers, Volume 1: Electrostatics and Magnetostatics, Dean James Friesen

Electromagnetic Modeling of Composite Metallic and Dielectric Structures, Branko M. Kolundzija and Antonije R. Djordjevic

Electromagnetic Scattering and Material Characterization, Abbas Omar

Electromagnetic Waves in Chiral and Bi-Isotropic Media, I. V. Lindell, et al.

Electromagnetic Diffraction Modeling and Simulation with MATLAB®, Gökhan Apaydin and Levent Sevgi

Engineering Applications of the Modulated Scatterer Technique, Jean-Charles Bolomey and Fred E. Gardiol

Far-Field Wireless Power Transfer and Energy Harvesting, Naoki Shinohara and Jiafeng Zhou, editors

Fast and Efficient Algorithms in Computational Electromagnetics, Weng Cho Chew, et al., editors

FDTD Modelling of Metamaterials: Theory and Applications, Yang Hao and Raj Mittra

Generalized Multipole Technique, Christian Hafner

Grid Computing for Electromagnetics, Luciano Tarricone and Alessandra Esposito

High Frequency Electromagnetic Dosimetry, David A. Sánchez-Hernández, editor

High-Power Electromagnetic Effects on Electronic Systems, D. V. Giri, Richard Hoad, and Frank Sabath

Integral Equation Methods for Electromagnetics, Joseph R. Mautz and Nagayoshi Morita

Intersystem EMC Analysis, Interference, and Solutions, Uri Vered

Introduction to Electromagnetic Wave Propagation, Paul Rohan

Introduction to the Uniform Geometrical Theory of Diffraction, Derek A. McNamara, C. W. I. Pistorius, and Jan Malherbe

Iterative and Self-Adaptive Finite-Elements in Electromagnetic Modeling, Magdalena Salazar-Palma, et al.

Machine Learning Applications in Electromagnetics and Antenna Array Processing, Manel Martínez-Ramón, Arjun Gupta, José Luis Rojo-Álvarez, and Christos Christodoulou

New Foundations for Applied Electromagnetics: The Spatial Structure of Fields, Said Mikki and Yahia Antar

Numerical Analysis for Electromagnetic Integral Equations, Karl F. Warnick

Parallel Finite-Difference Time-Domain Method, Wenhua Yu, et al.

Practical Applications of Asymptotic Techniques in Electromagnetics, Francisco Saez de Adana, et al.

A Practical Guide to EMC Engineering, Levent Sevgi

Quick Finite Elements for Electromagnetic Waves, Second Edition, Giuseppe Pelosi, Roberto Coccioli, and Stefano Selleri

The Scattering of Electromagnetic Waves from Rough Surfaces, Peter Beckman and Andre Spizzichino

Shipboard Electromagnetics, Preston Law

Understanding Electromagnetic Scattering Using the Moment Method: A Practical Approach, Randy Bancroft

Wavelet Applications in Engineering Electromagnetics, Tapan K. Sarkar, Magdalena Salazar-Palma, and Michael C. Wicks

For further information on these and other Artech House titles, including previously considered out-of-print books now available through our In-Print-Forever® (IPF®) program, contact:

Artech House
685 Canton Street
Norwood, MA 02062
Phone: 781-769-9750
Fax: 781-769-6334
e-mail: artech@artechhouse.com

Artech House
16 Sussex Street
London SW1V 4RW UK
Phone: +44 (0)20 7596 8750
Fax: +44 (0)20 7630 0166
e-mail: artech-uk@artechhouse.com

Find us on the World Wide Web at: www.artechhouse.com